AF252038

A CABINET OF CURIOSITIES

The Myth, Magic and
Measure of Meteorites

A CABINET OF CURIOSITIES

The Myth, Magic and Measure of Meteorites

Martin Beech

Campion College, The University of Regina, Canada

World Scientific

NEW JERSEY · LONDON · SINGAPORE · BEIJING · SHANGHAI · HONG KONG · TAIPEI · CHENNAI · TOKYO

Published by

World Scientific Publishing Co. Pte. Ltd.
5 Toh Tuck Link, Singapore 596224
USA office: 27 Warren Street, Suite 401-402, Hackensack, NJ 07601
UK office: 57 Shelton Street, Covent Garden, London WC2H 9HE

Library of Congress Cataloging-in-Publication Data
Names: Beech, Martin, 1959–   author.
Title: A cabinet of curiosities : the myth, magic and measure of meteorites /
    Martin Beech, Campion College, The University of Regina, Canada.
Description: Hackensack, NJ : World Scientific, [2021] |
    Includes bibliographical references and index.
Identifiers: LCCN 2020053137 | ISBN 9789811224911 (hardcover)
Subjects: LCSH: Meteorites.
Classification: LCC QB755 .B44 2021 | DDC 523.5/1--dc23
LC record available at https://lccn.loc.gov/2020053137

British Library Cataloguing-in-Publication Data
A catalogue record for this book is available from the British Library.

For any available supplementary material, please visit
https://www.worldscientific.com/worldscibooks/10.1142/11954#t=suppl

Desk Editor: Ng Kah Fee

Typeset by Stallion Press
Email: enquiries@stallionpress.com

Prologue

Geologists have a saying — rocks remember

Neil Armstrong (1930–2012):
first man on the Moon,
July 21st, 1969

On the desk, to the right of my keyboard, sits a meteorite. It is a small fragment, just a few centimeters across, of North West Africa (NWA) 869. Not a particularly rare kind of meteorite, I purchased it, for a modest sum, from a dealer in rocks and minerals some 15 years ago. It looks very ordinary perched upon my desk, and yet NWA 869 is a product of the most amazing journey through time and space. The small rock on my desk, that I can cradle comfortably in one hand, is old, older, in fact, than the Earth — its constituent grains and minerals having condensed and accumulated some 4.56 billion years ago from the gas and dust within the solar nebula, the primordial disk of material from which the planets had yet to form. It is one of the originals, long incorporated within the greater body of an asteroid, it survived the turmoil and accretion of planet formation, and it escaped the random scattering and chaos associated with planetary migration. For millennia upon millennia it has relentlessly circled the Sun within the great expanse bounded by Mars and Jupiter — the main-belt asteroid region — until one fateful day, some 5 million years ago, its parent asteroid suffered a catastrophic collision with another minor planet. Shards of the colliding asteroids were ejected into the surrounding space, and one fragment, perhaps 4 to 5 meters across and weighing in at some 2–3 metric tonnes, set out on a long and tortuous journey, powered predominantly by a gravitational dance with the planet Jupiter, that ended through an encounter with Earth's atmosphere about 4500 years ago.

After careening through the upper atmosphere, undergoing an incredible retardation in velocity, NWA 869 raced across the sky as a brilliant fireball to land (perhaps entirely unseen by human eye) in southern Morocco within the northern reaches of the Sahara Desert. At this time, the fourth dynasty of Egypt was nearing its close, ending the Golden Age of the old kingdom, and in England, the construction of Stonehenge had yet to begin. The Assyrian Empire was in its early ascendency and Gilgamesh was soon to become King of the Sumerian city-state of Uruk. For us, NWA 869 fell into an ancient world, a world of myths and legends — indeed, the ancient world that ushered-in the beginnings of human civilization. Buried within the Saharan sands NWA 869 has been silent witness to virtually all that is known in written form of human history — it was here when all the major religions of today were new, it was here when the Romans ruled over Europe and northern Africa; it was here when the Vikings ruled northern Europe and bartered for gold with traders in Iraq. It was here when the Vikings first sailed to Iceland and Greenland, and then on to the North American Continent. It was here through the medieval period, and the bloodshed of the crusades; it was here during the French and American revolutions, and it was here during the great human-wasting of the 1st and 2nd World Wars. This one small rock, sitting somehow expectantly on my desk, has counted and marked the centuries of humanities' struggle to become what it is today, and one can only imagine what it might say if it could only speak. While stubbornly mute with respect to the history of human struggle, however, NWA 869 has yielded much scientific information about itself, and its long history has been deep searched by human imagination and technology. This is a rock that seems to transcend time — we come and we go in our all too fleeting lives, but it persists.

Meteorites from the NWA 869 strewn field are characterized by being highly wind-eroded, sporting surface features that have been polished through countless sand grain abrasions — these are rocks that have literally been shaped and shaded by the sands of time. The exposed surface has a greenish-grey tinge to it, and what was once a smooth fusion crust is now dimpled with a deep reddish-brown, waxy lustre. I do not know who originally found my fragment of NWA 869, but the first meteorites to be given the 869-designation appeared in the marketplaces of Morocco during the first few months of the 21st century. My fragment is far from being a singular find, and to-date some 2000 kgs of meteoritic material, numbering thousands of fragments, have been collected from the (still secret) fall location. In terms of specific classification NWA 869 is an L3-6 regolith breccia.[1] The L designation indicates that it has a relatively low abundance of nickel-iron and sulphides, and that it is a regolith breccia indicates that the parent body was composed of many distinct lithologies all of which were mixed and compacted together upon the surface of an ancient parent asteroid. Just 3% of the thousands of known L chondrite meteorites are classified as regolith breccias, and this makes NWA 869 extra special. Indeed, NWA 869 is a meteoritic mongrel; the product of the shock-compression and

Figure 1. My fragment of NWA 869. Inspiration on a desk — plucked from the sands of the Sahara Desert, the sample is some 6 cm across.

---

[1]A breccia is a rock composed of many broken fragments all cemented together by a fine-grained matrix. The form and composition of a brecciated meteorite, in principle, tells us about the collisional history of its parent asteroid.

thermal modification of regolith material. The meteorite sitting on my desk was forged by at least one, but probably many, asteroid upon asteroid collisions. Indeed, my desk companion had suffered much blunt-force trauma before finding its way to the Earth and its Saharan resting place.

My NWA 869 fragment seems all peaceful as it rests before me, but if I could zoom in and observe its interior and discern its atomic make-up then the apparent peace would be shattered. By no means all, but a goodly number of its constituent atoms are diseased — their nuclei bloated — in the form of radionuclides. These unstable atoms betray not only the origins of NWA 869, but they attest to its journey to the Earth, and its long strafing by cosmic rays. Indeed, the whole solar system is awash in a particulate sea of cosmic rays — charged particles, protons, and atomic nuclei, that have been accelerated to near the speed of light. Formed in distant supernovae explosions and channeled through vast light years of interstellar space on a conveyor-belt of galactic magnetic field lines, cosmic rays relentlessly plunge into the solar system and any object not shielded by a dense planetary atmosphere or thick layer of rock will begin to accrue cuts, scars, and atomic transformations. Indeed, cosmic rays have enough energy to penetrate through several meters of dense rock, and they can literally knock the stuffing out of (and into) otherwise stable atomic nuclei. Such extra stuffing and/or ejections can cause unstable atomic isotopes to form within a meteorite fragment as it journeys towards the Earth, and it is by analyzing such cosmogenic radionuclides that a measure of a meteorite's journey time to Earth (as well as upon Earth) can be obtained.

Key to extracting a time interval from radionuclides is a measure and understanding of their decay constant, or ticking rate, — this is usually expressed in terms of the so-called half-life: $T_{1/2}$. Indeed, the half-life is literally a measure of how long it takes the number of parent radionuclide atoms, at some time $T = 0$, to transform into the stable daughter state. This steady decay from the radionuclide parent to a stable daughter end-state is found to vary as an exponential law (discussed further in Chapter 4), with $N(T) = N(T = 0)\exp(-T/T_{1/2})$, where $N(T)$ is the number of unstable radionuclides at time T. It is the minus sign in the exponential term that accounts for the diminished number of unstable atoms. Given the conservation of nuclei, however, for every parent atom that decays, a stable daughter atom is formed. By measuring the abundance of the parent and daughter atoms, therefore, a measure of how long any specific radionuclide has been actively ticking can be obtained. Furthermore, since different radionuclides decay, or tick, at different rates, some undergoing rapid decay in a matter of days, or years, or centuries, with others taking eons, so a timeline picture of a meteorite's journey from the epoch of its formation, to the moment of is ejection into space, and even any subsequent collisions before encountering the Earth, can be built up.

A meteorite's age is based upon the presence and incorporation of long-lived radionuclides at the time of its crystallization, and these include such unstable atoms of uranium-238, rubidium-87, and argon-40. For these atoms, uranium-238 decays into lead-206 with a half-life of 4.47 billion years; rubidium-87 decays into strontium-87 with a half-life of 49.23 billion years, and argon-40 decays into argon-39 with a half-life of 1.25 billion years. It is by measuring the parent to daughter concentrations of such atomic nuclei that the bulk material of (nearly[2]) all meteorites are consistently found to be more than 4.56 billion years old — and this corresponds to the time at which planet building began within the Sun's natal dust-disk. Analysis of many different fragments of NWA 869, by the argon-40 $\rightarrow$ argon-39 decay method, indicates a formation age of 4402 ± 7 million years. Some fragments, containing what are called impact melt rocks (IMRs), show younger formation ages, indicating that the parent body, from which the meteorite is derived, must have suffered at least two distinct impact events, and probably more, when the IMRs were formed at times 2.2 and 1.8 billion years ago. Measurements based upon beryllium-10 radionuclide (with a half-life of 1.3 million years) and neon-21 (which is actually a stable atom) accumulation indicate that the material that was to become NWA 869 was ejected

---

[2]The term 'nearly' is used here since some meteorites, especially the lunar and Martian meteorites, can have more recent formation ages.

from its larger parent body 5 ± 1 million years ago — presumably as a result of some catastrophic collision and break-up event in the main-belt asteroid region. A comparison of carbon-12 dating and cosmogenic beryllium-10 abundance indicates a terrestrial age of 4.4 ± 0.7 thousand years — this is the epoch (circa 2400 BC) at which the NWA 869 fragments fell to the ground, and where, thereafter, its constituent atoms were shielded by the Earth's atmosphere from further cosmic ray alteration.

Having typed the above, I am left marveling at the deep history packed into and then extracted (by great scientific alchemy) from the fragments that accompanied the fall of my humble sample of NWA 869. If ever magic, myth and measure existed, it is there, in that small rock, to the right of my keyboard, resting quietly on my desk.

Contents

Introduction: The Cabinet of Curiosities

*In the end, perfection is just a concept — an impossibility we use
to torture ourselves and that contradicts nature*
> Guillermo del Toro, *Cabinet of Curiosities:*
> *My Notebooks, Collections,*
> *and Other Obsessions* (2013)

The Compact Oxford English dictionary contains 16 descriptions under the word *curiosity*. Its earliest known written usage in the English language, curiously, dates back to the 14th century. Indeed, writing circa 1391, in the very first English language astronomy text, *A Treatise on the Astrolabe*, Geoffrey Chaucer provides us with the following: "to know the degree of the sonne by thy reit, for a manner curiosite." This word usage, in fact, is applied to a heading under which Chaucer explains to "lyte Lowys my sone"[1] how to measure the altitude of the Sun with an astrolabe. Curiously, and in contrast to Chaucer, several hundred years later, in 1586, Thomas Bowes in his translation of Pierre de La Primaudaye's *L'Academie Française*, was to inform us that, "Their subtitles and bold curiosities, who have sought to plucke ... out of heaven the secrets hid from the angels." Indeed, while curiosity is a good trait it can sometimes be a bad preoccupation. For all this, however, curiosity is all about investigation and questioning, and we are curious creatures[2] — creatures,

---

[1] Curiously, Chaucer didn't actually have a son called Lewis — or Lowys — small or otherwise.

[2] Indeed, as we all know, curiosity killed the cat — a saying that is curiously modern in its introduction; its first recorded (approximate) usage dating to 1909, when William Sidney Porter under the pseudonym of O. Henry wrote in his short story *Schools and Schools* that, "curiosity can do more things than kill a cat."

that is, who are enthralled by the rare and novel; thirsting to collect, collate and understand (at least in some small part) the world around us. The Cabinet of Curiosities is the quintessential physical embodiment of that human desire to seek out new knowledge and wonderment, and whether in formal cabinet form, a box under the bed, a book on a shelf, or as some grand-edified national building and institution, the *Wunderkammer* has always been with us.

There are many purposes and forms to a cabinet of curiosity, they can be macabre and consist, as in the present day museum of Viktor Wynd in England, of the abnormalities of nature — a two-headed snake, the skeleton of a conjoined human child, a frog with no eyes (the possibilities and horrors are near endless). They can be refined, scientific and logically organized, consisting of many trays of shells or fossils, multi-coloured beetles or glimmering rocks. And, they can be just a collection of odds and ends — the organized jumble of things that have caught the collector's eye. The cabinet of curiosities in this book is made up of the formal, that is, the scientific, the mythical, and what might reasonably be called the magical. Firstly, the cabinet, the physical container, contains words and images, the inked-simulacra of real three-dimensional objects, and it contains a history of the curiosity that has long accompanied the fall of material from the sky. Second, this textual cabinet presents an outline of the deduced accumulated knowledge, science in its purest form, of meteorites — fragments of rock and iron that are as old as the Sun and solar system. And thirdly, this cabinet contains the intangible, the myths and misunderstandings that have developed in parallel, but apart from, the scientific study of meteorites — these are the infamous meteor-wrongs. The inclusion of this latter part perhaps needs a little more justification in the modern era, where the scientific approach is held supreme. The myths and magic (magic being used here to mean shear wonder) associated with the history of meteorites has no scientific value *per se*, but they very much enrich the human story that has become the narrative of meteoritics — they provide colour and context. For indeed, the history of science, and the development of scientific explanations, is all about struggle and perseverance, imagination and curiosity. Furthermore, the history of science tells us that the individual scientist most often gets things wrong, but importantly that mythical *thing* called *the scientific method* eventually finds a correct and collective viewpoint: from the myths, magic, and measure emerges a picture that is something close to reality. Scientists place great value on the measure part of their work, but it is the magic and myth, imagination and curiosity that provide the overarching theory and historical context.

One of the first renowned cabinet of curiosities was that maintained and constructed by the Danish physician Ole Worm during the early to mid-17th century. This cabinet, while not the first nor the largest, is principally known today for the text, *Museum Wormianum*, published in 1654 (after Worm's death) which described its contents in detail (Figure 1). Apart from being an eclectic collection of curiosities gathered during his extensive travels, Worm's

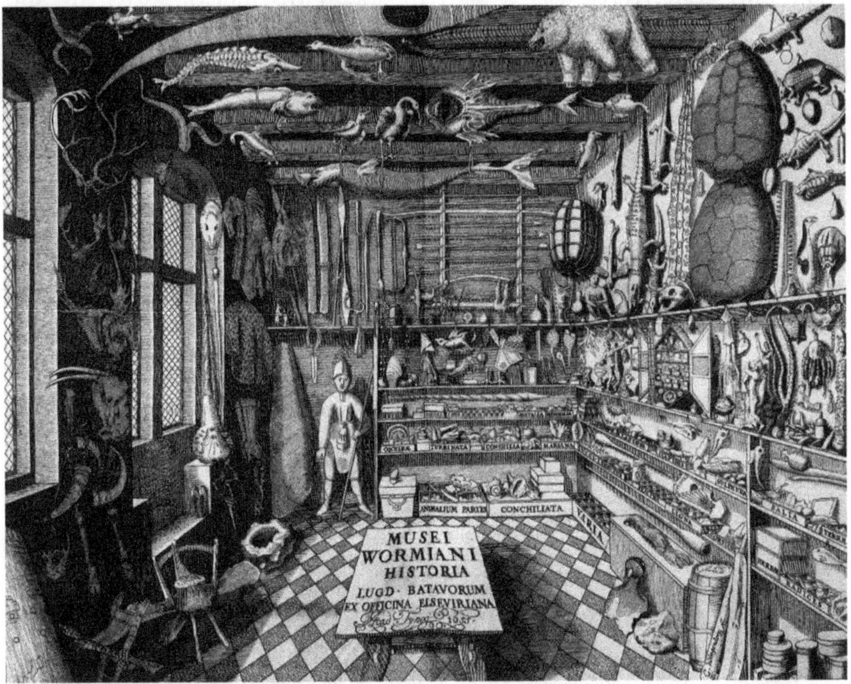

Figure 1. *Musei Wormiani Historia.* Wood cut from 1654 showing the cabinet of curiosities and *naturalia* assembled by Ole Worm.

Wunderkammer also contained contemporary wonders — perhaps the most incredible of which is the *statua librata pondere mobilis*, a human automaton that could mimic hand movements and gestures via internal strings and gearing. This early robot is seen in Figure 1 where it is dressed, spear in hand, as an imagined native American. The contents of Worm's Wunderkammer were arranged according to similarities — environment, kind, mass, animal, vegetable, and mineral, and while the cabinet contained many rocks and fossils, it contained no meteorites — for indeed, no such thing as a meteorite (as an extraterrestrial object) was recognized at that time.

Cabinets, such as that assembled by Ole Worm, were a direct outcome of the erstwhile Age of Exploration. Starting in the 16th century it was realized that there was a vast world *out there* to discover, and exotica, of one form or another, was poring into Europe. The ownership of a cabinet, literally a small room, filled with curios (real and contrived) was a sign of social distinction, wealth, and intellectual depth. These cabinets, however, were not for the public or the *hoi palloi*, nor were they typically for scientific enquiry; rather they were opulent showpieces for the well-connected to display to their well-connected visitors and friends. Cabinets only later, during the 18th century, became what might be called grand-amateur, pre-institutional museums, constructed to actually foster scientific enquiry. These latter cabinets often becoming so vast that they took on national importance. The first museum open to the public in England was The Ark in Lambeth, London. Developed, stocked, and owned by John Tradescant the elder, and later expanded upon by his son John Tradescant the younger, The Ark was a collection of the strange, the practical, and the marvelous. Indeed, it was a collection of fact and fiction, and we learn from a visit to The Ark, by German traveller and diarist George Stirn, in 1638 that:

> *"In the museum of Mr. John Tradescant are the following things: first in the courtyard there lie two ribs of a whale, also a very ingenious little boat of bark; then in the garden all kinds of foreign plants, which are to be found in a special little book which Mr. Tradescant has had printed about them. In the museum itself we saw a salamander, a chameleon, a pelican, a remora, a lanhado*

from Africa, a white partridge, a goose which has grown in Scotland on a tree, a flying squirrel, another squirrel like a fish, all kinds of bright colored birds from India, a number of things changed into stone, amongst others a piece of human flesh on a bone, gourds, olives, a piece of wood, an ape's head, a cheese, etc.; all kinds of shells, the hand of a mermaid, the hand of a mummy, a very natural wax hand under glass, all kinds of precious stones, coins, a picture wrought in feathers, [and] a small piece of wood from the cross of Christ."

The Tradescant Ark, full as it was of magic, myth, and make-believe, was eventually gifted to (although some have suggested stolen by) Elias Ashmole in the early 1660s, where after it became, in the 1680s, the foundation collection of the Ashmolean Museum at the University of Oxford. A similar such story of a private cabinet becoming the foundational component of a national institution is that surrounding the collection by physician, botanist and long-time President of the Royal Society of London, Sir Hans Sloane. Indeed, Sloane amassed one of the largest collections of curiosities in the world, eventually containing more than 71,000 items. Following an illustrious career and completing a 14-year stint as President of the Royal Society of London, Sloane, upon his death in 1753, bequeathed his collection of treasures to King George II and the British Nation — thereby establishing the core collections behind what are now the grand institutions of the British Museum, the Natural History Museum, and the British Library.

Most of the early cabinets of curiosities contained an eclectic rag-tag of objects — anything that was strange, rare, or miraculous could be included. More specialized and focused collections containing just plants, or minerals, or shells did exist, however. The Duchess of Beaufort, in the 1600s, for example, amassed a collection of some 1500 plant specimens gathered from around the word, while Hans Sloane's collection contained an estimated 5843 seashells. Geologist and obstetrician Gideon Mantell (who described the first Iguanodon dinosaur bones) amassed some 20,000 fossils in his private collection and museum — before selling them in later life, under straitened circumstances, to the British Museum in 1838.

The first meteorite collections to be housed in National Museums were generally secondary developments, typically founded upon a relatively small number of meteorites acquired in the purchases of much larger mineralogical collections and cabinets. Indeed, at a time when national prestige and vigor were rated according to ownership of *things*, and especially curious *things*, most of the world's largest collections of meteorites can trace their first major purchases and larger donations to the early to mid-19th century. Typically, it was not until the early to mid-20th century before *bona fide* research departments, dedicated to the study of meteorites as objects in their own right, first began to appear.

It has been estimated that some sixty-thousand distinct meteorites have been collected and catalogued over human history — the vast majority, of course, having been collected in the past several hundred years. Each of these multitudes of meteorites has its own story to tell, and each has a place in the human measure of the solar system. Some meteorites are famed because of when and where they fell, others are renowned because of the scientific insight that they provided. Others are perhaps run-of-the-mill in terms of their terrestrial history, but they are all witnesses to the long history of the solar system. Indeed, as will be explored in Chapter 1, meteorites are primordial, and they contain within their compass the very story of planetary creation. There is deep science buried within the chemistry and mineralogy of meteorites, and the task of the meteoriticist is to decode and unweave the complex messages that meteorites contain.

Not only do meteorites have immense scientific value, however, they also have a fascinating hold upon the human imagination — people have marveled and fainted at their unexpected appearance throughout recorded history, and many strange myths, legends and down-right strange ideas and associations have arisen over time — and some of these have even propagated to the modern era. Sometimes, in contrast to the norm, it is not so much the meteorite's own story that turns out to be interesting, but it is the narrative of the human adventure and activity (and occasionally the lies, theft, and stupidity) that makes a meteorite noteworthy. One classic example of such associated human adventure and sub-defuse is that surrounding the Cape York meteorites stolen (for this appears to be the appropriate word) by Robert E. Peary (of the US Navy) in the late 1890s. Not only did Peary simply take several massive meteorites from the indigenous Inuit, but he technically stole them from another sovereign nation — Denmark. Peary was a major figure and *personality* in the 19th century business of Arctic exploration, although his achievements and actions can be, and indeed have been, heavily criticized by modern commentators. Indeed, Peary claimed to have reached the north geographic pole on April 6th, 1909, but this, by various avenues of reasoning, has been clearly disputed — he was either mistaken in his claim or he wilfully lied. Peary additionally fathered (at least) several Inuit children outside of his marriage, and then happily (for so it seems) abandon them. He was a larger than life figure, consumed by an inner self-desire for fame and fortune, and little else beyond his own ego appeared to matter.

The existence of at least some iron meteorites on the west coast of Greenland (internationally recognized as a dependency of Denmark since 1814) was reveled during the Arctic explorations and search for the famed northwest passage by John Ross in the early 1800s. Indeed, Ross found various metal artifacts being used by Inuit hunters, and knowing that there was no tradition of iron smelting, reasoned that the metal was most likely meteoric in origin (Figure 2). The exact

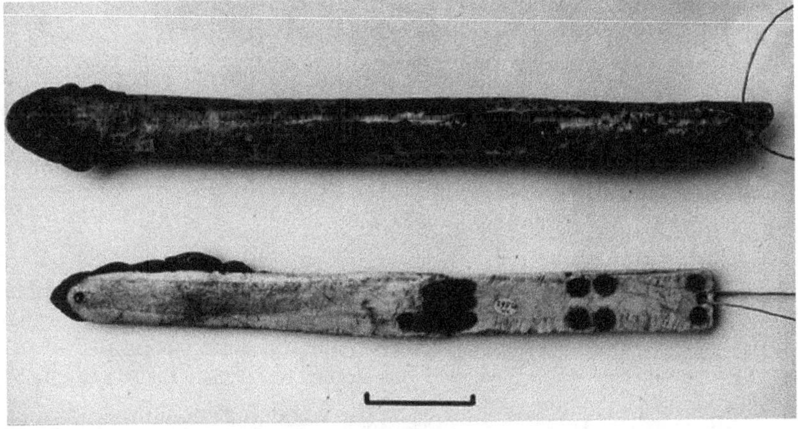

Figure 2. Inuit hunting spears tipped with slices of Cape York meteoritic iron.

location of the meteorite(s), however, was a closely guarded secret —
that is, until Peary, in 1894, bribed a local hunter (on promise of
receiving a rifle) to reveal its whereabouts. Indeed, Peary had set
out upon the Greenland Expedition in 1893, with 14 companions and
his pregnant wife. At this time his goals were to establish a farthest
north record and survey northeast Greenland. Due to various factors
he achieved neither of these goals, and accordingly, in an attempt
to redeem himself (and the expedition) he decided to look for the
fabled Iron Mountain described by Ross in 1818. Peary recounted
his adventures in volume II of his *Northward Over the Great Ice*
(published in 1898). "*On the 16th of May, I left the lodge again with
[Hugh] Lee, my iron-runner sledge, and ten dogs, in search of the Iron
Mountain of Melville Bay.*" His Inuit guide Tellakoteah had explained
to Peary that, "*there were three saviksue (great irons) of varying
sizes, the smallest about the size of a mikkie (a dog), indicating a
dog curled up, the second considerably larger, and the third still larger
than the second.*" These three meteorites are now known under the
names of the *Dog*, the *Woman* and the *Tent*.[3] The discovery of the
first of the iron meteorites took place on Sunday morning, May 27th,
1894, and Peary writes,

> "*the brown mass, rudely awakened from its winter sleep, found for
> the first time in its cycles of existence, the eyes of a white man
> gazing upon it .... In addition to the thick blanket of snow, the
> meteorite was completely covered with a half-inch thick coating of
> ice ... Tellakoteah tells me that the Innuits [sic] call the meteorite
> a woman in a sitting position, and he says it used to be much larger
> and higher that it is now, but his people had gradually worn it down,
> and many years ago natives from Peterahwik broke off the head and
> carried it away.*"[4]

---

[3]This meteorite is also known under the name *Ahnighito* — which is the middle
name of Peary's daughter Marie, "the snow baby" born during the 1893 Greenland
Expedition. It does not actually mean anything in the Inuit language.
[4]Danish mineralogist Vagn Buchwald has argued that the *Dog* is actually the
former head of the *Woman*.

Having been brought to the locality of the iron meteorites by his guide, Peary set about establishing his ownership: *"I scratched a rough P on the surface of the metal, as an indisputable proof of my having found the meteorite."* With ownership thus established, Peary thereafter blatantly, over the next several years, appropriated the meteorites for his own financial gain, after what might generously be called a series of heroic engineering efforts. His initial efforts in 1894 were unsuccessful, but returning the following year he was able to maneuver the 3000-kg *Woman* and 400-kg *Dog* aboard his ship. An attempt to retrieve the 30,900-kg *Tent* was made in 1896, but again it was not until the following year that it could be safely stored aboard his transport vessel (Figure 3). The *Tent* arrived at Brooklyn Naval Yard in October 1896, and there it stayed until 1904, when, with much fanfare, it was transported aboard a carriage pulled by 28 horses to the American Museum of Natural History. Indeed, once transported to New York, claiming that they had been a gift from

Figure 3. The 30.9-tonne *Ahnighito* (the *Tent*) meteorite (draped by "old glory") being loaded aboard Peary's ship the *Hope*.

her husband, Josephine Peary set about selling the meteorites to the highest bidder, and eventually the Museum acquired them for the tidy sum of $40,000[5] generously donated by Museum President, banker and philanthropist Morris K. Jesup. No money, apology or even a by-your-leave, has ever been issued to the Inuit of Havighivic in compensation for their cultural and material loss. The Peary's, of course, were a product of their time, and their glaringly colonialist attitudes were largely, in one form or another, emulated by all other explorers of that era.[6] The fact that Peary had to self-finance his expeditions probably had much to do with his decisions making as well. The Cape York meteorites offer but one example of a cultural theft made in the name of private gain weakly disguised as scientific research. From their very first appearance at the American Museum of Natural History, however, *Ahnighito*, the *Dog* and the *Woman* have attracted the wonder and imagination of the visiting public. They are truly impressive objects, and to this day, *Ahnighito* is still the largest meteorite on public display in any museum in the world.[7]

Eight major fragments from the Cape York region have now been identified, and a distinct meteorite strewn field, perhaps some 100 km in length, is beginning to emerge, although given the harsh conditions much of the fall locality has not been mapped or even systematically surveyed. This being said, Danish mineralogist Vagn Buchwald has conducted a number of surveys in the area, finding, in 1963, the 20-metric-tonne *Agpalilik* (*the Man*) meteorite. The most recent find in the Cape York peninsula was that of the 250-kg *Tunorput* meteorite in 1984. Of recent note it has been suggested that the meteorites may be linked to the Hiawatha crater, only recently discovered in 2018 — this connection has not as yet been fully established, but if true then the Cape York meteorites were fragments associated with a 2-km or so diameter asteroid that produced the 31-km diameter Hiawatha crater (Figure 4).

---

[5]In 2020 dollars this would be a sum in excess of 1 million dollars.

[6]It has been noted by various commentators that the Inuit he encountered regarded Peary with "fear," and that he was, "hated for his cruelty."

[7]Indeed, it is also the 2nd largest iron meteorite that has ever been discovered, bested only by the Hoba Meteorite in Namibia.

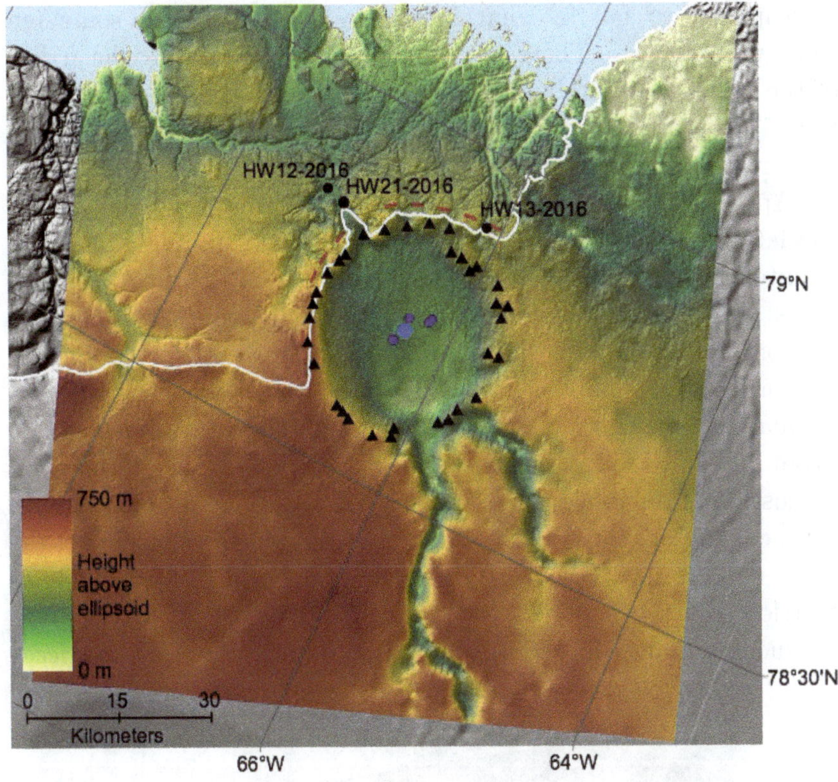

Figure 4. The Hiawatha crater in northern Greenland. The map is from an airborne radar survey. Image courtesy of NASA.

The Cape York meteorites are classified as IIIAB irons and are noteworthy because of their chemically inhomogeneous nature. This group of irons, the type IIIABs, are thought to have crystalized from the fluid melt accumulated within the central core of a large, so-called differentiated, asteroid. The first solids to crystalize in the outer regions of such a molten core will be depleted in nickel, but as the cooling and crystallization proceeds, so the solids will become increasingly nickel rich — the Cape York meteorites appear to be composed of crystallization fragments from across the core of their parent asteroid. It has been estimated that the parent asteroid to the Cape York meteorites underwent catastrophic break-up 95 million years ago and that the meteorites themselves crashed through Earth's atmosphere, to fall in what is now called Greenland, some ten thousand years ago. There is no Inuit tradition recounting the arrival of the meteorites, although knowledge of them stretches back more than 1000 years into the past, and they are accordingly classified as *finds* rather than *falls*.

Apart from being a meteorite, which automatically makes it a special object, there is nothing especially outstanding about the Launton meteorite (Figure 5). While it is one of the most common types of stony meteorite (an H4 chondrite — see Chapter 1), it is more interestingly a member of the more exclusive group of meteorites that qualify as a *fall*. That is, it was observed as a bright, detonating fireball as it descended through Earth's atmosphere at about 7:30 PM on Monday February 15th, 1830. Creating loud percussions, "resembling the rapid discharge of three ordinary guns," the incoming meteorite both excited and amazed the residents of the small Oxfordshire (UK) village of Launton. Local resident Thomas Marriot was particularly startled at the meteorite's arrival, and commented that he looked up and saw an object "the size of a cricket

Figure 5.   The Launton meteorite. Image courtesy of The Natural History Museum, London.

ball" hurtling towards him. Fearing that it might strike him, "he instinctively lowered his head to avoid it." The meteorite came to rest in the garden of John Bucknell, burying itself approximately 30 cm ("nearly a foot") below ground in newly dug topsoil. The recovered meteorite weighed in at 2 pounds 5 ounces (about 1 kg), and, being a single mass, the meteorite could sit comfortably in the outstretched palm of one hand. While it is likely that more than one fragment struck the ground in the area surrounding Launton that February evening, no other masses have been recovered.

It was to be nearly one whole year before the first written account of the Launton meteorite appeared in print. Compiled by William Stowe (a surgeon working in Buckinghamshire) the account appeared in the January 1831 issue of *The Magazine of Natural History* — a new journal catering to the grand-amateur reader and the diffusion of knowledge that first saw print in 1829. Immediately upon finding this learned magazine, the modern reader finds a curiosity. Even though the magazine has the subtitle, "journal of zoology, botany, minerology, geology and meteorology," it is within the meteorology section that the Launton fall is indexed. Although very definitely a solid rock, the magazine editors did not feel compelled to place the account under either the geology or mineralogy titles — why was this? The reasons behind the editor's placement are the historical grist behind the beginning chapters of this very book (see Chapter 2), and they are reflected in the common word origin behind meteorite and meteorology. As will be discussed in Chapter 2 the etymology of meteorite and meteorology is based upon the Greek word *meteoron*, meaning *something high up*. That is, the words are taken to be related to phenomenon pertaining of the Earth's upper atmosphere. Indeed, one of the curiosities encountered by the modern reader when reading over William Stowe's account of the Launton meteorite is that he takes particular pains to investigate the meteorological conditions on the day of the fall — "I find that the day had been foggy, with the wind in the north; and the barometer was unusually high, being at 10 AM, 30.9, and at 10 PM 30.8; the thermometer being at the same times respectively 43° and 26°. There was nothing like thunder in the atmosphere at the time mentioned [7:30 PM]." By this detailed weather account Stowe is essentially building upon the concept that meteorites are objects formed within Earth's upper atmosphere. Indeed, he writes that, "there has been great diversity of opinion among scientific men as to the origin of these bodies," but then goes on to argue that, "it requires no great stretch of credulity or imagination to attribute their formation to the transmuting powers of electricity." Specifically, Stowe argues that, "in the great laboratory of the atmosphere, electricity being the chemist, changes may possibly occur attended by the formation of iron and other metals from the consolidation of simple elementary substances." Writing, as he was, in 1830 Stowe was not greatly behind in his thinking with respect to the origins of meteorites, and he was essentially tapping into the then still relatively new *wonder topic* of electricity, concluding, in fact, that, "can we doubt the possibility of more astonishing productions being formed with the great battery of Nature?" Astonishing, indeed, and from a modern perspective absolutely wrong — but that, as we shall see, is very much the story behind the myth and magic of meteorites.

In some sense, with the appearance of Stowe's account, the story of the Launton meteorite goes quiet, not to reappear again until 1916. For all this, however, the narrative takes us in new directions. Indeed, Stowe notes in his account that shortly after the meteorite fell it was acquired by the Reverend Doctor John Fiot Lee of Hartwell House (Figure 6). To this announcement, in slightly critical tone, Stowe comments that, "Had I been so fortunate as to have obtained it, I should have deposited it in the Ashmolean Museum at Oxford, as being a *county* curiosity, worthy of a place in that valuable, and scientifically arranged collection." Certainly, Stowe's sentiments are sound and indeed, the incredible curiosities that meteorites are, they should be properly curated in museum collections. But, for all this, who was the Reverend Lee, and what did he want with such an object as the Launton meteorite? Fortunately, as historians, we need not look to far in order to find the relevance of Lee's interest. Indeed, the Reverend Lee was at the center of an eclectic clique of grand-amateur astronomers and men of science, and the cultural

Figure 6. The Reverend Doctor John Lee of Hartwell House.

hub of these committed, but non-tenured, scientists[8] was Hartwell House. Lee in essence was the grand instigator, and while he did not spend large amounts of time conducting astronomical research, he encouraged and facilitated others in doing so. Foremost among his *famulus* was the well-respected astronomer Captain W. H. Smyth. Indeed, Smyth is still remembered among amateur astronomers to this very day for his book the *Cycle of Celestial Objects* (first published in 1844). Smyth was awarded the gold medal of the Royal Astronomical Society in 1845 for this work, and he later became President of this learned (and still thriving) society. Very much a product of the Victorian era, in addition to his military career, Smyth was a Fellow of the Royal Society (elected in 1826), a Fellow of the Society of Antiquaries of London (elected 1821) and a founding member of the Royal Geographical Society (1830) as well as the Royal

---

[8]It is perhaps worth pointing out that the term scientist was first coined (in public) by William Whewell at the 1833 meeting of the British Association, although the word appears to have seen earlier use in various reviews and letters.

Numismatic Society (1836). The Reverend Lee was also a Fellow of the same societies supported by Smyth, and in addition he was a Fellow of the Philological Society (1831) and a founding member of the Royal Meteorological Society (in 1850). With his attachment to so many different societies, Lee excited and moved within the various orbits attached to the scientific elite of Victorian England, and at Hartwell House he established one of the finest Wunderkammer's in Great Britain.

Orphaned at a young age, Lee inherited Hartwell House and a considerable fortune from his uncle William Lee Antonie, and he used his wealth and leisure time to amass a vast collection of curiosities, antiquities and treasures. Writing in his *Ædes Hartwellianæ* (notices of the manor and mansion of Hartwell), published in 1851, Captain Smyth noted that, "the museum is appropriated to a miscellaneous collection of articles culled from the animal, vegetable, and mineral kingdoms." To this Smyth notes with respect to the mineral cabinets that, "the specimen which perhaps deserves to head the list in the mineral kingdom, as extra-mundane, is the black meteoritic stone that fell at Launton.... Because it long constituted one of the very few whose time and place of descent are well authenticated." The Launton meteorite, therefore, was the pride of Lee's mineralogical collection, but this did not mean that it was an easy object to study.

In August 1854 William and Robert Chambers, writing in their *Chambers's Journal of Popular Literature, Sciences and Arts* described a visit to Hartwell House. Using the term pilgrimage to describe their visit, the Chambers commented that, "nothing can be conceived more beautiful of its kind than the park which unfolds Hartwell House within its umbrageous arms." Reluctantly, for so it seems, the Chambers crossed the surrounding parkland and entered into the manor house proper. The glowing terms of appreciation were no less panegyrical than those used for the parkland, and their narrative begins with a brief description of the various paintings (by Vandyck, and Reynolds, amongst others) that decorated the house's walls and corridors. The observatory wing is described as being "filled with curious and valuable astronomical instruments, besides containing a rich collection of works treating of that science.... The number of books is indeed so great, that some are to be found in almost every room of the house." To reach the museum at Hartwell, one has to climb to the first floor, "the great staircase conducting to it is a stately oaken structure of easy ascent and great breadth." Indeed, the visitor to the museum is swept up towards an opening vista that contains "a vast assemblage of all kinds of curiosities, collected with great diligence and at much expense by Dr. Lee while he was a travelling bachelor for the University of Cambridge." "The Grecian and Egyptian antiquities are particularly numerous," the Chambers continue in their account, "and the geological and mineralogical department singularly rich. All the articles are named and described, and thus the visitor is instructed as well as interested while examining the collection."

It is within this latter comment by the Chambers that we hit upon the problem hinted at by Thomas Stowe in his account of the Launton meteorite. Indeed, the point is that Lee's collection was private, and it was only upon a vetted request that one might get to see its contents. It was not a display open to the general public, but a display only accessible to a select few with the right social bearing and connections — and, indeed, many luminaries of Victorian science are known to have visited and studied at Hartwell House. In the Ashmolean Museum, at least in principle, the Launton meteorite could have been viewed by an interested public. Indeed, it appears that Lee guarded his prize meteorite sample with some zeal, refusing even to grant the requests from qualified scientists applying to study its properties. Stowe comments, correctly in his 1831 account that, "it [the meteorite] appears to contain nickel, which has long been considered the constant associate of iron in meteoric productions," although he is less sure about the meteorite containing chromium (see Chapter 1). In the hope of solving the issue of its chemical composition, Stowe sent "a small fragment" to the editor (J. C. Laudon) of the *Magazine of Natural History* in the hope that it might be forwarded to "Mr. Farraday, or some other competent person, for minute analysis and conformation of its celestial origin." The sample was eventually given to Edward Turner,

Professor of Chemistry in the London University (now University College, London), but the fragment forwarded by Stowe proved too small to yield any meaningful results. The only account concerning the Launton meteorite, while in the Hartwell House museum, is that given by chemist W. H. Miller (University of Cambridge) in 1841 when he briefly noted a visit, along with Johann Franz Encke (who we shall encounter again later) in August 1840. After much delay, however, Lee eventually presented the British Museum (now the Natural History Museum, London) with a cast of the meteorite in 1853.

It was not until the analysis by George Thurland Prior in 1916 that the chemical make-up of the Launton meteorite was to be described, and the presence of nickel-iron and chromium were confirmed. Prior provides an original mass of 1028 grams for the Launton meteorite and gives its dimensions as $4^1/_2 \times 3 \times 2^1/_2$ inches. A 25-gram slice was cut from the main mass (see Figure 5) and it was this smaller piece that was exposed to chemical analysis. Even the analysis by Prior turned out to be long delayed, although much of the delay time was due to the apparent loss of the meteorite.

Lee's death in 1866 resulted in the partial dispersal of his long and lovingly acquired collection, and it is at this time that the Launton meteorite, although a prize of his collection, seemingly disappears. That is, however, until enquiries were made of the Hartwell House estate by Sir Arthur Church in 1889. Church was at that time the curator of the meteorite collection at the British Museum and was interested in expanding the museum's collection. It was the spring of 1895, however, before Sir Lazarus Fletcher (then the new curator of meteorites at the British Museum) was able to visit Hartwell House and successfully identify and collect the meteorite for the national archive. It transpires that the Launton meteorite had been incorrectly labelled as being a fragment of the Esnandes meteorite — the correct reidentification being confirmed via the cast of the meteorite that Lee had sent to the British Museum some 42 years earlier. The Esnandes meteorite is interesting in its own right, having been an observed meteorite fall in France in 1837. This particular meteorite is somewhat more exotic than the Launton meteorite, being a stony

L6 chondrite (see Section 1.2), and it is singled out through the account that the meteorite was hot to the touch, and burnt the finder's fingers when first picked up. In addition to the Esnandes meteorite, Lee also acquired for his museum fragments of the Cold Bokkeveld meteorite, which fell at the Cape of Good Hope on October 13th, 1838. From a modern perspective it is surprising that Smyth (writing, recall, in 1851) would consider the Launton meteorite to be the prize of the Hartwell House collection, when in fact the Cold Bokkeveld meteorite is, scientifically speaking, by far the more important object — it being a CM2 carbonaceous chondrite — but we, of course, have the advantage of a greater period of study and scholarship behind our gaze.

Prior used the story concerning the apparent loss (and subsequent reidentification) of the Launton meteorite as an argument against meteorites being housed within private collections, and this is an argument that continues to simmer even to this very day, with scientific institutions, private collectors, and meteorite dealers regularly squaring off against each other, not only in biding wars, but in the basic stance of who should have access to such rare astronomical treasures: scientist at national research institutions, or entrepreneurs just trying to make an honest living in an open and unrestricted market? This being said there have been attempts to produce mathematical formula to gauge (see Chapter 2) the value of a given type of meteorite, although free-market rules tend to apply today, with the person offering the largest pot of money generally taking home the prize of the meteorite. There is additionally the long-running and often highly contentious legal debate to consider — who actual owns the meteorite (and who can therefore legally sell it) once it has been found: is it the finder, the land owner, the entity that owns the mineral rites below the surface soil, or the state? The answer to this legal question, in spite of there often being clear legislation, is, "it depends."

The 5th edition of the *Catalogue of Meteorites* (edited by Monica Grady, 2000) indicates that the entire extant mass of the Launton meteorite (some 999 grams) is held within the collection of the Natural History Museum in London. In spite of this, the closeted-life of this particular meteorite, cut off from public wonder and gaze, has continued through to this very day — such, of course, is the fate of many (actually, the majority of) curated and cared for meteorites. Lee's 1853 cast of the meteorite can be seen, however, at the Natural History Museum, Oxford. The most recent public airing of the Launton meteorite occurred in 2011, when it formed part of an art installation, *The Robinson Institute*, created by Patrick Keiller at the Tate Gallery in London. Keiller's Tate installation was developed from an earlier film project, *Robinson in Ruins*, which followed the ramblings of an "erratic scholar on a journey through landscapes in the south of England." *The Robinson Institute* featured two meteorites — that of Wold Cottage, which fell on December 13th, 1795, and the Launton meteorite. The inclusion of these meteorites was predicated upon the conclusion by Robinson (the unidentified protagonist behind the *Institution* and the film), "that a meteorite fall necessarily coincides with an event of major historical significance." The Wold Cottage fall, according to Robinson, marked the appearance of the 1795 amendment to the Settlement Act — this amendment allowing for the free movement of the poor and disenfranchised to go where jobs were available, and this, "enabled industrial capitalism to develop." As for Launton, Robinson (through Keiller) explains, "1830 was a year of revolutions, in France and Belgium, and in England, where the Captain Swing[9] riots began at the end of August and continued through the autumn." Meteorites (and meteors) have long been interpreted as the harbingers of both good and ill (depending upon one's biases), but there is, of course, no causal relationship between the deeds and actions of human society and the appearance of extraterrestrial material upon Earth's surface. For all its quiet, hidden, and un-exceptional terrestrial residency, the Launton meteorite hardly seems a good prop for stratifying a time of revolution, riots, strife, and death.

---

[9]Captain Swing was the name appended to various pamphlets and letters produced during the rural riots that erupted in England when laborers protested the introduction of new threshing machines — machines which threatened their very trade and livelihood.

In this introduction I have specifically focused attention on two very different meteorites, one massive and (in)famously found in the frozen Arctic, with a less than praiseworthy provenience of theft and private profiteering; the other small, hardly known, genteel (perhaps), and for a large fraction of its time on Earth rarely seen. Neither meteorite has been at the center of any sustained scientific research effort (indeed, they are both somewhat ordinary as meteorites go), but they are fascinating, nonetheless, for the whirlwind of stories, attributions, and historical connections that surround and emanate from them. They highlight, each in their own way (even though they are far from unique and many similar such examples could have been chosen), some of the tensions, triumphs, and tribulations that pervade the world of meteoritics. And, these same tensions, triumphs, and tribulations will reappear, at various times, throughout the remainder of this book.

Meteorites — they have been a part of human thought and activity for millennia. They are treasured objects — both culturally venerated and scientifically valuable; they are objects of desire and obsession and occasionally they are objects of villainous greed. But, most of all, they are instigators of narrative and knowledge. Their magic is that of engagement — they can hold the human imagination like few other objects can, and their measure is the discovery of the origins of the solar system itself.

In terms of the story ahead, the remainder of this text is divided into two main sections. The first section offers a modern-day introduction to meteorite classification — what they are, how they are classified, and what this *measure* tells us about their potential origins. The next section turns the gaze backwards and looks to explore the long history of meteorites. This section will examine some of the myths and magic that our ancestors bestowed upon meteors and meteorites. Following the historical review, there are several sections that explore the astronomical details of meteoroid dynamics and the physics of a meteoroid's interaction with the Earth's atmosphere. Mathematical details and developments will be provided in these sections (and others), but for those not wishing to explore such numerical matters the text will also guide the understanding. The ultimate story that emerges from these technical sections is that concerning the origin of the solar system, and specifically, how meteorites, as pioneering components of the solar system, inform us about our origins. With the broad narrative on meteorites thus in place, in the second part of this book, we enter into a selection of cabinets of curiosities. The final sections cover the strange, and the wonderful, the matter of fact, the oddballs, and the final sections look at a number of topics which remain deeply mysterious and not fully explained, even in the modern era. The final sections do not need to be read in any specific order, and the reader is invited to pick and choose their way through the topics. Most of all, however, the fundamental property of any cabinet of curiosities is wonderment and marvel, and it is my hope that this text will act to inspire the reader to learn more about the myth, magic, and measure of meteorites.

Additional reading

M. Beech. *Meteors and Meteorites: Origins and Observations.* The Crowwood Press, Ramsbury (2006).

A. Bevan and J. De Laeter. *Meteorites — A Journey through Space and Time.* Smithsonian Institution Press, Washington (2002).

J. Burke. *Cosmic Debris: Meteorites in History.* University of California Press, Berkeley (1986).

W. Cassidy. *Meteorites, Ice and Antarctica — A Personal Account.* Cambridge University Press, Cambridge (2003).

A. Chapman. *The Victorian Amateur Astronomer.* Gracewing, Leominster (2017).

R. Dodd. *Thunderstones and Shooting Stars — The Meaning of Meteorites.* Harvard University Press, Cambridge (1986).

M. Golia. *Meteorite — Nature and Culture.* Reakton Books, London (2015).

F. Heide and F. Wlotzka. *Meteorites — Messengers from Space.* Springer-Verlag, Berlin (1995).

P. Keiller. *The Possibility of Life's Survival on the Planet.* Tate Publishing, London (2012).

J. Larsen. *On the Tail of Stardust.* Voyager Press, Beverley (2019).

P. Mauries. *Cabinets of Curiosities.* Thames and Hudson, London (2019).

H. McSween, Jr. *Meteorites and their Parent Planets.* Cambridge University Press, Cambridge (1999).

O. Norton. *The Cambridge Encyclopedia of Meteorites.* Cambridge University Press, Cambridge (2002).

G. Notkin. *Rock Star — Adventures of a Meteorite Man*. Stanegate Press, Tucson (2012).

R. Olson and J. Pasachoff. *Fire in the Sky: Comets and Meteors, the Decisive Centuries, in British Art and Science*. Cambridge University Press, Cambridge (1998).

S. Russell and M. Grady. *Meteorites*. The Natural History Museum, London (2002).

D. Steel. *Target Earth — The Search for Rogue Asteroids and Doomsday Comets that Threaten our Earth*. Readers Digest, Pleasantville (2000).

L. Weschler. *Mr. Wilson's Cabinet of Wonder*. Vintage Books, New York (1995).

V. Wynd. *The Unnatural History Museum*. Prestel Verlag, Munich (2020).

B. Zanada and M. Rotaru. *Meteorites — Their Impact on Science and History*. Cambridge University Press, Cambridge (2001).

Chapter 1

The End

Given the nature of this text, it seems appropriate to begin at the end of the story, rather than at the beginning. This reversal allows for the upfront introduction of meteorites, as they are presently understood and classified — leaving the story of how such information was gleaned to later chapters and sections. Accordingly, the first section will provide an outline of what a meteorite is, and how such objects are identified. The second section will examine the meteorite classification scheme, and then a final section will consider the circumstances of meteorite falls and finds, presenting in its close an in-depth analysis of the fall of the Buzzard Coulee meteorite in Saskatchewan, Canada.

1.1 What is a meteorite?

Commission F1 of the International Astronomical Union provides the following useful set of definitions:

- **Meteor** is the light and associated physical phenomena (heat, shock, ionization), which result from the high-speed entry of a solid object from space into a gaseous atmosphere.
- **Meteoroid** is a solid natural object of a size roughly between 30 micrometers and 1 meter moving in, or coming from, interplanetary space.
- **Dust** (interplanetary) is finely divided solid matter, with particle sizes in general smaller than meteoroids, moving in, or coming from, interplanetary space.
- **Meteorite** is any natural solid object that survived the meteor phase in a gaseous atmosphere without being completely vaporized.

A number of qualifications are added to these definitions, but the main ones of interest to us are that: "a meteor brighter than absolute visual magnitude (distance of 100 km) -4 is also termed a **bolide** or a **fireball** (Figure 1.1). A meteor brighter than absolute visual magnitude -17 is also called a **superbolide**".[1] Additionally, "a meteoroid in the atmosphere becomes a meteorite after the

---

[1]The magnitude scale is based upon the logarithm value of the measured energy flux in the visible part of the electromagnetic spectrum. That is, if the light emitted between the red and blue wavelengths of light produces f joules worth of energy per square meter per second at the location of a detector (the eye, or more usefully in terms of recording a measurement a CCD camera) then the apparent magnitude m is defined as: $m = -2.5 \log f + K$, where K is a calibration constant. The absolute magnitude M is defined in terms of the flux that would be measured at some fixed distance away from the meteor (generally taken to be 100 km). In this manner, $M = m + 5 \log(100/d)$, where d is the actual distance to the meteor (measured in kilometers) at the time of the flux measurement. Given the negative sign in the definition, the brighter an object is, so the more negative is its associated apparent magnitude. The apparent magnitude of Sirius, the brightest star, is -1.46; the apparent magnitude of planet Venus (from Earth) is -4.14 at its maximum brightness; and the Moon, at full phase, has an apparent magnitude of -12.5. Accordingly, a fireball or bolide is a meteor that appears to be brighter than the planet Venus and a superbolide is a meteor that appears brighter than the full moon.

Figure 1.1. A bright, fragmenting fireball observed over San Francisco Bay Area on October 17th, 2012. Image courtesy of NASA/Robert P. Moreno Jr.

ablation stops and the object continues on **dark flight** to the ground. A meteorite smaller than 1 millimeter can be called a **micrometeorite**. Micrometeorites do not have the typical structure of a fresh meteorite — unaffected interior and fusion crust. Foreign objects on the surfaces of atmosphere-less bodies are not called meteorites (i.e. there is no meteorite without a meteor). They can be called **impact debris**."

These definitions and qualifications are all based upon observable phenomena and generally measurable characteristics, but they don't tell us much about meteorites — other than they are objects that have arrived externally to be found on the surface of a planet that supports an atmosphere (effectively meaning Mars, Earth, and Venus, and one moon — Titan). So, we can certainly talk of meteorites on the surface of Mars (see Section 7.11), and, of course, we can talk about meteorites on the Earth (Venus presents some additional problems for meteorite survival — see Chapter 6). By passing through and interacting with the gasses in an atmosphere, the solid body of a meteoroid produces a meteor — the light phenomenon associated with atmospheric interaction — and finally, if the conditions are right, the meteoroid once reaching a dark flight phase (that is, a phase in which mass loss via ablation is no longer active) can deliver a meteorite to the ground — and this meteorite can (at least in principle) be collected and taken to a laboratory for analysis. There is no reference in the definition for a meteorite that pins it to a specific origin — it need only be a natural solid object. In this manner, meteorites, in principle, could be derived from any of the terrestrial planets, cometary nuclei, asteroids, and even Kuiper Belt Objects and Jovian planet moons — although, as ever, there are qualifications that apply to all of these potential parent bodies.

Definitions, of course, only get us so far, and in this book a parochial, Earth-O-centric, view will largely be adopted, and it is the arrival of extraterrestrial material to the Earth's surface that is of primary interest. Even with all of these modern-day definitions in place, the phenomenon of a meteorite fall is as dramatic in the present era as it was to our distant ancestors many thousands of years ago. For indeed, meteorites have been observed to fall from the sky for as long as there have been eyes around to see them — indeed, meteorites have landed on Earth since it first took on its near final form some 4.5 billion years ago. To see a fireball or superbolide (Figure 1.2) is a life-defining event — there is literally nothing similar to such a phenomenon in everyday life, and eyewitnesses are usually left in a state of shock and disbelief upon seeing such a brilliant meteor move through the sky. Such witnesses, and the vast majority of people on Earth, at any one epoch, never get to see such an event in their entire lifetime. Very, very few people get to actually witness a meteorite (at the end of its dark flight stage) hit the ground. Our hapless and frightened witness, Thomas Marriot, in Launton, Oxfordshire, throwing himself to the ground (as described in the introduction) was one such rare and lucky individual that witnessed an actual fall. Experiencing such an event is perturbing to the psyche, and one's inner-world compass struggles with the thought that stones can actually fall out of the sky. The conviction, as future sections will explore, that stones simply cannot fall from the sky, without having been first launched by a great explosion or some volcanic eruption, historically drove investigations of meteorite dropping events in entirely the wrong direction. In spite of thousands of eyewitness reports and trustworthy plebeian accounts, the scientific acceptance of meteorites, as being a true astronomical phenomenon, is of remarkably recent genesis (see Chapter 2). Ancient or modern,

Figure 1.2. The meteor train associated with the Chelyabinsk superbolide. This image was captured about 200 kilometers from the meteor's path, and about one minute after the parent meteoroid entered Earth's atmosphere. Image courtesy of Alex Alishevskikh and NASA.

however, the startled witnesses of a fireball and the thundering fall of a meteorite, stumble for the right words to describe what they have seen. There is typically no parallel upon which to draw a simile.

Eyewitness accounts can yield approximate atmospheric flight data, but to truly get to grips with the typical numbers relating to fireball characteristics and meteoroid ablation, some form of calibrated instrumentation system is required. This might be a system of widely spaced cameras, video recorders, or a radar transmitter and receiver network. A number of photographic surveys have been conducted since the mid-20th century, including the Prairie Meteorite Network in the United States (which operated from 1964 to 1975), the Meteorite Observation and Recovery Project (MORP) in Canada (which operated from 1971 to 1985), and the European Fireball Network, which has operated continuously since 1959. Canadian astronomers Ian Halliday, Alan Blackwell, and Arthur Griffin, of

Table 1.1. The characteristic properties of meteorite-producing fireballs. Data from Halliday *et al.*, *Meteoritics*, **24**, 65–72 (1989).

Quantity	Median Value
Duration	4.2 seconds
Zenith angle	51.0°
Beginning height	72.0 km
End height	31.0 km
Initial speed	15.2 km/s
Terminal speed	8.2 km/s
Maximum brightness	Mag. −9.2
Height of maximum brightness	47.5 km

the Hertzsberg Institute of Astrophysics in Ottawa, analyzed the characteristics of fireballs photographed with the MORP cameras and derived the characteristics of a typical meteorite-producing fireball, with their results being shown in Table 1.1. The meaning behind most of the quantities listed in Table 1.1 are self-evident, but their relevance and/or definitions will be explored in further chapters.

While it is truly remarkable that, if one is very lucky, it is possible to see a meteorite fall to the ground, it is equally as remarkable that one can also stumble across a meteorite at random. Meteorites can and do fall to the Earth all the time (see Chapter 18), and no location is excluded. Stand still long enough and in principle you will eventually be hit by a meteorite. Admittedly, you will have to stand still for a very, very long time, but something like 20,000 meteorites, with a mass larger than 0.1 kg (the mass of a 4-cm sphere of rock), fall on the Earth every year — that's about 54 such meteorites every day.[2] Another way of looking at such numbers is to imagine that you

---

[2]Writing in the journal *Nature* (**318**, 317, 1985) Ian Halliday (Herzberg Institute of Astrophysics, Ottawa) and coworkers estimated that an annual hit rate of 0.0055 impacts on people per year for the population of the United States and Canada. That's about one person being hit every 180 years. Extrapolating this to a global scale suggests that the human–meteorite encounter rate is about one hit every nine years. Given the record of meteorite hits and near misses (see Chapter 18), this number seems to be on the high side — hits being, in fact, much rarer. Of course, the analysis of human hits is strongly influenced by how long people

are a farmer who owns and regularly patrols a sector of land with a total area of $10\,km^2$. The expected number of 0.1-kg meteorites to fall on your patch of land is accordingly, $10/(510 \times 10^6) \times 20,000 = 0.0004$ per year (the large divisor in this calculation is the surface area of the Earth). So, in your lifetime of say 100 years, there is about a 1 in 25 chance that a meteorite will actually fall in your own backyard. If, however, the land you own has been accumulating meteorites since the end of the last glaciation, which ended some 10,000 years ago, then your patch of land will already have been hit by 4 meteorites. On average then, your patch of land will contain 1 meteorite per 2.5-square-kilometer sector. Now, if you are prepared to walk $10\,km$ a day, and scanning 1 meter either side of your path as you systematically survey your sector, then within 125 days the whole sector will have been scanned, and you should have found one 0.1-kg meteorite. Over a lifetime, therefore, of intermittently searching your land, the odds are now much more in your favor of finding at least a few meteorites. Sadly, finding meteorites is not quite so straightforward, but the exercise indicates that you are much more likely to find a meteorite, in a systematic search of a given area, than you are to see one fall in that same area. Key to finding a meteorite, however, is knowing what to look for, and accordingly, the next several sections will discuss the very basics of meteorite classification and identification.

1.2 Classification

Classification is all about finding the similarities and differences between objects that can be observed, measured, and quantified. The most basic division of meteorites is that which splits them into being either stony or iron. But such a division fails to take into account many distinctive properties of meteorites (that distinguish them from terrestrial stones and iron) and their thermal/chemical

---

actually spend outside of a protective building or structure on any given day. For some people this might be many hours, for others it could be just minutes. Additionally, the calculation assumes people are spread across the land evenly when, in fact, most people live in village, town, and city clusters.

Figure 1.3.   Chondrules in the NWA 5731 (LL3.2) chondrite meteorite. The largest chondrules visible are about 1 mm in size.

evolution. Instead, the basic meteorite classification scheme proceeds according to the division of chondritic or achondritic (that is, without chondrules). The chondritic meteorites are those stony meteorites that are predominantly composed of small spherical grains called chondrules (Figure 1.3) — the word chondrule being derived from the Greek for *seed*. The chondrules are characteristically composed of minerals such as olivine, pyroxene, and feldspar, and are typically between 0.1 and 1 mm in diameter. The chondrules are all held in place by a fine-grained matrix composed of broken pieces of chondrules, nickel-iron, and various mineral assemblages (such as troilite and apatite), all with characteristic sizes smaller than about 0.1 mm. Importantly, chondritic meteorites are composed of both low and high-temperature melting point components and are accordingly called undifferentiated — meaning that the whole assemblage has not been transformed by being appreciably heated. The other term for this is that chondritic meteorites are primitive, and that they represent the basic course material of the early solar system (see Chapter 5).

The chondrite meteorites are divided into three main classes: the ordinary chondrites, carbonaceous chondrites, and the enstatite chondrites. The ordinary chondrites are the most common form of meteorite, either as a fall or a find (see below) and they are further subdivided into the H, L, and LL groups. Here the H indicates a high total iron content, L indicates a low total iron content, and LL indicates a low total iron and low metal content. This latter division has evolved from what is known as Prior's rule (after George Prior, who was Keeper of Mineralogy at the Natural History Museum in London). This latter rule, introduced in 1916, indicates that, the smaller the amount of nickel-iron in a chondrite meteorite, the richer it is in nickel, and the richer in iron are the magnesium silicates. The basic point, however, is that differences in the amount of metal present is used to divide the ordinary chondrites into three groups, H, L, and LL. Furthermore, the enstatite meteorites have significantly more metal than the ordinary chondrites, and the carbonaceous chondrites have significantly less. In addition to the metal content, ordinary chondrites are additionally distinguished according to petrological type. This scale is based upon the observed characteristics of the chondrules, and upon the characteristics of their constituent mineralogy. The enstatite chondrites have more total iron content than the H-chondrites, but characteristically most of their iron is in the form of a metal or sulfide rather than an oxide — they tend to be rich in the mineral enstatite, which is the reason for their name. The silicates in enstatite meteorites are additionally the most oxygen-poor within the solar system. Carbonaceous chondrites have a dark, fine-grained matrix rich in organic compounds and hydrated minerals. They are the most primitive, that is, least thermally altered since formation, of all the meteorites, but they do indicate a mineralogy consistent with aqueous alteration.

Achondrites, as the label indicates, contain no chondrites and they are differentiated. That is, the achondrites are derived from more primitive, primary material that has been chemically and thermally altered. This class of meteorite has many subgroups, including the iron meteorites, the stony-iron pallasites, and magmatic groups that have separated out from a liquid melt (see Figure 1.4). The angrites largely consist of the mineral augite and are among the oldest igneous rocks known. The aubrite achondrite meteorites are largely composed of iron-free magnesium pyroxene (enstatite),

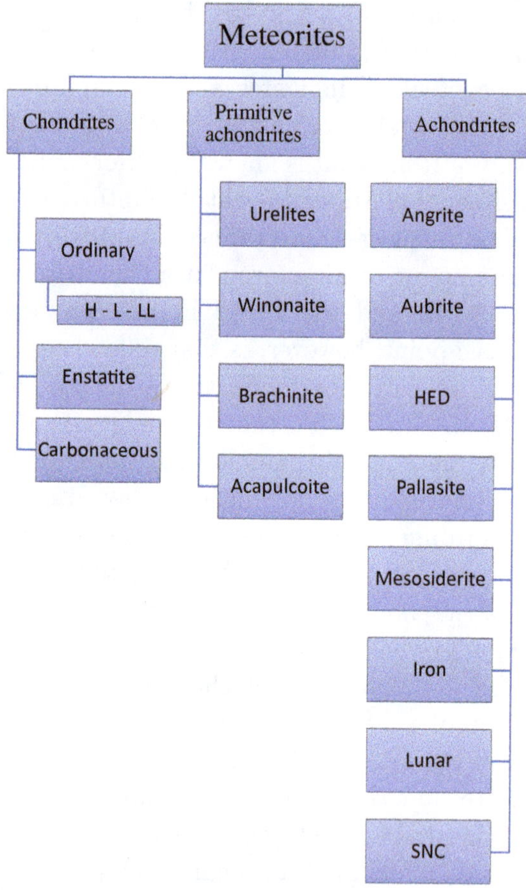

Figure 1.4.   A basic meteorite classification scheme showing the major meteorite classes.

and they formed by cooling from a magmatic melt under highly reducing conditions. The HED meteorites are the most abundant, in terms of collected numbers, of all the achondrites, and they formed from basaltic magmas (see Chapter 13). The SNC meteorites are basalts and are derived from planet Mars (see Chapter 16); the lunar achondrite meteorites, as the name implies, are derived from the Moon. The pallasites are distinguished as being iron meteorites containing large cavities filled with olivine crystals (Figure 1.5). The mesosiderites are also composed of silicates and metals, but their

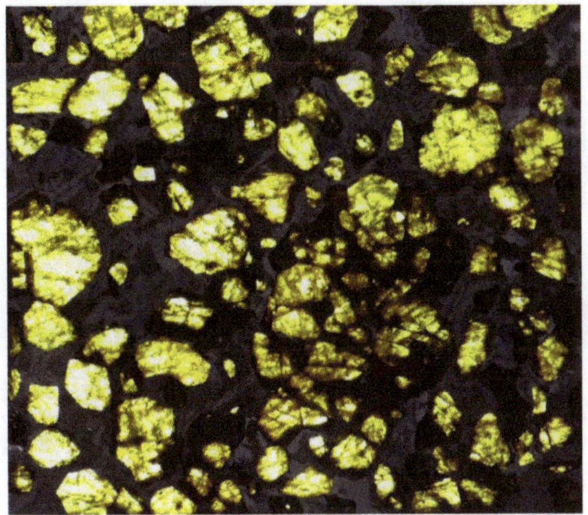

Figure 1.5. Backlit image of a thin slice through the Fukang stony-iron (pallasite) meteorite. The yellow olivine crystals are set within a nickel-iron matrix. Image courtesy of Natural History Museum, Vienna.

texture is much finer grained and more irregular than that exhibited by pallasites.

The iron meteorites are differentiated meteorites consisting almost entirely of nickel-iron. Many subgroups of iron meteorites are recognized based upon their physical structure and the variation of nickel and trace elements of iridium, gallium and germanium. One of the unique and defining features displayed by iron meteorites is the Widmanstätten pattern (Figure 1.6). First described by William Thomson in 1804, the pattern is more generally attributed to Count Aloys de Widmanstätten who noticed it in 1808 after experimenting on the Hraschina meteorite.[3] Generally, however, following Thomson, who worked on the Krasnorjarsk meteorite (this is the original pallasite found by Peter Pallas in 1772), the Widmanstätten pattern is revealed by polishing and then acid etching the cut surface of an iron meteorite. The pattern appears as a criss-cross of lines of kamacite and taenite — the two crystallization forms of nickel-iron.

---

[3]This, in fact, was the first witnessed fall of an iron meteorite — it fell in Croatia on May 26th, 1751.

Figure 1.6. Widmanstätten lines in the Gibeon IVA iron meteorite. Image courtesy of Wikipedia Commons.

This is a very specific pattern that comes about due to the diffusion of nickel atoms under very slow cooling conditions of just a few degrees per million years. The classification scheme for irons has been described (more than once) as being confusing but serviceable, and in all it recognized some 13 different groups. Four groups of irons were initially distinguished according to trace elements of gallium and germanium, and these were identified by a Roman numeral I to IV. More refined analysis, using other trace elements, split the original 4 groups into 12 new ones, and these are distinguished according to an additional capital letter that runs from A to F. The 13th iron classification group is that of "anomalous" (the catch-all bin for all those irons not fitting neatly into the other 12 groups).

The primitive achondrites have characteristics that fall between those of the chondrites and the achondrites. They show clear signs of thermal metamorphism, partial melting, and recrystallization, but have a bulk composition similar to that of chondrites, and they also contain relic chondrules. The winonaites and acapulcoites, for example, show clear evidence of partial melting and contain relic chondrules, but have compositions that fall between the H and enstatite chondrites. The brachintes are odd, however, in the sense of being highly olivine rich in comparison to ordinary chondrites. The urelites have a seemingly unique composition and are distinguished by their high carbon content, and by containing micron-sized diamonds. The diamonds were probably produced by collision induced shock waves that propagated through the urelite parent body. The texture of these meteorites reflects that expected of silicate grains that have accumulated on the floor of a magma chamber.

There are many, many more intricacies to meteorite classification than those indicated above — but this is not the book to discuss, nor describe, the chemical and petrographic machinations. Suffice to say, there are numerous meteorite groups that can be distinguished within the chondrite and achondrite divisions. Of interest, here, however, is the identification of source regions — where did the various types of meteorite form, and what are the potential connections between the various types? That meteorites are indeed derived from different parent bodies, and are not terrestrial in origin, is betrayed by the measurement of their oxygen isotope ratios.

1.2.1 *Isotopic anomalies*

After hydrogen and helium, the most abundant element in the Sun and the solar system is oxygen, and importantly, oxygen can be found in three isotopic forms: ^{16}O, ^{17}O and ^{18}O. Study of terrestrial rocks indicates that the ratios of $^{17}O/^{16}O$ and $^{18}O/^{16}O$ vary in a systematic manner. Indeed, there are a number of mass-dependent physical, as well as chemical processes, that can drive changes in the ^{17}O and ^{18}O abundances relative to ^{16}O — evaporation, for example, will preferentially drive away ^{17}O because it is lighter, and concentrate ^{18}O because it is heavier. For terrestrial and Moon rocks, the fractionation processes have played out in such a manner that all material plots on a line of slope $1/2$ in the $^{17}O/^{16}O$ versus $^{18}O/^{16}O$ diagram[4] (Figure 1.7).

---

[4]The fractionation diagram is constructed according to the comparative ratios of $^{17}O/^{16}O$ and $^{18}O/^{16}O$ to values derived from ocean water (standard mean ocean water, SMOW) giving a relative value δ in parts per thousand (‰).

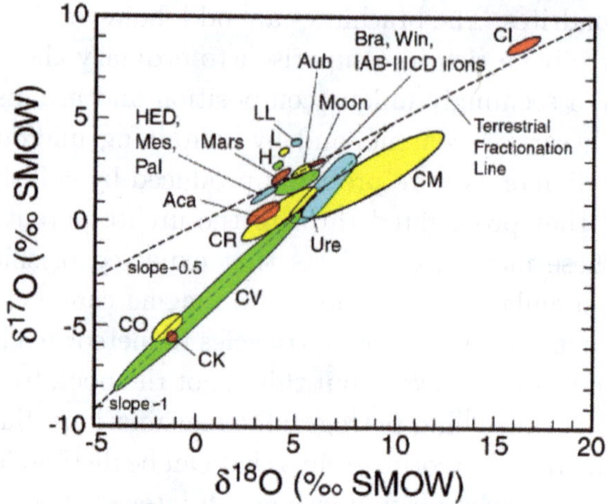

Figure 1.7. Isotope ratios of oxygen for different meteorite classes. Key: Aca = acapulcoite, Aub = aubrite, Bra = brachinite, Mes = mesosiderite, Pal = pallasite, Ure = urelites, Win = winonaite. The carbonaceous chondrites are distinguished by a C followed by a classification letter. H, L, and LL indicate the location of the ordinary chondrites. HED corresponds to the howardite, eucrite and diogenite meteorites (see Chapter 13). The Moon and Mars labels indicate the lunar and SNC meteorites (see Chapter 16).

The first oxygen isotope values for meteorites were measured in the mid-1960s, by research groups located at the California Institute of Technology and the University of Chicago, and it was immediately clear that most meteorite classes do not plot upon the terrestrial fractionation line. While meteorite groups such as the enstatite chondrites and the aubrite achondrites straddle the terrestrial fractionation line, the H, L and LL chondrites plot in regions above the line, with the primitive achondrites and iron achondrites plotting below. The isotopic values for the refractory inclusions and chondrules in carbonaceous chondrites (especially the CVs) show the greatest deviation with respect to the terrestrial fractionation line — indeed they fall on a line with a slope of order 1 rather than $1/2$. The reasons for this variance are not universally agreed upon, but one theory argues that the material incorporated into the carbonaceous chondrites formed in a region of the solar nebula that was enhanced in ^{16}O — this extra ^{16}O being derived from a supernova explosion that, by chance, was located close to the young solar nebula. The fact that different meteorite classes plot in distinct regions, however, indicates that they must have formed in different, unmixed regions (that is, in specific reservoirs with their own distinct initial isotope ratios) of the solar nebula. In addition, the differences between the oxygen isotope values, for distinct meteorite classes, can be used to estimate the number of parent bodies that are represented within the known meteorite record.

A recent study of high-precision oxygen isotopic vales by Richard Greenwood (The Open University, UK) and coworkers found that somewhere between 120 and 132 asteroids are implicated as the parent bodies to all known meteorite types. Of these about 60 are responsible for the iron meteorites, 35 to 40 for the achondrites and 25 to 32 for the chondrites. Of the achondrites, the pallasites appear to be derived from 6 distinct parent bodies, while the primitive achondrites are derived from just 2 parent bodies. Likewise, the aubrites appear to be derived from 2 parent bodies. The H, L and LL chondrites appear to be derived from 3 to 5 parent bodies, with the carbonaceous chondrites being derived from some 10 to 15 parent bodies. With the exception of the HED meteorites, which have been linked to one specific asteroid (asteroid 4 Vesta — see Chapter 10), it has not been possible to identify, within the main-belt asteroid region, the specific parent bodies to the various meteorite classes. Given the relatively small number of parent bodies required to account for the known meteorite classes, it is clear that the meteorites that have been collected and sampled on Earth are but a small fraction of the meteorite classes that actually exist. Indeed, some researchers have suggested that there should be a continuum of chondrite meteorite classes (rather than the distinct groups seen in the oxygen fractionation diagram), but the dynamics of meteorite delivery are such that we only get to sample a small subset of the available total on Earth (see Chapter 3.1).

1.2.2 *The asteroid connection*

That there are different types of meteorite, from the chondrites to the achondrites, informs us that at least some of the primordial asteroids must have undergone some considerable amount of internal

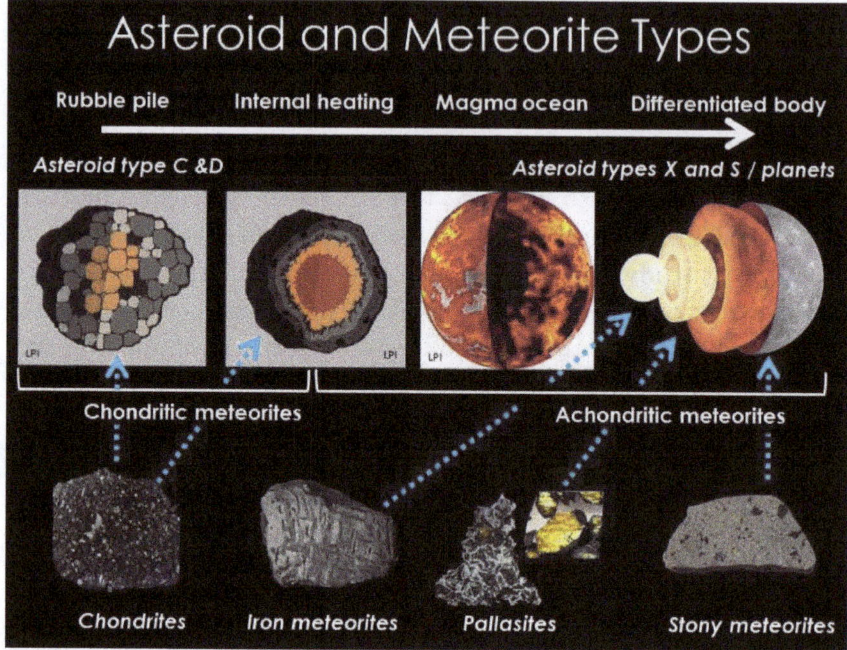

Figure 1.8.   The association of various meteorite types with different asteroid formation/source locations. The arrow at the top of the diagram (and image insets) indicates the structure and metamorphism variation from a primordial, rubble pile asteroid through to a fully differentiated asteroid. Asteroid structure and asteroid types are described in Chapter 12, and the internal heating mechanism is discussed in Chapter 13. Image courtesy of https://ukantarcticmeteorites.com/meteorite-science/.

thermal alteration. The key mechanism here is differentiation which is a process that allows for the heat metamorphism and physical migration of material in an initially homogeneous body (Figure 1.8). The least heat-altered carbonaceous chondrite meteorites are taken to be derived from primitive, rubble pile asteroids. These bodies are remnant planetesimals that have not undergone any significant internal alteration since they first formed (see Chapter 5). The achondrites and ordinary chondrites, in contrast, are derived from different regions of asteroids that have undergone internal differentiation, and which have then been fragmented and dispersed by collisions with other asteroids. The internal differentiation results in the formation of a core–mantle–crust layering, with the higher density material residing in the core. In this manner, some of the iron meteorites are thought to be derived from the iron-rich cores of differentiated asteroids. The pallasites are thought to be derived from the material that solidifies at the core–mantle boundary, while the stony achondrites are derived from the mantle region. The least heat-processed ordinary (stony) chondrites are derived from the outermost crustal layers.

1.2.3 *Recognition: Is this one?*

If past experience, over the last 35 years, has taught the author anything about meteorites, it is that whenever anyone says that they have found a meteorite, they are most probably wrong. Experience gained during the Prairie Meteorite Search,[5] for example, indicates that approximately 1 strange or odd-looking rock in about 400 being touted as a meteorite, was actually a meteorite. The odds are indeed very much against the casual finding of a meteorite, but fortunately it does happen, and there are some basic meteorite/non-meteorite

---

[5]Instigated by A. Hildebrand (University of Calgary) in 2000, and supported in later years by M. Beech (Campion College at the University of Regina) and P. Brown (Western University), the Prairie Meteorite Search (PMS) ran until 2007. The project was based upon the proven meteorite-finding principle initiated by Harvey Nininger in the 1930s, and that principle is to talk and be available to those people who are most likely going to find meteorites. In this manner, farmers and land-owners, who are out in their fields day after day, and often generation after generation, are by far more likely to discover a meteorite than someone who lives in a city or a person out for a Sunday afternoon stroll. And, the statistics of finds in Canada bear this principle out. The PMS proceeded by hiring a summer student who was then tasked to travel across the Prairie provinces of Alberta, Saskatchewan, and Manitoba. Their modus operandi was to make contact with rural communities, using local radio and newspaper interviews, and set up meeting dates and locations were people could bring in their strange and odd rocks. In the case of the Prairie provinces the land surface was effectively scoured clean during the last ice age, but has since accumulated meteorite falls for the past 10,000 years. The land has additionally undergone extensive agricultural development during the past century, and it is this factor that makes the prairies an especially rich area for meteorite finds.

characteristics, depicted below, that can be looked for, before seeking further advice about a potential find.

(i) Not all meteorites are magnetic, but the majority typically will be and accordingly they will influence a magnetic compass or attract a small magnet. If your find is uniformly magnetic, over its entire surface, then it could be a meteorite. Most terrestrial rocks, but again not all, will fail this test.

(ii) If you can see the interior, is it, in the case of a stony rock, composed of chondrules, sub-millimeter-sized grains, intermixed with metallic veins and/or blebs? Technically, no Earth rock has the same chondritic make-up as a meteorite, but unless you really know what you are looking for this hardly helps. However, a grainy, grayish colored interior is a sign of a rock possibly being a meteorite. If the interior has a uniformly smooth texture, showing no constituent grains, then it is not a meteorite. If you can see veins of quartz crystals, then it is not a meteorite.

(iii) Is the outer surface of the rock composed of a thin, dark brown/blackish layer no more than 1 millimeter in thickness? If so, then this could be a fusion crust. This thin outer layer is the cooled-off molten surface produced by ablation as the meteoroid descends through the atmosphere (see Chapter 6). Many types of weathered terrestrial rocks can develop an outer rind, and this can be easily confused with a fusion crust. The fusion crust is actually a metal-rich glass and it is easily chipped from the surface, but a weathered rind is typically more difficult to remove.

(iv) A weathered meteorite might have no visible fusion crust but have a distinct surface rusting. This, however, is something that many terrestrial rocks can also show, and it is accordingly not a definitive trait of a rock being a meteorite.

(v) Are there thumbprint-like indentations on the rocks surface, or distinctive molten flow lines? These dimples, called regmaglypts, and flow lines are a product of the ablation process and are unique in formation to meteorites (Figure 1.9). Many terrestrial rocks can mimic these features, however.

Figure 1.9.   One of the many Pultusk (Poland) chondrite meteorites that fell on January 30th, 1868. Note the thin, dark black/brown fusion crust, the regmaglypt *dimples*, and the oxidization (rusting) evident on the fractured (front right) face. Image courtesy of the Geological Museum of the Institute of Geological Sciences, Polish Academy of Sciences.

(vi) Meteorites are not heated throughout their interiors, nor are they completely molten as they descend through the atmosphere, so any rock that has a bubbly appearance, or has clearly been completely melted, or is full of small holes (technically called vesicles), it is not a meteorite. One of the most common, so-called, meteor-wrongs is that of industrial slag. This material is typically light in weight, full of holes, and may contain isolated magnetic blebs and glass like inclusions. Sometimes, however, slag can be uniformly black and smooth. In general, if you find an odd-looking rock or iron mass close to railway lines, or an industrial site, or in an old quarry, it is highly unlikely to be a meteorite.

Just because a rock looks odd, or wasn't in the driveway the other day, or has an unusual color, doesn't mean it has to be a meteorite. Likewise, just because you have found a lump of rusted iron in the corner of a field, or in a forest stand does not mean that it has to be a meteorite. Indeed, the definitive test for the identification of an iron meteorite is a chemical one that determines its nickel content. Iron meteorites do technically have a fusion crust, although it is generally not at all obvious, they will be strongly magnetic over their entire surface, and they can show regmaglypts and material flow lines. Key to identifying meteorites is experience — so educate yourself by looking at and handling as many actual meteorites as you can. The internet is not always going to be helpful in identifying a rock — a picture paints a thousand words, and it can implant many a misguided diagnosis; multiple factors, not just external look, are required to make a correct diagnosis. The determination of what specific type of meteorite that you might have found is something that will have to be performed by a professional researcher in a laboratory setting. In the modern era there are many institutions and/or dealers that will test and perform a preliminary analysis on any finds — usually for free. A recommended first step towards the identification of any find, however, is to consult the webpages hosted by Washington University in St. Louis: https://sites.wustl.edu/meteoritesite/identification/.

1.3   Finds, falls and strewn fields

The heritage of a meteorite is often based upon whether it is a fall or a find. These two terms are essentially self explanatory, with a meteorite being a fall if it has a known date of arrival on Earth and can be definitively linked to some specific fireball event. A find, in contrast, is a meteorite that was found by chance, and has no known fall date. Falls tend to be scientifically more interesting since they have a distinct provenance with respect to fall location, fall time, specific fireball, and potentially from the fireball observations a specific orbit, which in turn can be linked to a specific region of the asteroid belt and then to a potential parental asteroid type. These links and associations will be explored in the following chapters. Finds are additionally important, and even if their origins and arrival times are unknown, they delineate and define the compositional range of meteorite types. It is also the case that many meteorite finds have wonderfully complicated stories to accompany their discovery[6] (see Figure 1.10 and also Section 7.1).

The Catalogue of Meteorites, compiled and edited by Monica Grady [2], provides a detailed account of known meteorite falls and finds from the earliest records through to the end of 1999, and it accordingly allows for an analysis of fall against find statistics. Table 1.2 is a breakdown of the number of falls and finds, as derived from the *Catalogue*, in terms of stony, stony-iron, and iron types. In each case, falls and finds, the number of stony meteorites is considerably larger than the number of iron meteorites, and the number of irons is considerable larger than the number of stony-iron meteorites. In terms of meteorite falls, 94% have been determined to be stony, with iron meteorites accounting for another 5%. This distribution changes but little with respect to finds, with stony meteorites being slightly more common at 96%, with irons being

---

[6]The Willamette meteorite story provides a classic example of past cultural insensitivity, individual enterprise, and corporate greed. Known to the Clackamas Chinook peoples of Oregon as *Tomanowas* (meaning *visitor from the sky*), the meteorite has been venerated and endowed with magical properties since prehistory. Indeed, the iron is believed to be a glacial erratic, transported, during the last ice age, from its fall location in southern Alberta, Canada, by ice rafting driven by the catastrophic flooding associated with the collapse of ancient Lake Missoula. The Clackamas peoples were forcibly removed from their ancestral lands in 1855 but knowledge of *Tomanowas* remained within the tribal memory. In 1902, homesteader Elis Hughes rediscovered the meteorite and recognized its value. Knowing, however, the meteorite was not on his land, he set about moving it the several kilometers to his home (recall Figure 1.10 — Hughes is seen to the left and behind the meteorite). This herculean task took many months, but once completed, Hughes built a shed around the iron and then charged 25 cents to view it. Realizing that the meteorite had been moved from their land, the Oregon Iron and Steel Company (OISC) filed a lawsuit against Hughes in 1905, and they were successfully awarded ownership. In 1906 businessman and politician William Dodge Sr. bought the meteorite from the OISC for $26,000. Shortly thereafter, Dodge donated the meteorite to the American Museum of Natural History in New York, where it has been on display ever since. In 1999, the Confederated Tribes of the Grand Ronde Community of Oregon (CTGRC) successfully sued the museum for ownership and access under the Native American Graves Protection and Repatriation Act. A compromise deal was struck, however, with the meteorite remaining on display at the museum, but tribal members being allowed private access to the meteorite in order to perform their cultural ceremonies.

Figure 1.10.   The Willamette iron meteorite — weighing in at 14,150 kg this impressive mass is the largest meteorite find from North America.

Table 1.2.   Deduced numbers of falls and finds for stony, stony-iron and iron meteorites. Data from [2].

Class	Fall #	Fall %	Find #	Find %	~Find/Fall	Total
Stony	940	94	20574	96	22	21514
Stony-iron	12	1.2	104	0.5	9	116
Iron	48	4.8	817	3.5	17	865
Total	1000		21495			22495

slightly less common at just 3.5%. Whether falls or finds, stony-irons account for about 1% of the totals. For any given class, however, finds always outnumber falls by a factor of at least ten. This essentially means that if you find a meteorite, or chance to see one fall, the odds are such that it is more than likely that it will be a stony meteorite. Irrespective of being a fall or a find, the most common type of stony meteorite is L6, and this is closely followed by H5. The most common iron meteorites, either as a fall or a find, are IIIAB and IAB.

Most falls produce more than one meteorite on the ground, and this results in the formation of a strewn field. The shape of a strewn field will depend upon the fireball angle of incidence, with a vertical path producing a near circular distribution of meteorites, and a fireball on a very shallow angle to the horizon producing a long elliptical distribution of meteorites. Mass sorting of meteorites is observed to take place along the strewn field's major axis, with larger meteorites traveling further along the trajectory (down range) than smaller mass meteorites, which fall up range. This effect is explained according to the ballistic coefficient B, which is a measure of a body's ability to overcome air resistance in flight. The ballistic coefficient is defined in terms of the inverse of the negative deceleration (giving in units of kg/m^2), such that,

$$B = \frac{M}{\Gamma S} = \frac{2}{3}\frac{\rho}{\Gamma} L \qquad (1.1)$$

where M is the mass of the body, S is the cross-section area of the body, ρ is the density of the body and L its diameter (assuming the body is a sphere), and Γ is the drag coefficient (see Chapter 6). On the basis that all of the fragments have the same density, then the ballistic coefficient will vary simply according to physical size L, and, of course, larger objects will be more massive than smaller ones. The ballistic coefficient is such that the larger its value, so the smaller is the deceleration, and the further will the given body tend to travel. It is for this reason that the meteorites within a strewn field are mass sorted, with the larger (in size and mass) meteorites travelling further down range than the smaller ones.

The strewn field need not always be orientated along the direction of the fireball flight, since the dynamics of smaller mass meteoroids can be strongly influenced by atmospheric winds. Depending on atmospheric wind strength and direction, meteorites can land many kilometers away from the extrapolated ground track of the parent fireball (see Section 1.4 below and specifically Figure 1.19). Figure 1.11 shows the strewn field deduced for the 1868 Pultusk

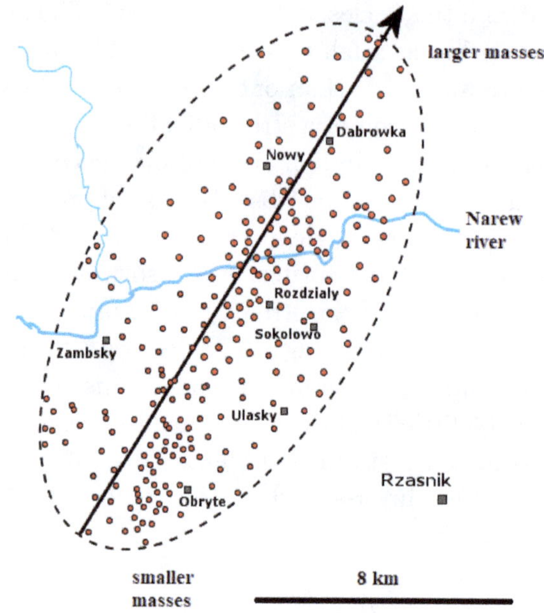

Figure 1.11. The Pultusk meteorite strewn field. The arrow marks the direction of flight, with larger-mass meteorites being found down range and smaller-mass meteorites up range. The town of Pultusk is approximately 10 km to the west of Obryte. North is to the top in the diagram. Image adapted from Wikipedia Commons.

meteorite fall. In this case the strewn field corresponds to an elongated ellipse about 18-km long and 9-km wide, giving a strewn field area of 127 square kilometers. The strewn field for the Homestead meteorite is shown in Figure 1.12, and in this case the ellipse is approximately 12-km long and 5-km wide, giving an area of about 47 square kilometers.

1.4   The fall of the Buzzard Coulee meteorite

The extremely bright fireball that presaged the arrival of the Buzzard Coulee meteorite first became luminous at UT 00:26:40, November 21st, 2008. Lasting for about 6 seconds the fireball was probably observed by many thousands of eyewitnesses in the Canadian Prairie Provinces of Alberta, Saskatchewan, and Manitoba. The appearance

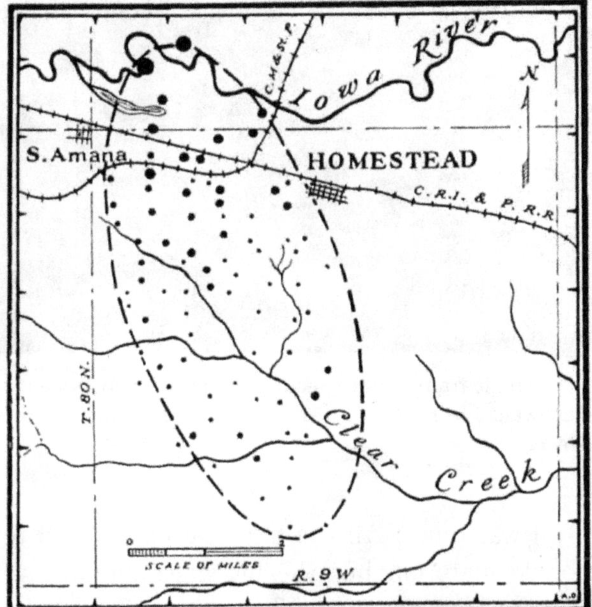

Figure 1.12.   The strewn field for the Homestead (L5) meteorite that fell on February 12th, 1875. Close on 100 meteorites have been found in this strewn field which has an estimated area of 47 square kilometers. The direction of flight was from south to north (bottom to top of image), with the larger-mass meteorites (large dots) falling down range of the smaller mass meteorites (small dots). The largest (main) mass corresponds to a 34-kg fragment, with approximately 230 kg of meteoritic material being recovered in total. The scale bar at bottom left in the image corresponds to a distance of 2 miles (3.3 km). Image adapted from Wikipedia Commons.

of the fireball (Figure 1.13) generated almost unprecedented media attention, and several hundreds of eyewitnesses submitted reports on what they saw and heard to the Canadian Fireball Reporting website. The story and discovery that unfolded in the wake of these fireball reports is fairly typical of any meteorite fall — the investigators having to deal with a mass of initial data, some reports being confusing and contradictory, out of which a coherent picture has to be formed before a reasonable ground search can begin. The following is an overview of how the events concerning the Buzzard Coulee meteorite fall unfolded.

Figure 1.13.   A single frame from a security camera video sequence captured at Biggar, Saskatchewan. The range to the fireball is of order 160 km. Image courtesy of R and D. Meger.

Initially, all was quiet. All the information pointed towards the fall of meteoritic material, but the big question was, where is it? One week after the passage of the fireball, the first meteorites were successfully found, after an extensive search and interview campaign, by Dr. Alan Hildebrand (University of Calgary, Alberta) and (then) Masters student Ellen Milley on a frozen pond in Buzzard Coulee, located about 40 km southeast of Lloydminster, in Saskatchewan [3]. Subsequent systematic ground searches, conducted over several years, have resulted in the recovery of more than two thousand meteorite fragments from the Buzzard Coulee strewn field, with the masses found ranging from just a few grams (see Figure 1.14 inset) to an impressive 13.1 kg.

Infrasound detection (see Chapter 21) data from instruments located in Manitoba, Ontario, Greenland, Washington, and Utah provided an equivalent energy estimate for the fireball of 0.32 ± 0.09 kilotons of TNT, suggestive of an initial entry mass of order 10 to 15 tonnes. Figure 1.14 shows the light variations associated with the fireball as recorded at Campion College, at the University of Regina (at a range of some 520 km). Rapid, short-duration light variations clearly indicate multiple fragmentation events occurring after the attainment of peak brightness, and a large terminal fragmentation event is apparent about 0.8 seconds after the time of peak brightness.

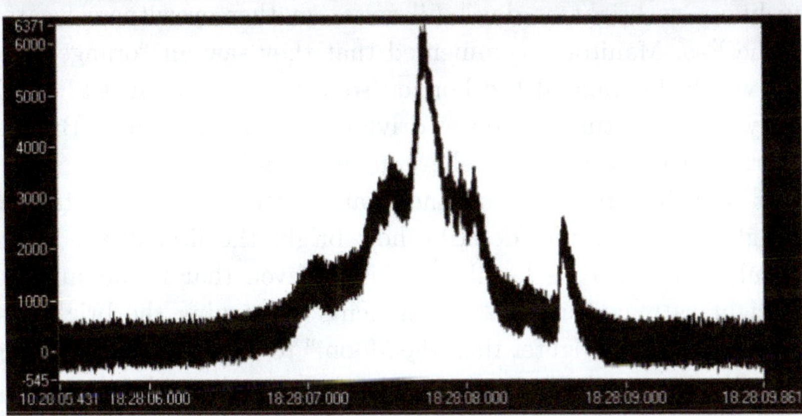

Figure 1.14.   A montage sequence of images related to the fall of the Buzzard Coulee meteorite on November 20th, 2008. The lower diagram shows the light curve (brightness variation as a function of time) as recorded at a range of some 400 km. Distinctive flares are seen in the light curve and these correspond to individual fragmentation events. The main upper diagram shows an image of the fireball as recorded by an observer in Edmonton at a range of 200 km. The upper right inset shows an enhanced image from towards the end of the fireball phase, and a number of distinct fragments can be seen. The lower left inset shows the author's finger and thumb holding up a small meteorite fragment found in the Buzzard Coulee strewn field.

To the radiometer's detection threshold of approximately magnitude −4, the event lasted for a total of 2.5 seconds. Since the fireball occurred very low on the horizon from Regina a significant brightness correction is indicated. The standard correction tables indicate that for a zenith angle of order 90 degrees a correction of some 9 magnitudes is required. This sizeable correction places the peak brightness at about magnitude −20 — a magnitude comparable to that of the Sun (which has an apparent magnitude of −26.75).

That the Buzzard Coulee meteorite fireball was a truly spectacular sight is easily gleaned from the general comments provided by eyewitness reports: it "looked like a hole was burnt right through the ozone" wrote one observer from Edmonton, Alberta — giving a descriptive image, but a physically incorrect interpretation of the event. Another observer from Millet, Alberta, commented that it looked like a "giant white ball ... [which] disintegrated into black nothingness moments before it hit the Earth." From Dore Lake, Saskatchewan, an eyewitness commented that the "whole western sky lit up as bright as day." Likewise, another eyewitness located in The Pas, Manitoba, commented that they saw an "orange color trail which illuminated the horizon somewhat like a sunset." And, finally, one eyewitness, who was driving through the town of Brooks, Alberta, wrote "I couldn't believe my eyeballs."

It is readily apparent from the email reports that all eyewitnesses struggled to accurately describe how bright the fireball was. This difficulty, of course, is hardly surprising given that it was at least as bright as magnitude −20. Comments concerning the brightness varied from "way brighter than the Moon," to "some sort of lightning storm," "like sunrise," "it lit up the area nearly as bright as the Sun," "like a white flare trailing golden sparkles," and, like "a welder's spark ... which lit up the sky as if it were daytime." While many eyewitnesses were fortunate enough to catch the entire display, others experienced just a fleeting and peripheral glimpse of the fireball. Accordingly, one eyewitness from Edmonton described the event as "a large, quick, bright flash of light," while another eyewitness from Kindersley, Saskatchewan, simply wrote that he saw a "huge flash in the sky." The descriptions of the fireball's general

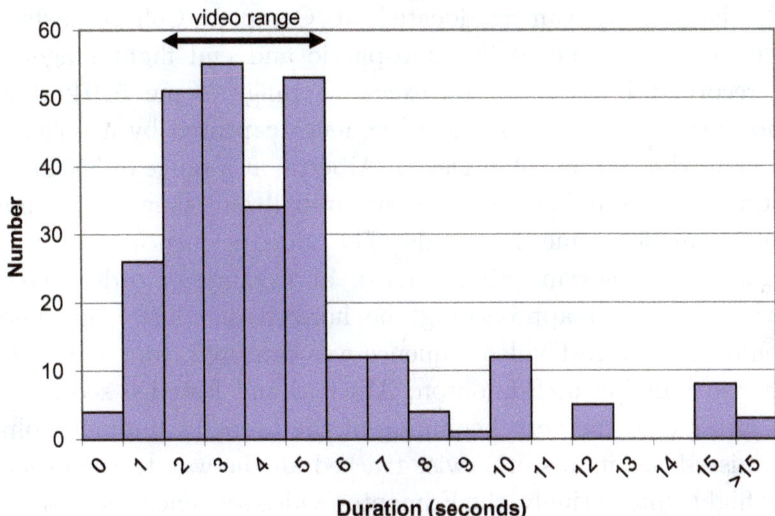

Figure 1.15. Eyewitness estimates of the fireball duration. The modal time estimate is 3 seconds. The arrowed horizontal line indicates the range in durations as revealed by video captures recorded at ranges varying between 150 and 500 km.

appearance echo those given for its brightness: an observer from Hardisty, Alberta, commented that "the fireball and tail appeared to consume most of the sky ... around the head of the fireball I saw various colors which included orange, red, and greens." Another eyewitness driving near Vermillion, Alberta, described the display as being, "like a million lights turned on at once." Eyewitness estimates of the fireball's duration vary, of course, according to the individual viewing circumstances. Figure 1.15 shows the range of duration estimates. Clearly most observers place the duration somewhere between 2 and 5 seconds (recall Table 1.1), although a good number of observers suggested times well in excess of 10 seconds; three observers even suggested that the fireball lasted for 30 seconds. While not articulated within their reports those observers indicating that the fireball duration was in excess of 10 seconds may well have been referring to the time over which they could see the fireball and its remnant train.

The various video captures of the fireball suggest a probable duration of 6 to perhaps 7 seconds. The video sequence captured with the all sky camera located at Campion College lasts for 2.5 seconds, but the early atmospheric and end flight stages are not recorded because of the excessive range (some 520 km) and atmospheric extinction. A video sequence captured by a police car dashcam while on patrol in Devon, Alberta, at a range of 350 km did capture the beginning phase of the fireball and shows a luminous trail lasting for some 4 seconds. The video sequence captured in Biggar, Saskatchewan (Figure 1.13), at a range of order 160 km, shows the fireball approaching the horizon and lasts for some 5 seconds. The longest video sequence was captured from a 10th floor high rise building in Edmonton, Alberta, and lasted 6 seconds. In this latter case the very beginning of the fireball's luminous phase was missed but it was otherwise tracked all the way to the onset of dark flight. Interestingly, the Edmonton video sequence also indicates that three, and possibly more, very large fragments survived the ravages of atmospheric flight (Figure 1.14). The largest Buzzard Coulee meteorite fragment found to date has a mass of 13.1 kg, and there is accordingly every reason to believe that several similar mass (if not more massive) meteorites have yet to be found in the strewn field.

In total 424 email reports directly relating to the observation of the Buzzard Coulee fireball were received at the Canadian Fireball Reporting website. The first eyewitness report arrived at the fireball reporting website within 15 minutes of the event having occurred, and over the ensuing 6 hours a further 84 reports were submitted. In the 24-hour interval from midnight November 20th to midnight November 21st (CST) an additional 242 eyewitness reports were sent in. Figure 1.16 shows the number of reports received during the time interval from November 20th (starting at 18:42 CST) to midnight November 30th (CST). A flurry of great initial activity is clearly present in the reporting distribution, with about one-third of all the reports being received within 24 hours of the event having taken place. A smattering of eyewitness reports continued to arrive up to a week after the event had occurred. The locations from which the various eyewitnesses saw the fireball are illustrated in Figure 1.17.

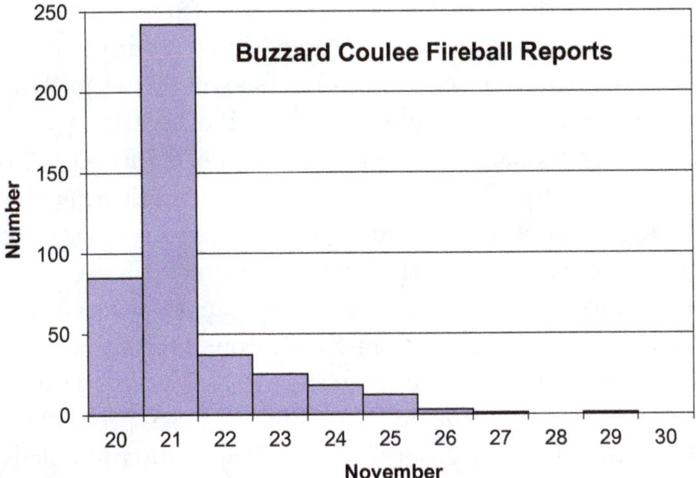

Figure 1.16. Number of eyewitness reports relating to the Buzzard Coulee meteorite fireball received per day from November 20th to midnight November 30th.

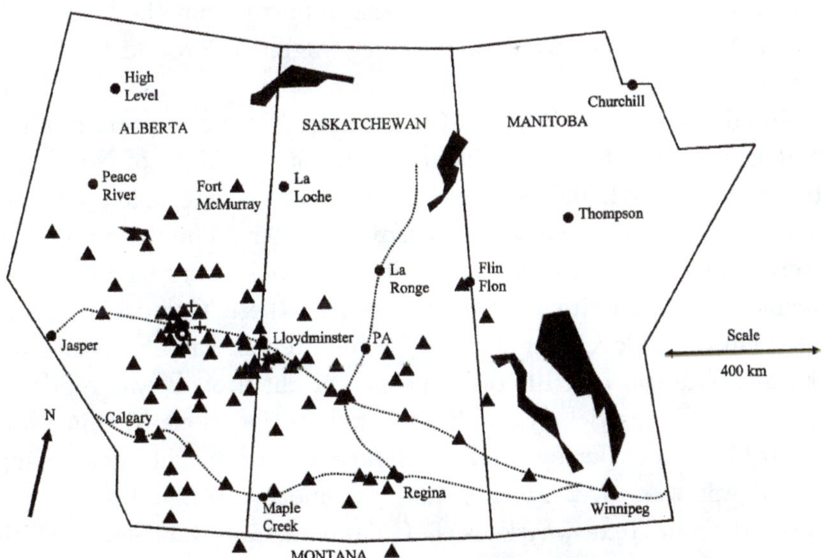

Figure 1.17. The spatial distribution of observers (triangles) who reported seeing the Buzzard Coulee meteorite fireball. Not every eyewitness location is shown since many were received from the same city or town — some 84 reports (about 20% of the total number), for example, were submitted from Edmonton, Alberta, alone.

Within Canada, the fireball was seen from as far away as Winnipeg, Manitoba (at a range of some 930 km), but the majority of eyewitness reports were received from residents living in Alberta — the fireball occurred there at 17:30 MST and accordingly was witnessed by many people traveling home from work. Indeed 172 reports (40.6% of the total) were received from people indicating that they were sitting within a motor vehicle. Five reports were received from pilots who witnessed the fireball while their aircraft were in flight; one report was received from a passenger in an aircraft that was in the process of taking off from Edmonton International Airport. Relatively few reports concerning the fireball were received from observers north of latitude 55°N — this is most probably due to a combination of cloud cover and the low population density in northern Alberta and Saskatchewan. Observations of the fireball were also reported from eyewitnesses in the northern United States. From Scobey, Montana, for example, at a range of order 600 km, one observer reported that they "saw a flash like lightning," while other observers near Kremlin, Montana, reported that the "main ball was accompanied by numerous smaller fragments around it" (recall Figure 1.14 inset).

In addition to the Buzzard Coulee fireball, email reports relating to at least eight other fireballs appearing on the night of November 20th were received. Indeed, this caused some considerable confusion to a number of eyewitnesses and media reporters. The various emails received indicate that at least two bright fireballs were seen from the Pacific Coast of British Columbia and northern California during the evening of November 20th. A bright fireball was also seen over southern Ontario and Illinois. Another bright fireball was reported from New Brunswick, Canada, as well as locations within New Hampshire, New Jersey, and Vermont. A very bright, detonating fireball was seen over Florida, and yet another bright fireball was reported from Texas, Arkansas, Oklahoma, and Louisiana. With respect to this latter event, an email report was received from a radio station in Little Rock, Arkansas, indicating that they had received some 10 calls from listeners concerned about bright flashes that had been seen in the sky. Email reports of other fireballs seen on the night of November 20th were also received from as far a field as the island of Bermuda and the Orkney Islands in Scotland.

In addition to providing a once-in-a-lifetime visual display, some 59 (14% of the total) eyewitness reports of the Buzzard Coulee fireball indicated the presence of associated sounds (see Chapter 21). This percentage is actually fairly typical with respect to similar such bright fireball events. Three eyewitnesses reported sensing strange smells; one observer reported that his dog was spooked and greatly agitated just before and after the passage of the fireball; while another observer reported feeling heat upon the side of his face. Indeed, it was this sensation that caused the latter eyewitness to turn his head and thereafter see the fireball. The locations from which sound phenomena were reported are shown in Figure 1.18.

The larger ellipse in the figure indicates that sound phenomena were widely reported at ranges up to several hundred kilometers from the fireball's ground path. Two observers in Avonlea, Saskatchewan (southwest of Regina), reported hearing a "staccato" sound from what would be a range of 500 km. Many observers in Edmonton (range 230 km) also reported hearing staccato, popping, and sharp sounds at the same time that the fireball cut across their eastern horizon. Only electrophonic (also called anomalous) sounds are likely to be heard at such distances and witnessed at the same time as the fireball is seen (see Section 7.17). What are reasonably interpreted as electrophonic sounds were reported by people who were driving as well as observers who were out hunting or simply walking in the street. Interestingly there are reports from five drivers who had their radios on that they saw the fireball but heard no sounds. Likewise, two eyewitnesses were having telephone conversations, four were at home watching television, and one was using a computer when the light from the fireball was observed — and yet they heard no sounds. These observations are consistent with the argument that electrophonic sounds are generated via the propagation of very long wavelength radiation derived from a narrow bandwidth region in the electromagnetic spectrum. The three eyewitness reports of sensing odd smells, the spooked dog, and the eyewitness who felt a heat sensation may also be related to the electrophonic phenomena, and

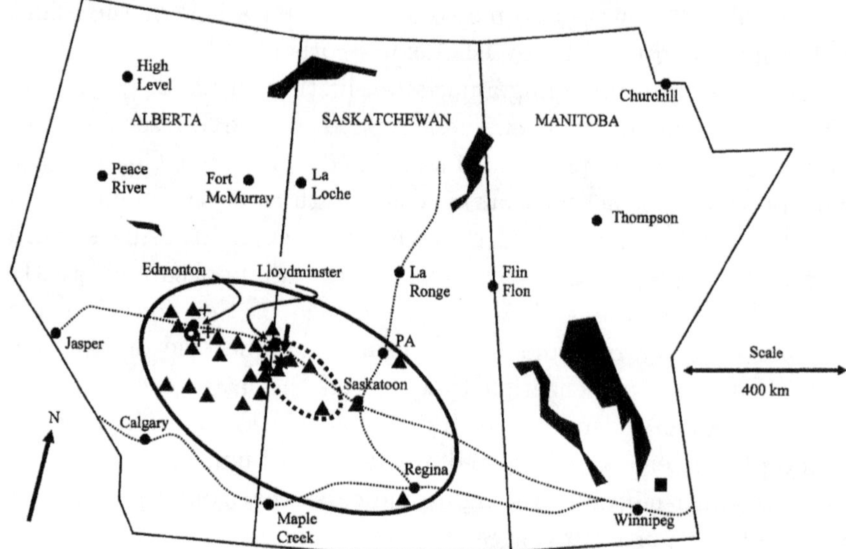

Figure 1.18. The distribution of eyewitnesses that reported sounds associated with the Buzzard Coulee fireball. Many of the triangles indicate multiple eyewitness locations. The smaller ellipse indicates the region in which observers reported sonic booms, while the larger ellipse corresponds to the zone in which electrophonic sounds were experienced. The filled square close to Winnipeg, Manitoba, indicates the location of the Lac du Bonnet infrasound detector. The crosses indicate the location of the three observers reporting smells, the circle indicates the location of the one observer who reported a heat sensation. The short arrow to the east of Lloydminster indicates the approximate north-south ground path of the fireball which entered Earth's atmosphere at a very steep angle (estimated to be ∼60 degrees to the horizon).

similar such experiences have been reported in relation to the passage of other very bright fireballs such as the Chelyabinsk fireball of February 15th, 2013.

Closer to the ground path of the fireball and the eventual strewn field, at ranges of perhaps 50 km and less, many eyewitnesses reported hearing sharp crackling sounds. These sounds are most probably associated with the arrival of closely spaced sonic booms — the result of multiple fragmentation events. Sonic booms and rumbling sounds were only reported by those eyewitnesses located at ranges less than about 150 km. At Biggar, Saskatchewan (recall Figure 4.13), a distinct rolling thunder-like sound was heard about five minutes after the fireball. In the Lloydminster area, in contrast, with ranges of order 50–100 km, various eyewitnesses reported very loud sonic booms and seismic related phenomena: "it sounded like a large freight train" commented one observer. Other eyewitnesses described the sounds as being "like a helicopter overhead," a "large rumbling sound," "like a jet engine" and that it "crackled like a rocket." Several eyewitnesses in the Lloydminster area also reported that their houses shook; one eyewitness even commented that they ran upstairs thinking a tree had fallen on their house. Perhaps the most remarkable sound report is that received from an observer located in the hamlet of Lone Rock, Saskatchewan. Their somewhat laconic report reads, "Saw the flash. Lit up the sky to daylight brightness. Stepped out on deck on south side of house. Looking almost directly up I saw the fireball and flame streak. At the same time heard a high-pitched whistle with throbbing base [sic] sound. Windows rattled in the house..." In this case it is clear that the observer actually heard the rotating meteorite fragments (at that time in their dark flight phase) passing overhead. Indeed, Lone Rock is located less than 10 km away from where the first meteorite fragments were eventually found.

The Buzzard Coulee meteorite strewn field is approximately 7-km long and 3-km wide (Figure 1.19) and is situated mostly within open pastureland. Multiple ground searches of the strewn field have revealed a rich collection of several thousand meteorites, and the surface density in the up-range region, where mostly small-mass meteorites are found, is estimated to be about 2000 per square kilometer. No animals were reported injured during the fall, nor were there any reports of buildings being damaged. The largest fragment, weighing in at some 14 kg, was found close to the side of the road at Gibbons Ridge at the southernmost end of the strewn field. While no injuries occurred during the fall, some eyewitnesses did report having near motoring accidents as a result of seeing the fireball. One driver near to the town of Ohaton, Alberta, commented, "saw a spark in my windshield [which] looked and felt like a train coming on top of my windshield — I put on the brakes and skidded to a stop." Another

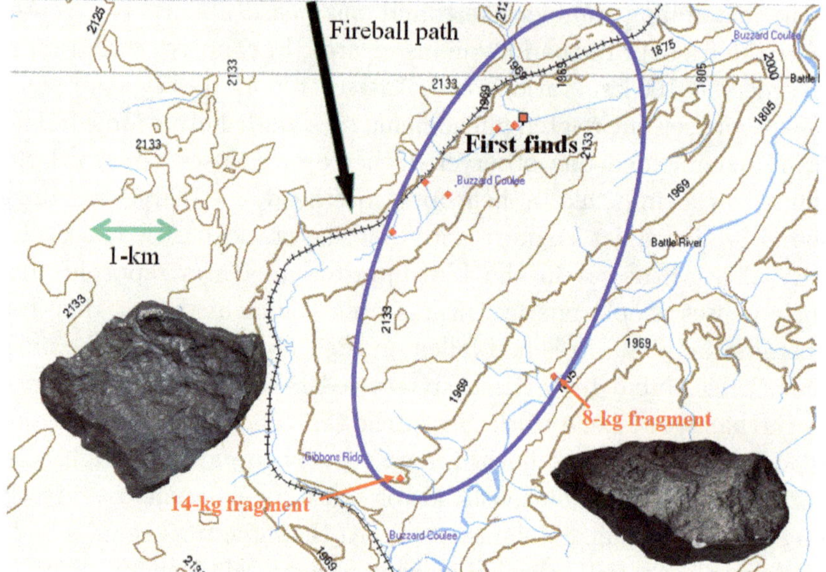

Figure 1.19.   The Buzzard Coulee meteorite strewn field. The strewn field is *twisted* with respect to the fireballs ground path (black arrow) as a result of smaller-mass meteoroids being preferentially deflected by atmospheric winds. The horizontal arrow indicates a scale of 1 km.

driver traveling near Elk Island National Park, Alberta, commented that upon seeing the fireball they "made an auto response to the brake."

Many eyewitnesses were deceived by the appearance of the fireball to think that it must have produced meteorites close to their viewing locations. Indeed, several observers reported that they actually set out to search for the "crash site," and were incredulous of the fact that the suggested meteorite search area was many tens, if not hundreds of kilometers away from where they were situated. In addition, two eyewitnesses from near Edgley, Saskatchewan (located to the east of Regina), who saw the sky flashes associated with the fireball set out to find the "impact crater," and by chance came across a field in which a farmer was burning off the remnants of an old building. Thinking that this was the devastation associated with the meteorite fall they informed the local media, who sent out a camera crew to record the story.

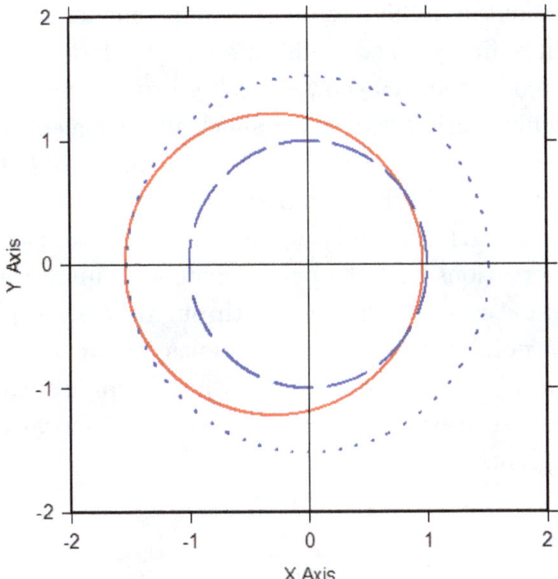

Figure 1.20.   The estimated orbit (red ellipse) for the parent meteoroid of the Buzzard Coulee meteorites as deduced by Milley and Hildebrand [3]. The long-dashed and short-dashed circles in blue correspond to the orbits of Earth and Mars respectively. The direction to the first point of Aries is vertically downward from the Sun in this diagram, and the Sun is at the (0, 0) point in the diagram, and each axis is labeled in astronomical units.

While instrumental data is available for the Buzzard Coulee fireball, a good derivation of the parent meteoroid's orbit has proved difficult to ascertain. Much effort was directed towards this problem, however, by Alan Hildebrand and Ellen Milley (University of Calgary) and Figure 1.20. shows the derived orbit. The data analysis indicates an orbit with a semi-major axis of $a = 1.25$ AU, an orbital eccentricity of $e = 0.23$, and an inclination of $i = 25.0$ degrees (see Chapter 3 for a discussion of orbital parameters). These orbital parameters make the parent body to the Buzzard Coulee meteorites a so-called Apollo asteroid (see Chapter 12). For all this, the deduced orbit is a little unusual in having an aphelion point not much greater than the orbital radius of Mars.

Chemical and petrographic analysis of Buzzard Coulee meteorites has revealed them to be H4 ordinary chondrites [3]. Chondrule sizes vary from about a millimeter across and smaller, and these are embedded in a fine-grained, uniformly gray colored matrix. Many of the collected meteorites showed freshly broken surfaces, but none revealed any brecciation textures or shock alteration effects. The density of measured samples varied from about 3260 to 3500 kg/m^3 — exactly as expected for H chondrites. One particularly interesting feature, however, is that relating to the fusion crust. It is clear from the observations that the parent meteoroid underwent multiple fragmentation events as it descended through the atmosphere (recall Figure 1.14), notably at altitudes as low as 12 and 2.5-km altitude, and reflective of this many meteorites are distinguished by having immature fusion crusts, with angular surfaces showing numerous small regmaglypts.

References

[1] M. Beech. "The Millman fireball archive." *Journal of the Royal Astronomical Society of Canada*, **97**, 71–77 (2003).

[2] M. Grady. *Catalogue of Meteorites*. Natural History Museum, London (2000).

[3] Much of this section is taken from an article by the author that appeared in the May 2009 issue of *Meteorite* — the international quarterly of meteorites and meteorite science. Preliminary results of the first field investigations are provided by A. Hildebrand *et al.*, "Characteristics of a bright fireball and meteorite fall at Buzzard Coulee, Saskatchewan, Canada, November 20th, 2008," paper presented at the 40th Lunar and Planetary Science Conferences (2009). Also presented at this meeting were papers by Erin Walton *et al.*, "Mineralogy, petrology, and cosmogenic radionuclide chemistry of the Buzzard Coulee H4 chondrite," and Hutson *et al.*, "A first look at the petrography of the Buzzard Coulee (H4) chondrite, a recently observed fall from Saskatchewan." Ellen Milley's masters thesis, presented to the Faculty of Graduate Studies at the University of Calgary in 2010: "Physical properties of fireball-producing Earth-impacting meteoroids and orbit determination through shadow calibration of the Buzzard Coulee meteorite," can be found at: https://prism.ucalgary.ca/handle/1880/47937. A review of the bulk physical characteristics of meteorite samples has been presented by Fry *et al.*, "Physical characterization of a suite of Buzzard Coulee H4 chondrite fragments." *Meteoritics and Planetary Science*, **48**, 1060–1073 (2013).

Chapter 2

The Beginning

History is not just one damn thing after another. In fact, history is the same old same old, a single grand and relentless process of adaptations to the world that always generates new problems that call for further adaptions.

Ian Morris. *Why the West Rules —*
For Now: The Patterns of History, and
What They Reveal About the Future (2010)

Astronomer Carl Sagan once remarked that to make an apple pie from scratch, you must first create a universe. The same could be said of meteorites. Indeed, all the ingredients to make a star, a planet, an asteroid, and a cometary nucleus must be formed somewhere and some when, and tracing all the steps backward to the original, basic raw material of everything, then we arrive at the Big Bang formative-creation moment of our universe — dated to some 13.8 billion years ago. Of course, starting with primordial nucleosynthesis takes us back far too far for present purposes, but technically that's where every story begins. Moving closer to the beginning of our story, we could start 4.56 billion years ago with the formation of the solar nebula (indeed, this is the topic of Chapter 5). This was the time at which the asteroids were first beginning to form, and importantly, the asteroids are the parent bodies to meteorites. Even this takes us too far back for the topic of this section. The history of meteorites is certainly long and has no real single moment of origin, but in terms of discourse and explanation, the history of meteorites can readily begin with the ancient Greeks some 2500 years ago. Indeed, it was the ancient Greek philosophers who realized that the world could be understood according to rational argument — the world and the heavens were not the simple play things of capricious gods, but, rather, the workings of the universe were constrained to obey fixed rules, and constrained to follow strict patterns of behavior. These rules were immutable (that is god interference proof) and understandable, and most importantly findable under that action of human thought and observation. Certainly, the ancient philosophers were not applying anything close to what would be recognized today as scientific principles, but they were performing the vital task of ordering observations and from these developing overarching connections and explanations. The fact that today virtually all their explanations are considered wrong is not the point. All science is provisional, and our knowledge at any one time builds upon the continued discussion of ideas and the continued collection of more and better data.

2.1   Something in the air

As with most scientific topics of interest today, the story of meteorites can be taken to begin with Aristotle. It is from this towering historical figure that a reasonable history of meteoritics can be traced. Aristotle didn't initiate the study of meteorites (as we recognize it today) but he did attempt to place the various phenomena we now associate with meteorites into a single comprehensive theory. The key historical text associated with Aristotle is *Meteorologica*, written circa 350 B.C. The title of Aristotle's work betrays its topic — that is, the atmosphere. The modern words meteorology, meteor, as well as meteorite, all stem from the ancient Greek *meterion*, meaning *high up*. In the modern era these words have adopted separate meanings and correspond to different fields of study, but to the ancient Greeks, as Aristotle reveals, the weather, wind and rain, lightning, meteors (shooting stars), meteorites, and even comets were all atmospheric phenomenon. The only distinction between these various entities was in their composition and their resultant tendency (or final cause). The meteorological theory outlined by Aristotle is based upon observations, rational ideas, and unified concepts, albeit that they are entirely wrong ideas by modern standards.

Taking a step back from *Meteorologica* for a few moments, Aristotle was responsible for developing a detailed cosmological model and a detailed theory for the structure and characteristic properties of matter. The cosmological model was Earth-centered and divided into two dynamically distinct domains. Surrounding a spherical Earth and lying below the sphere of the Moon was an envelope of air (with varying characteristic domains) — this envelope is what we would call the atmosphere. The sphere of the Moon marked a fundamental division in the cosmos. Below the realm of the Moon, Aristotle reasoned, all motion was rectilinear — that is objects moved in straight lines and change and transmutation was possible. From the top of the atmosphere outwards the cosmos changed. Motion was now entirely circular and eternal — the heavens from the Moon to Saturn (then the outermost known planet) were perfect in form and in motion, and this motion was driven by the outermost celestial sphere which contained the stars (the Prime Mover). Exactly how the motion was transferred from the outermost celestial sphere, which rotated around a stationary Earth once every 24 hours, to the various planetary spheres was not entirely clear to Aristotle, but the idea was that the various linking machinations were capable of explaining the observed planetary motions.[1]

Since phenomena such as lightning, shooting stars (meteors), and comets were short-lived, changeable, and occurred at unpredictable times, Aristotle reasoned that they must belong to the domain

---

[1]Foremost among the planetary observations to explain was that of retrograde motion. This motion is particularly obvious with respect to what are now called the superior planets (that is Mars outward). Retrograde motion is observed when a superior planet approaches opposition (that is when the Sun and planet approach 180 degrees apart on the sky as seen from the Earth). At the times of opposition a superior planet temporarily reverses its direction of motion with respect to the background stars, for several months, tracing out a distinctive loop-like motion on the sky. The problem for Aristotle and fellow Greek philosophers was to resolve the retrograde motion against an underlying requirement that all celestial motion was circular in form. While philosophers after Aristotle were able to develop exquisite mathematical models to account for retrograde motion (as summarized by Ptolemy in his famous *Almagest*) the observation was eventually given a natural (and correct) explanation by Copernicus in the mid-16th century.

below the first celestial domain that contained the Moon. They were phenomena that occurred in the upper most reaches of the atmosphere — above the Earth and below the Moon. It is worth noting at this stage that while Aristotle introduced various distinct domains and ordering within his cosmology, there was no specific scale to be applied. This being said, later philosophers did have a reasonably good measure for the physical size of the Earth's sphere and the distance to the Moon and Sun. Indeed, Erastothenes circa 240 B.C. derived an accurate measure for the Earth's diameter by comparing simultaneous Sun angles from two distant locations, and Hipparchus circa 150 B.C. determined that the Moon was at least 62 Earth radii away (it is actually on average 60.32 Earth radii away). Aristarchus of Samos circa 270 B.C. had additionally estimated that the Sun was 20 times further from the Earth than the Moon, but in this measurement he was also out by a factor of 20 times. Taking these measurements at face value, the atmosphere began at the Earth's surface and extended a further 62 Earth radii outward. This extended upper region was the realm of meteorology.

In addition to dividing the cosmos into two distinct regions — the lower region of change and straight-line motion, and the upper region of perfect, constant, and unchangeable circular motion — Aristotle also developed a detailed theory of matter. This theory was based upon the various ways in which four fundamental elements combined and upon the natural (that is fundamental) tendencies of those elements. The four fundamental elements were air, water, earth, and fire; of these elements earth and water had a natural tendency to move downwards while air and fire had a natural tendency to move upwards. Pure earth, if allowed, would move all the way to the center of the universe (which corresponded to the location of the center of the Earth); in contrast, if allowed, pure fire would ascend all the way to the top of the atmosphere and the bottom of the first celestial sphere for the Moon. Here, the 'if allowed' means if nothing else gets in the way. Indeed, Aristotle argued that the Earth was a sphere because this was the shape that allowed all the free elements of earth to get as close to the center of the universe as possible — each component, as it were, jostling with its surrounding components until it could move no closer to the center. Water and air had tendencies between those of earth and fire, and also allowed, according to the fractional mix, for the specific characteristics of all kinds of matter. A combination of earth and water would produce cold mud; a combination of water and fire would produce hot steam; a combination of water and air would produce clouds that hovered appropriately between the Earth's surface and the Moon's celestial realm. Add the elements of air and fire into the meteorological realm and ignite it — say by frictional heating — then you had the rapidly burning linear train that characterized lightning. If the train of fire and air burned slightly more slowly, then you had a meteor or shooting star. Take extra hot dry air and accumulate that vapor at the very top of the atmosphere, ignite the vapor-cloud so that it burned very slowly, then one had a long-lived slowly moving comet. If, on the other hand, the atmospheric elements combined in such a way as to produce something that was predominantly earth-like in composition then it would fall from the sky, following the earthly tendency to move downward, and a stone or iron meteorite would be formed.

Aristotle's cosmology and his theory of elements were flexible and fully capable of explaining, to the satisfaction of his followers, all of the observed atmospheric phenomena. It could account for what was observed in terms of its physical make-up (solid, wet, dry, cold, warm, etc.) along with its location and motion. Aristotle also introduced a fifth element, the element of quintessence, but this element was not to be found in the terrestrial or atmospheric realms. Indeed, this was the element of the celestial region — the planets were made of quintessence, and these objects were perfect in form, and unchangeable in their characteristics and motion.

Aristotle brought all his ideas on the properties of matter and the cosmos to bear in his *Meteorologica*, and in typical style, he offers an explanation for everything from fireballs, to comets, to the origin of the Milky Way, to thunder and lightning, rainbows and aurora. It was a grand synthesis that described everything, and, of course, it explained nothing in a specific or predictable manner. For all this, Aristotle distinguished between different kinds of fiery meteors.

If the light emanation was wide and extended in length (a fireball to us) it was called a *beam*. If the light trail was narrow but extended in length, it was called a *faces* or *lampadas* (a torch) or *caprae* (a goat). The latter name was applied if the emanation threw off sparks (that is it showed distinct fragments). If the emanations were scattered and small, then they were called *stellae trascurrentes* or *stellae trajicentes* (shooting stars). Many Latin derived names have been invoked with respect to bright meteors, including *globi scintillarum* (sparkling globes), *serpens* (snake), *stella ardens* (blazing star), *volatus draconis* (flaying dragon). The term *Draco Volans*, literally flying dragon, is a medieval invention, and some authors suggest that it may have originally been a term applied to aurora — but now it is taken to refer to a very bright meteor (or fireball). The origin behind the association of fireballs with snakes and dragons may well lie behind the observation of the long-lived luminous trains that can remain along the path of a fireball — these trains being twisted and deformed into snake-like shapes by the differential action of atmospheric winds (Figure 2.1). The term *bolis* has also long been used in the description of bright meteors, and this is derived from the Latin *bolidas* which literally means missile.

Not only were the Greek philosophers highly imaginative and intellectually capable, they, like present-day scientists, were not simply prepared to take just any one's explanation for phenomenon[2] — although the writings of Aristotle and Plato generally tended to dominate. So while certain aspects are common to the many theories that the ancient philosophers developed in order to explain the world, each would have some unique feature, foundation, or outlook. Perhaps the strongest alternative to Aristotle's fiery vapors origin for meteors was that introduced by Epicurus in the 3rd century B.C., and championed by Diogenes Laertes in the 3rd century A.D. This

---

[2]This outlook is enshrined in the motto *Nullius in Verba* adopted by the Royal Society of London when first formed in 1662. This motto translates from the Latin to "take nobodies word for it" and it underscores the scientific dictate of rejecting, out of hand, dogma as an acceptable means of explanation.

Figure 2.1.   A persistent, luminous train left along the path of a bright Leonid fireball seen on November 17th, 1998. The trail was visible for some 20 minutes, continuously evolving into newly contorted and twisted shapes due to wind speed variations at different atmospheric heights. Image courtesy of the ROTSE project.

atomic theory explained fiery meteors as the chance meeting of atoms capable of generating fire.

The only meteorite that Aristotle described in his *Meteorologica* was that of the fall in Aegospotami in 467 B.C. Perhaps the most renowned of ancient falls, Aristotle actually rejects the idea that it was a heavenly object, arguing that it was a stone blown from the top of a mountain by a strong wind. Plutarch in the first century A.D. discussed the fall (and criticizes Aristotle) in his biographical account of the Spartan admiral Lysander. Indeed, Lysander oversaw the naval defeat of the Athenians at Aegospotami in 405 B.C. Plutarch explains that according to local belief a stone of vast size had fallen from the sky, and that its arrival was predicted by Anaxagoras.[3] If a prediction was actually made, then Anaxagoras was very lucky that something

---

[3] Anaxagoras (born circa 500 B.C.) was the first philosopher to correctly explain lunar and solar eclipses, and he additionally argued that the Earth could be in

actually occurred, but apparently he had argued that heavenly bodies (in Aristotle's otherwise perfect realm) could be loosened by some slip or shaking of their circular motion, and accordingly plunge downward to the Earth as a shooting star. Plutarch also notes that should such objects be falling from heaven, then most would fall in the sea (Oceania), and this explains why they are only rarely observed and found on *terra firma*. In support of this extraterrestrial notion, Plutarch also reports that Daimachus had observed a "fiery body of vast size" (presumably a comet) in the sky for 75 days straight before the meteorite fall took place.

While perhaps not fully convincing to everyone, Aristotle's meteorological ideas rippled down the centuries as the go-to theory for meteors. Pliny the Elder in his *Naturalis Historia* reiterated Aristotle's explanation for fiery meteors in the 1st century A.D., and so too did Seneca in his *Naturalis Quaestiones* written at about the same time — their only criticisms being the naming of the various forms. One-and-a-half thousand years forward in time, the Paracelsian and Oxford scholar Robert Fludd happily adopted Aristotle's ideas into his *Meteorologia Cosmica* published in 1626 (Figure 2.2), and Francis Bacon in his *The New Atlantis* (published posthumously in 1627) was to describe high towers, and a spacious experimental chamber, where study was made of "diverse meteors; as wind, rain, snow, hail; and some of the fiery meteors." Located on the imagined island of Bensalem in the Pacific Ocean, Bacon's passionately described Salomon's House is the utopian university, and it is there that the Bensalemites study, measure, and catalog all nature: "we have also precious stones of all kinds ... a number of fossils and imperfect minerals ... likewise loadstones of prodigious virtue, and other rare stones, both natural and artificial" — but, apparently, no stones that fell from the sky.

Even 75 years later, John Harris in his *Lexicon Technicum* (published in 1704) was to write under the term *meteors*, that they are "various impressions made upon the elements, exhibiting them

---

motion about the Sun, but this latter idea gained no strong attraction until later revived by Copernicus in the 16th century.

Figure 2.2. Adapting and expanding Aristotle's *Meteoroligica*, Fludd divided the atmosphere into a series of distinct zones. The outermost region contained the stars and the Milky Way. Below this was the region containing comets (of which several distinct tail-types are shown) and then aurora. Below this zone are the clouds, where rain, snow, hail stones, and thunderbolts are produced (even a rain of frogs is shown — at image center). This is also the zone where meteors are produced. In the lowest realm such phenomena as lunar halos, rainbows, and parahelia are produced.

in different forms — fiery meteors are such as consists of a fat, sulphurous kindled smoke." As for stones falling from the sky, Rene Descartes (in 1843) advocated an origin from within clouds,[4] and Diderot in his *Encyclopedie* (printed between 1751 and 1766)

---

[4]Descartes in fact attacked Aristotle's concept of hylomorphism in which the properties and tendencies of physical matter result from a combination of matter (elements) and form. In his 1637 review of meteorology Descartes argued that meteorites were a form of imperfect bodies, formed by the action of lightning upon atmospheric dust grains. He also offered an experimental proof, noting that rainwater collected in jars always contained earthly sediments, and if to these one added sulphur, saltpeter, and heat then a stone will form.

indicated that thunderstones[5] were nothing more than mineral matter fused together by atmospheric lightning. Not only were meteors and meteorites atmospheric in origin, Robert Hooke in his acclaimed *Micrographia* (published in 1665) repeated the prescribed wisdom that no small bodies existed in interplanetary space, and Isaac Newton in his famed *Principia* (first published in 1678) pushed the idea that only planetary-sized objects could exist in the celestial realm — in this manner comets were inflated to gargantuan sizes in comparison to the kilometer-sized objects we know them to be today. Indeed, comets had been separated out from Aristotle's meteorological scheme through the parallax observations of Tycho Brahe in 1577. This remarkable and unexpected observation placed at least one comet in a location well beyond the Moon's sphere, and signalled, along with the cosmological shift brought into play by Copernicus in 1553, the beginning of the end of Aristotelian cosmology.

2.2   The investigations of Edmund Halley

The 18th century saw the appearance of many bright, well-observed fireballs and meteorite falls, and while the temperament for scientific investigation was in its ascendency, the question as to the origins of such phenomena remained unclear. Edmund Halley saw in the beginning of the century by introducing a new epoch of measurement and theoretical reasoning, and specifically published three important papers concerning the atmosphere and meteors. The first paper, published in the Proceedings of the Royal Society of London in 1686, concerned Earth's atmosphere, and Halley reasoned that not only must its density decrease with height above the Earth's surface, but that it could not extend much more than 50 miles in altitude. This latter deduction set an upper bound upon the domain in

---

[5]The term thunderstones has been applied to many objects from ancient arrowheads, to fossils, meteorites, and indeed, almost any oddly shaped rock. The origin of such objects is attributed to lightning strikes into the ground. Lightning can, under the right circumstances, produce fulgurites, which are melted and fused soil and sand grains.

which fireballs might be observed, and this is exactly what Halley investigated in the wake of a brilliant fireball witnessed on March 6th, 1716.

Analysis of the eyewitness accounts indicated the fireball could be located at a height of some 40 kilometers in altitude, and Halley reasoned, "'Tis hard to conceive what sort of exhalations should rise from the Earth, either by the action of the Sun or subterranean heat, so as to surmount the extreme cold and rareness of the air in those upper regions." Here Halley is attacking Aristotle head-on, and suggests as an alternative model that, "it must be some collection of matter form'd in the aether, as it were by some fortuitous concourse of atoms, and that the Earth met with it as it past along its orb, then but newly formed, and before it had conceived any great impetus of descent towards the Sun." This is a bold hypothesis based upon the atomistic theory of Epicurus, which is consistent with gravitational theory, but does invoke the notion of spontaneous formation of matter within space. This latter notion may seem odd, but when Halley was writing it was still generally believed that life could spontaneously arise from decaying matter. Additionally, sometimes, new theories simply need to make extraordinary claims — the steady state cosmology developed by astrophysicist Fred Hoyle, Herman Bondi, and Thomas Gold beginning in the 1950s, for example, invoked the spontaneous generation of hydrogen atoms in order to keep the density of the universe constant. Halley left the issue of what material form the fireball might have taken as being open to debate, but very little debate was forthcoming. And, indeed, Halley himself began to waver in his conviction.

Following the investigation of another brilliant fireball, seen on March 19th, 1718, he once again endorsed the idea that its path was set by a sulphurous train of vapor, although he adds a twist and suggests that the vapors were the same as those responsible for causing earthquakes and the aurora. This latter capitulation was, to a certain extent, a falling into step with Isaac Newton. Indeed, Newton had argued in his *Opticks* (published in 1704) that, "sulphurous streams, at all times when the Earth is dry, ascending into the air, ferment therewith nitrous acids, and sometimes taking fire cause lightning and thunder and fiery meteors." It is worth pointing out at this stage that the sulphur and nitre that Newton and others at that time were invoking had very different meanings to those of today — indeed, sulphur essentially invoked the idea of combustibility and nitre was taken to be any volatile substance. Interestingly, Halley in his analysis of the 1718 fireball draws an analogy between its appearance and experiments conducted by John Whiteside that he had observed at Elias Ashmole's Museum in Oxford. These experiments concerned the heating of gunpowder in an evacuated cylinder (a risky experiment one imagines). The vapors produced were apparently observed to "shine in the dark, and ascend to the top of the receiver [though exhausted]." So, while Halley eventually formulated a revised (even updated) Aristotelian outlook for fireball origins, he had additionally revived the alternative idea that they might be produced by some form of extraterrestrial matter.[6]

The idea that meteors were some form of electrical phenomenon operating in the upper atmosphere began to take hold towards the mid-18th century. Indeed, Benjamin Franklin's correspondent and friend John Perken in Boston suggested in a 1753 letter that shooting stars were electrical fires. There was no specific reason to associate fiery meteors and fireballs with some form of electrical action, other than a vague similitude of appearance with sparks and lightning. Rather it was a form of the herding effect that often occurs in the sciences when a new phenomenon is opened up to study: all of a sudden the new innovation explains all that was previously unexplained. The great power of science, however, is to keep testing the ability of a theory to explain the available observations. John Pringle in England, for example, rejected the electrical effluvia idea in his review for the Royal Society of London of the November 26th, 1758, fireball. Indeed, Pringle returned to a

---

[6]An additional reason that the atomic theory of Epicurus was not looked upon favorably in Halley's day is that it is an essential materialistic philosophy. This philosophy has been heavily criticized by church philosophers during the later part of the 17th century, and ran counter to the then dominant doctrine of continuous matter introduced by Descartes.

meteorological explanation, suggesting that the fireball was caused by the impingement of dense air upon the lower atmosphere — drawing an analogy with a cannon ball that can be made to skip across water if fired at a very low angle of incidence. Pringle further rejected outright the idea that meteorites might fall to the ground.

As the 18th century began to close, however, the association between fireballs and some form of electrical discharge strengthened. Le Roy, who investigated a bright fireball seen on July 17th, 1771, concluded that it was some form of electrical manifestation. Likewise Charles Blagden suggested that the bright fireball of August 18th, 1783, was caused by the motion of a mass of electrical fluid. Just a few years before this, in 1781, however, a posthumous publication by Thomas Clap (a past President of Yale College) appeared in which it was suggested that fireballs were caused by Earth-orbiting comets dipping into the lower atmosphere. This idea built upon the observation that the derived speeds of fireballs were similar to that of the Earth's orbit about the Sun (about 30 km/s) — an argument that had previously been invoked as a clear reason why fireballs could not be the result of some extraterrestrial matter interaction. Clap's innovation was with respect to the dynamics of fireballs, rather than their actual composition — and this is an idea, in fact, that has more recently been reworked (see Section 20.4). Towards the close of the 18th century, the accepted wisdom was such that fireballs were the result of some electrical phenomenon and that meteorites could not possibly have an extraterrestrial origin — but all this, however, was about to change, and within a time span of some 10 years centered upon 1800, not only was the extraterrestrial origin of meteorites established, but so too was a clear link between the fall of meteorites and the passage of bright fireballs through the atmosphere. Key to this new revolution was an important publication by Ernst Chladni in 1794, and the fall of the L'Aigle meteorites in 1803.

2.3 The investigations of Ernst Chladni

Ernst Chladni is one of those unsung heroes of science, not only can he be reasonably labeled as the father of modern meteoritics, but he was also the founder of the physical science of acoustics. He traveled widely and was connected with all the major thinkers in Europe. His investigations into meteorites and fireballs followed a conversation with and suggestion by George Lichtenberg in 1792–93. His thesis, with the rather long-winded and generally uninformative title, *Concerning the origin of the mass of iron discovered by Pallas and others similar to it, and concerning a few natural phenomena connected therewith*, appeared in 1794, and within its first few pages he outlines his main argument that meteorites and fireballs are related, and that meteorites have an extraterrestrial origin. The mass of iron discovered by Pallas in Chladni's title refers to the prototype pallasite meteorite, found and described by Peter Pallas in 1772. To defend his thesis Chladni hammers away at the weaknesses of the prevailing theories for fireballs (the electrical and the Aristotelian) and draws strongly upon the eyewitness accounts of numerous meteorite falls. He accordingly argued that the fireball phenomenon is the result of a frictional deceleration action working on a solid mass moving through the atmosphere. Chladni also brought shooting stars into his theory, arguing that they were just lesser fireballs. Chladni also took on Newton in his thesis, arguing that it was just as reasonable to argue that space was full of small masses, than it was not — the latter situation being advocated by Newton. Indeed, there was no direct proof that planetary space was empty, and Chladni suggested that small fragments could easily be produced via explosions and/or collisions, and that they were simply a population of objects that had not been incorporated into the planets.

Chladni's book appeared at what was a transitional time in the study of meteorites, although it is also the case that the full acceptance of his thesis took many decades to come about. Importantly, shortly after his book appeared, a number of dramatic meteorite producing events occurred. Within a year of Chladni's book appearing there was a spectacular fall recorded at Wold Cottage in northern England. This event of December 13th, 1795, occurred in daylight, at 3:30 in the afternoon, and the meteorite fell just a few meters away from several farmhands. In the following year, the meteorite was put on display in London, where interested parties could pay 1 shilling to hear and read about the stone's remarkable arrival from the sky. Further to the Wold Cottage fall, the L'Agile meteorite fall, as investigated by Jean-Baptiste Biot, occurred in 1803, and Biot reported upon the veracity of the fall and spoke to its extraterrestrial origin. All of a sudden meteorites were in the news and on the minds of researchers.

Inspired by the appearance of Chladni's book, two students of Lichtenberg, Heinrich Brandes and Johann Benzenberg, decided to put the extraterrestrial origin idea to the test, and accordingly set out in 1798 to record the paths traced out by shooting stars on the sky from two separate locations a known distance apart. The idea being that such parallax measurements would reveal beginning and end heights — if meteoroids entered the atmosphere from space then the end heights should be consistently lower than beginning heights. As with many experiments, the idea is straightforward enough, but the actual measurements are hard to make and the data collected is far from easy to interpret [1]. Between September 11th and November 4th, 1798, Brandes and Benzenberg recorded some 21 meteors that they deemed to have been recorded simultaneously. For these they calculated an average beginning height of 50.1 ± 29.1 miles (about 91 kilometers), although there was a wide spread between the extremes — the lowest beginning height being a totally implausible 6 miles high and the highest beginning height being 132 miles. Only 4 of the 21 meteors observed were deemed to be sufficiently well recorded to enable beginning and end height determinations. And, the numbers, once reduced, indicated that 2 of these had end heights lower than their beginning heights (that is they moved downward through the atmosphere), while the other 2 had end heights higher than their beginning heights (that is moved upwards through the atmosphere). It was a 50/50 split and Chladni's thesis had not been proven, but then, it had not been negated either. The reason for the large range in the beginning heights and the uncertainty in end heights was a result of the baseline distance between the two observers being far too small — just a few kilometers — and this compromised the accuracy of the parallax determinations. Brandes returned, this time with his own students, to the determination of meteor heights in 1823, and this time collected simultaneous observations data on 63 meteors. Some 36 of these meteors had their beginning and end heights determined, and of these 26 were deemed to be moving downwards in the atmosphere (1 appeared to be moving horizontally, and 9 appeared to be moving upwards). While these later results are more supportive of the thesis outlined by Chladni, the methodology at play was not sufficiently accurate to provide a definitive result — this would only be achieved once instrumental techniques were developed in the early 20th century.

Inspired by the London showing of the Wold Cottage meteorite, then President of the Royal Society, Sir Joseph Banks decided it was high time that the physical and chemical nature of meteorites be systematically investigated. And to this end, he had just the right chemist in mind: Edward Howard. Elected a Fellow of the Royal Society of London in 1799, Howard was awarded its highest honor, the Copley Medal, for his work on isolating mercury fulminate (a so-called primary explosive that was later used in percussion caps) in 1800. Working with French émigré and renowned mineralogist Count Jacques-Louis Bournon, Howard set about the systematic study of 4 stony meteorites (Wold Cottage: L6, Siena: LL5, Benares: LL4 and Tabor: H5) and Steinbach: IVA-an, the Pallas iron, Campo del Cielo: IAB and Siratik: IIAB. Rather than perform a bulk analysis Howard and Bournon disaggregated the meteorites to study their physical make-up. In Howard's 1802 final report to the Royal Society it was noted that stony meteorites are composed of millimeter-sized chondrules (made of olivine and pyroxene), that they contained troilite (iron-sulphide), and that they have numerous small inclusions of iron. Importantly as well, Howard showed that the iron was an alloy containing nickel. Howard also drew attention to the black iron-oxide (fusion) crust sported by the stony meteorites, and performed experiments with an electrical arc to show that it arose through an intense surface heating effect. The importance of this latter finding was in the fact that it pointed to an association between meteorites and a fireball heating phase. Turning to the iron meteorites, Howard importantly found that it had the same chemical properties as the iron grains found in stony meteorites, suggesting that there must be some commonality between the two meteorite types. As to the physical connection between meteorites and fireballs, Howard ends his report in indecision, and calls for more observations.

With his 1802 report presented to the Royal Society, Howard left the study of meteorites for other topics (specifically, the refinement of sugar), but the science of meteoritics can now, reasonably, be said to have been founded. As an indication of where popular opinion lay at this time, however, we can turn to *Kirby's Wonderful and Eccentric Museum*, a magazine of "remarkable characters, including all the curiosities of Art and Nature" published by Ralph Kirby between 1803 and 1920. Indeed, sandwiched between stories of a cat that lived for 25 months without drinking, a chicken born with 4 legs, a curious account of mermaids, the salt mines of Poland, and a description of the great Tun at Heidelberg, we find an account of Howard's report to the Royal Society. The scientific audience may well have begun to appreciate the extraterrestrial origin of meteorites by the beginning of the 19th century, but the public was going to take a lot more convincing.

Along with the occurrence of several memorable falls and the first chemical and mineralogical analysis of meteorites, within 6 years of Chladni's thesis appearing, an important astronomical discovery was made. This was the discovery of 1 Ceres, the first of the asteroids, by Charles Piazzi. This new celestial body was first thought to be the fabled planet that had long been speculated to reside between Mars and Jupiter, but the rapid discovery of three more asteroids, Pallas, Juno, and Vesta indicated that the true picture was much more complicated (see Chapter 13). While the discovery of the four asteroids between Mars and Jupiter was at least suggestive of their being the remnants of an exploded planet, Chladni was not convinced that they had any connection with meteorites. The problem being that the asteroids moved along near circular orbits about the Sun, while meteorite parent bodies clearly did not. The problem was one of dynamics, and its resolution was still many decades away.

For Chladni, the timing of the fall of the L'Aigle meteorite in France, on April 26th, 1803, at high noon, was perfect. Indeed, the fall at L'Aigle marks a pivotal moment in the history of meteoritics. It was an event that couldn't be ignored — having arrived amidst multiple detonations and showering stones upon the ground, the authorities had been informed and answers were required. The orders came from on high, and Napoleon Bonaparte is rumoured to have instructed his Minister of the Interior to investigate the matter. Accordingly, the rising star of the French Academy of Sciences, Jean-Baptiste Biot,[7] then aged 29 years, was dispatched to Normandy. Biot completed a thorough review of the situation, collecting meteorites, investigating the extent of the strewn field (Figure 2.3), and collecting eyewitness reports. Many of the reports are dramatic in their detail, with accounts of a terrifying noise of thunder and the starling thuds of stones hitting the ground. One eyewitness reported that while he was working outside, a stone glanced along his arm and fell at his feet. Bending to pick it up, the stone was so hot to the touch that he let it go in fear. Biot's engaging and thorough report was eventually read to his Fellows at the Institute National on July 18th, 1803, and it was there that Biot confirmed the meteoritic nature of the stones, and argued that the stones were directly related to the passage of the fireball seen just prior to the fall. Furthermore, in accord with Chladni's thesis, Biot asserted that the meteorites were derived from a single mass of material entering into Earth's atmosphere from extraterrestrial space.

In spite of the publication of Chladni's text and Biot's investigation into the fall at L'Aigle, no clear consensus as to the true origin of meteorites appeared. The Aristotelian rising vapors theory was still being invoked as late as the 1850s, but this was a last hurrah. Additionally, by mid-century the hypothesis that meteorites were derived from the Moon had begun to take hold. This latter idea was in harmony with the then prevalent idea that lunar craters were actually volcanic calderas. Indeed, many observers (including Lichtenberg in 1775 and William Herschel in 1787 and others thereafter) had

---

[7]Biot is perhaps better known in the modern era for his formulation, with Felix Savart, of the Biot–Savart law of magnetostatics, which describes the magnetic field pattern generated by a constant electric current.

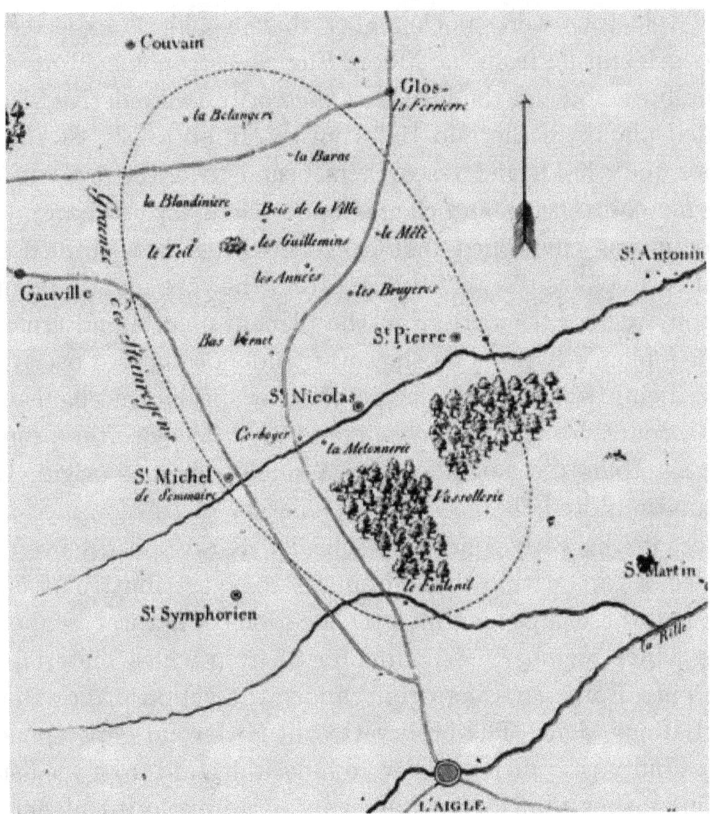

Figure 2.3. The strewn field of the L'Aigle meteorite fall as deduced by Jean-Baptiste Biot. The image scale is approximately 10 km across. The strewn field is enclosed by the dotted ellipse, and north is to the top of the image. Over 3000 fragments have been collected from the strewn field (to a total known mass of 37 kg), with the larger mass meteoroids being found in the southern end of the fall ellipse, indicating a fireball trajectory from the northwest.

reported seeing lunar volcanoes in actual eruption.[8] For a while, in the early 1800s, Chladni was supportive of the lunar volcano hypothesis, only to eventually reject it on the basis that observed

---

[8]The Moon has historically been a rich source of optical oddities. Transitory flashes are commonly reported to this very day (see Chapter 16), and luminous mists (commonly called transient lunar phenomena, TLPs) have been described on many occasions. The transitory flashes are probably meteoroid strikes into the Moon's regolith, but the origin of TLPs is still debated [2].

fireball velocities were much higher than would be expected from objects originating from the Moon. English astronomer Robert Greg eventually revitalized the idea that meteorites were derived from an exploded planet in the mid-19th century. Importantly, at this time the idea took hold, and researchers began to develop models that linked meteorite types and chemistry to planetary structure. In this manner it was envisioned that iron meteorites were formed in the central core, pallasites were derived from the core mantle boundary, and stony meteorites were from the planets mantle and crust. The exploded planet hypothesis, however, did not survive overly long, it soon being realized that there was no specific mechanism that could account for the cataclysm and that the orbital properties of the asteroids did not allow for a single point of origin. Daniel Kirkwood removed the need for a planetary disruption mechanism, however, in the early 1860s. Specifically, Kirkwood discovered the existence of gaps within the semi-major axis distribution of asteroid orbits, and these gaps identified the locations of natural escape zones (or resonance regions — see Chapter 3) from which material could evolve onto Earth-crossing orbits under the action of gravitational perturbations alone. The observation of Kirkwood gaps opened up the possibility that meteorites were indeed derived from the asteroid belt, but rather than being fragments of an exploded planet, they were fragments produced in the process of asteroid-upon-asteroid collisions.

By the close of the 19th century it was generally taken that meteorites were derived from the asteroid region, although no good instrumental data on orbits was available. This latter condition was only to change in the 1950s when the first photographic surveys, using multi-camera triangulation observations of fireball trails, were conducted.

2.4   Inspired by the Leonids

While the source region of meteorites was not to be fully understood until the early to mid-20th century, the origin of smaller meteoroids, i.e., those responsible for producing shooting stars and meteor showers, was identified in the early to mid-1800s. Meteor science literally burst onto the scene in 1833 when on the night of the 12th of November what is now known as the Leonid meteor shower produced a dramatic display, with many thousands of shooting stars being seen by observers located across North America [3]. An observer in Nashville described the event as follows: "about an hour before daylight I was called to see the falling stars, it was the most sublime and brilliant sight I have ever witnessed." Another observer reported that, "I should think, without exaggeration, that several hundreds of these beautiful stars [meteors] were visible at the same time, all falling in the same direction, and leaving in their wake a long stream of fire." This was by no means the first meteor storm to be recorded in history, but it was the first to be widely reported. The first important result to be extracted from of the 1833 storm was described by Dennison Olmstead, who noticed that all the shooting stars could be traced back to a small region in the sky — the radiant point — in the constellation of Leo[9] (see Figure 2.4). This observation, as Olmstead and others realized, indicated that the meteoroids that generated the shooting stars must have encountered the Earth along parallel paths — that is they all had the same orbit about the Sun.

The second key result was first articulated by Yale Professor Herbert Newton in 1863. Newton had conducted a survey of historical records and found that other meteor storms had been observed in November, and that such storms tended to appear at time intervals corresponding to multiples of 33 years. Newton also drew attention to other distinctive meteor showers that appeared in April (the Lyrid meteor shower) and August (the Perseid meteor shower) each year — this latter yearly association, along with the radiant condition, indicated that meteor showers must occur when the Earth passes through a stream of meteoroids that were independently moving about the Sun along similar orbits. With these results in hand,

---

[9]Meteor showers are typically named according to the constellation in which the radiant is located at the time of shower maximum. Some showers, however, are named according to a specific star in the constellation that the radiant is found at the time of maximum activity.

Figure 2.4.   Artistic illustration of the radiant point of a meteor shower. All the shower meteors trace back to a common origin on the sky, which indicates that the meteoroids must have been moving along parallel paths.

Newton predicted that another Leonid meteor storm should be seen in 1866. True to the prediction another dramatic meteor storm was observed on the night of November 13th.

Another important result from this time was published by Daniel Kirkwood in 1861, at which time he suggested that the meteoroid streams responsible for producing annual meteor showers might be the debris of ancient disintegrated comets — there was no observational evidence to support this idea (and it had been suggested by others) but the timing was right. The point being that Giovanni Schiaparelli had observed the 1866 Leonid storm and deduced the orbital parameters for the meteoroids. Fortuitously, at about the same time Guillaume Tempel (in Marseilles) and Horace Tuttle (in the US) independently discovered a new comet — comet 1866 I (now comet 55P/Tempel-Tuttle) — and Schiaparelli realized that the comet's orbit was the same as that derived for the Leonid meteoroid stream. In 1867 Schiaparelli also determined an orbit for the August (Perseid) meteoroid stream, and noted that it was almost the same as that derived for comet 1862 III (comet 109P/Swift-Tuttle).

Other meteoroid stream–comet orbital connections were subsequently made, but it should be remembered in the late 1800s it was still not clear what comets actually where — indeed, it was generally assumed that they were relatively loose aggregations of meteoroids. This latter idea was enhanced as a consequence of another set of meteor storms in 1872 and 1885, this time associated with the Andromedid meteor shower, which had in turn been linked to a comet first seen in 1772 (comet 3D/Biela). This comet was observed to split into two components during its 1846 perihelion return — both nuclei were observed at the next 1852 perihelion return, but have not been seen since (this is why it has a D designation). It was when the Earth passed particularly close to the orbit of the comet in 1872 and 1885 that the meteor storms occurred. The 'gravel-bank' model for cometary nuclei held sway until well into the 20th century, only being superseded by the 'dirty snow ball' model, introduced by American astronomer Fred Whipple, in 1950 [5]. Whipple's model interprets cometary nuclei as being largely composed of ice with an intermixing of refractory grains. As the nucleus approaches the Sun, so the ice sublimates and the gas outflow carries the small silicate inclusions (now technically meteoroids) along with it and places them into an orbit similar to that of the parent comet Figure 2.5).

By the close of the 19th century it was generally agreed that meteoroid streams not only existed in the solar system, but also within interstellar space. The ultimate expression of this idea was articulated by British astronomer (Sir) Joseph Norman Lockyer in his 1890 *The Meteoritic Hypothesis*. Not only are comets in Lockyer's book described as aggregates of meteorites, but stars are considered

Figure 2.5. Gas and dust jets emanating from the surface of comet 67P/Churyumov-Gerasimenko as it rounded perihelion. The twin-lobe nucleus is some 4 km across. Image courtesy of ESA.

to be the visual phenomena (heat and light resulting from collisions) associated with the intersection points of vast, looping meteorite streams — supernova where the collisions of meteoroid clusters. Lockyer's text was a grand thesis, explaining all astronomical phenomena, but it never proved particularly influential — this being said, the idea that stars might generate some (or even all) of their energy through the surface accretion of meteoroids was actively discussed well into the first quarter of the 20th century.[10]

The 1885 Andromedid meteor storm had an important role to play with respect to the discussion concerning the origins of meteorites. Chladni had argued in his 1794 text that the meteoroid

---

[10]The generation of energy via surface accretion was no longer required once the idea of fusion reactions operating within a hot central core was introduced by Arthur Eddington in the mid-1920s.

grains that produced shooting stars were no different to the larger meteoroids that produced fireballs and meteorites. If meteoroids were derived from comets, then why not meteorites too? The connection with the Andromedid shower, and by association with comet Biela, was the fact that on the night of the 1885 storm, the Mazapil meteorite fell to the ground in Mexico [4]. Indeed, using Mazapil as the paradigm, it was suggested in the early 1900s that meteorite falls could be linked to the times at which the Perseid and Leonid meteor showers were active. The question, however, is one of probability — indeed, what are the odds that a meteorite will fall at the same time that some meteor shower is active? The probability for this, of course, is almost certainty, since meteorites are continuously falling somewhere or other on the Earth (see Chapter 18) — it is just a case of whether the meteorite is observed to fall. A simple time of fall argument, however, must be tempered, as noted by W. H. S. Monck in his 1904 *Catalog of Aerolites,* by observations clearly linking any meteorite that might chance to fall, to an active shower radiant, indicating that the meteorite had the same orbit as the shower meteoroids. Furthermore, Monck noted, there is an additional problem of meteorite survival, since shower meteoroids tend to enter the Earth's atmosphere with high velocities (60 km/s in the case of the Perseids, 72 km/s in the case of the Leonids, and 16.5 km/s in the case of the Andromedids) and this dictates intensive mass loss via ablation (see Chapter 6). For all this, strong arguments were presented that attempted to link iron meteorites (in particular) to cometary meteoroid streams. As late as 1919, for example, American astronomer William Pickering was to argue, not unreasonably at the time, that while iron meteorites were derived from comets, stony meteorites were actually terrestrial in origin, the latter having been placed into Earth orbit (and thereafter gradually falling back again) during the cataclysm that formed the Moon.

The problem with arguing that meteorites are derived from the main-belt asteroid region is how to get the meteoroids into an Earth-crossing orbit. Asteroids move along essentially circular paths and while collisions do occur between asteroids (as we now know, but was previously assumed) the problem is the collision speed is such that any fragments produced will travel at speeds too low to place them on Earth-intercepting orbits — this observation was employed by the formidable Ernst Opik (Armagh Observatory, Northern Ireland) in a 1961 research paper to argue for a "meteorites-mixed-in" cometary model. Building upon ideas introduced by Opik a decade earlier, however, Lames Arnold (University of California at San Diego) produced a series of detailed numerical models to show that meteorites could evolve onto Earth-crossing orbits without the aid of high velocity increments — it just took time, a long time, for the orbital evolution to take place. Importantly, Arnold's numerical simulations were consistent with then available cosmic-ray exposure ages of iron meteorites (see Chapter 4). It was not clear from Arnold's calculations that the orbital evolution timescale was consistent with the available cosmic-ray exposure ages for stony meteorites, but he did suggest that they too were derived from the asteroid belt. Following the work of Arnold, the solution to the meteorite source region problem was essentially one of technology development — the technology being that of the electronic computer and the ability to perform high-speed numerical calculations. Such detailed sets of calculations, considering orbital evolution over vast intervals of time, were eventually presented in fundamental works by George Wetherill (Carnegie Institution of Washington), in the mid-1970s, and Jack Wisdom (MIT), in the mid-1980s.

2.5 Collections

Meteorites have been collected, studied, and venerated in one form or another throughout recorded history [5,6]. For most of this history, however, it has been either an individual fall or a specific find that has attracted interest, but this began to change towards the close of the 19th century. It was at this time, and moving into the early 20th century, that the great modern museums were both founded and began to develop specific collections of meteorites (Table 2.1). Most of these museums amassed their initial collections through the purchase or donation of private collections — the cabinets of curiosities lovingly developed by eccentrics, landed gentry, and

Table 2.1. A survey of selected present-day meteorite collections. Columns one and two indicate the institute and its location. Column three indicates the approximate date at which the institution was founded, while columns four and five indicate the approximate (present day) extent of their collections divided according to the total number of meteorite samples and specific individual meteorites represented. (Webpage data on collection information accessed January 2020. See also [6].)

Location	Institution	Founded	Total	Individual
Vienna	Naturhistorische Museum	1748	10,300	2,500
Berlin	Museum fur Naturkunde	1781	6,000	4,100
Moscow	Russian Academy of Sciences	1749	25,000	1,230
Paris	Museum National d'Histoire Naturelle	1843	4,000	1,500
London	Museum of Natural History	1803	5,000	2,000
Castel Gandolfo	Vatican collection	1907	1,100	500
Washington D.C.	Smithsonian	1846	45,000	16,850
New York	American Museum of Natural History	1872	5,000	1,255
Chicago	Field Museum	1893	10,000	1,500
Freemantle	Western Australian Museum	1881	14,000	750

dedicated individuals. The new museums, as their collections grew, changed the way in which meteoritic research was practized, with researchers having access to many different kinds of meteorites and well-equipped laboratories in which to conduct their studies. Not only this, however, the new museums were also charged with making their collections accessible to the general public — at least to the eye. Museums, then and now, are key institutions with respect to education and the inspiration of wonder and awe. To this point it is a delicate path that museum directors and staff must tread, since more than often it is a complex research topic that requires explanation to a non-specialist visitor. Indeed, poet John Keats had argued in 1817 that Isaac Newton had destroyed the beauty of the rainbow by reducing it to prismatic colors, and that empirical reasoning "will clip an angel's wings." Not everyone will agree, then and now, with Keats — beauty is in the eye of the beholder, and certainly

Albert Einstein was to argue that, "who can no longer pause and wonder and stand in awe, is as good as dead: his eyes are closed." Museums not only open eyes, and inspire a sense of wonder, they play a vital role in bringing the stories and the science of meteorites to the public.

The acquisition of meteorites, whether by the individual or by a museum, has always been problematic — they are rare, sought-after by many, and accordingly expensive. In modern times one can readily purchase meteorite samples from numerous dealers (many of which have media-star and super-explorer status) who display their inventory on the internet. A common iron meteorite will typically sell for between 0.5 to 5 dollars per gram; a common stony meteorite will typically cost 2 to 20 dollars per gram. But prices, of course, are set by market forces —— rare kinds of meteorites, meteorites that have hit something, meteorites from the Moon or Mars, meteorites with a special historical connection, or some specific aesthetic shape, all fetch a premium price, making them accessible to the tenacious and wealthy buyer only (see below). The museum curator (who rarely, if ever, has a big enough budget) is caught up in this market pressure, and collections tend to grow through either generous private donation (of money and/or meteorites) or trades with other institutions or dealers, where a fair swap and exchange between parties is negotiated. In the past, various attempts have been made to develop a pricing formula for the exchange of meteorites, although how useful this has been is largely unknown — ultimately, in a bidding war, the person with the most money gets the prize.

The classic formula for the exchange value of a meteorite was presented by Ernst Wülfing in 1897. The thinking behind his formula is certainly naïve, but it had a scientific rational and its development was an attempt to bring some order to an otherwise unordered domain [5, 7]. Wülfing's formula considers three main factors with respect to the exchange value of a meteorite V, these are: W the weight of preserved material in a specific meteorite class, T the total weight of material in the specific meteorite class, and N the number of owners of the specific meteorite being valued. Accordingly, the relative value is given as

$$V = \sqrt[3]{\frac{1}{W\,T\,N}}$$

(2.1)

The cubic root in Equation (2.1) was introduced in an arbitrary manner to reduce the range of values between large and small-mass meteorites. The formula is constructed so that the relative value decreases according to the available mass of the specific meteorite available, the relative abundance of meteorites in the entire given meteorite class, and according to the number of owners — this number effectively tries to reduce the resale value of a meteorite, with the more owners that it has had, so the cheaper it should become. This N term, in the modern era, is either ignored entirely or it effectively takes on a value smaller than 1.

Warren Foote reviewed the exchange values derived from Wülfing's formula to the prices that actually applied to sales circa 1910, and he provides an interesting tabulation of meteorite values (for us) from a century ago. In the case of the Canyon Diablo iron, for example, Foote gives a value of 0.03 dollars per gram; the 1982 catalog by pioneering meteorite dealer Robert Hagg indicates the same meteorite selling at 0.22 dollars per gram. A web search of dealers in 2020 reveals samples being offered for between 0.5 to 2.0 dollars per gram. At the upper end of the scale, at a 2014 Christies auction, a pristine Canyon Diablo sample sold for an equivalent value of 1.6 dollars per gram. The value of the Campo del Cielo iron can also be tracked, being rated by Foote at 0.5 dollars per gram in 1912, but selling for about 1 dollar per gram by dealers at the present time — with an impressive 9 dollars per gram being realized for a pristine sample at a 2018 Christies auction. Famous falls have increased dramatically in value during the past century, with, for example, the L'Aigle meteorite being valued at 0.2 dollars per gram in 1912, but a pristine sample realizing an equivalent value of 396 dollars per gram, again, in a 2018 Christies auction. Wold Cottage has also shown a dramatic increase in value over the years. Edward Topham, who owned the farmland upon which the meteorite fell, sold the entire specimen to British naturalist James Sowerby in 1804 for

Figure 2.6.   Portrait of James Sowerby showing the Wold Cottage meteorite (middle right). The story of the meteorite's fall was extensively covered in Sowerby's *British Mineralogy* (published in 1806). Image courtesy of Wikimedia Commons.

10 Guineas (Figure 2.6), equivalent to the cost of a horse, or about 55 dollars at that time (or about 1200 dollars in modern funds), giving an initial value of about 0.002 dollars per gram. In 1837 Sowerby sold the meteorite to the British Museum for 250 pounds, equivalent to some 1300 dollars at that time, giving a healthy increase in value to some 0.06 dollars per gram. Foote gives a value of 2.6 dollars per gram for Wold Cottage in 1912, but one hundred years latter, a 5.5-gm specimen sold for the equivalent of 1591 dollars per gram at a 2014 Christies auction. Since being found, the value of the Wold Cottage meteorite per gram has increased by a factor of some 800,000.

In more recent times, even the high-end prices paid out at auctions are dwarfed by the value of the very rare Martian and lunar meteorites, where the prices will be at least several thousand dollars per gram. For all this economics, however, the important point to remember is that the value of a meteorite is not in its artificial fiscal loading, but in its history and its power to inspire wonderment and awe.

References

[1] M. Beech. "The makings of meteor astronomy: Part X." *WGN, Journal of the International Meteor Organization*, **23**(4), 135–140 (1995).

[2] W. Cameron. "Lunar transient phenomena." *Sky and Telescope* Magazine, March (1991). D. W. Hughes, "Transient Lunar Phenomena." *Nature*, **285**, 438 (1980). M. Beech and D. W. Hughes, "Seeing the impossible: Meteors in the moon." *Journal of Astronomical History and Heritage*, **3**(1), 13–22 (2000).

[3] D. W. Hughes. "The history of meteors and meteor showers." *Vistas in Astronomy*, **26**, 325–345 (1982).

[4] M. Beech. "The fall of the mazapil meteorite: From paradigm to periphery." *Meteoritics and Planetary Science*, **37**, 649–660 (2002).

[5] J. Burke. *Cosmic Debris: Meteorites in History.* University of California Press, Berkeley (1986).

[6] G. McCall, A. Bowden, and R. Howarth (Eds.). *The History of Meteorites and Key Meteorite Collections: Fireballs, Falls and Finds.* Geological Society, London (2006).

[7] F. Heide and F. Wlotzka. *Meteorites: Messengers from Space.* Springer-Verlag, Berlin (1995).

Chapter 3

Origins

Facts cannot be observed without the guidance of some theory. Without such guidance, our facts would be desultory and fruitless; we could not retain them: for the most part we could not even perceive them.

Auguste Comte. The Course in
Positive Philosophy (1830–1842)

3.1 Around the Sun

Throughout the vast majority of human history, the heavens have exemplified perfection and dependability, with the Sun, Moon, and planets moving against the backdrop of fixed-star constellations in a timely and (mostly) predictable manner. A sense of 'everything in its place' has prevailed in human thinking about the heavens, and with this came an implicit sense of stasis, not only was 'everything in its place,' but that place was where it always had been and always would be. The planets and the Moon moved in their appropriate orbs and the idea that such orbs might change seemed inconceivable. But, as the past several hundred years of mathematical research and ever more detailed observations have revealed, the heavens are far from constant, and the solar system is in a continuous state of change. Change, indeed, driven by myriad interactions, permutations, and rhythmic gyrations. There is certainly order on many levels, but there is also chaos and disorder, and it is the disorder that is responsible for meteorites arriving on Earth.

Let us begin our discussion on solar system dynamics by looking at the ordered component — remembering that the order has more to do with mathematical abstractions than being an aspect of reality.

To begin then, any object in orbit about the Sun must satisfy Kepler's three, empirically derived, laws of planetary motion. These laws were introduced by Kepler in 1609 and 1618 and they marked the end of classical Greek indoctrination. Kepler, by carefully analyzing the positional data related to planet Mars, as gathered by Tycho Brahe, discovered that planetary orbits are not circular and the planets do not move within their orbits at constant speed. Kepler's three laws of planetary motion are as follows:

K1: The orbit of a planet about the Sun is in the shape of an ellipse, and the Sun is located at one of the focal points.

K2: The Sun–planet line sweeps out equal areas (inside the elliptical orbit) in equal intervals of time.

K3: The square of the orbital period P is proportional to the cube of the orbital semi-major axis a, such that:

$$P^2 = Ka^3 \tag{3.1}$$

where K is a constant. If the period is measured in days and the semi-major axis in kilometers, then $K = 3.98 \times 10^{-20}$; more conveniently, however, if the period is measured in years and the semi-major axis in astronomical units,[1] then $K = 1$.

The physical reason behind the workings of his laws was never revealed to Kepler. This was only apparent some half-century after his death and with the 1687 publication of Isaac Newton's masterpiece *Philosophiae Naturalis Principia Mathematica* [1]. In particular Newton reveled that Kepler's laws were a direct consequence of the inverse square law behavior of the gravitational force, and the fact that there is a quantity of motion (the angular momentum) that never changes (that is, it is a conserved quantity). With the publication of Newton's *Principia*, the ways of *doing* science changed: an emphasis was placed upon mathematical development, generality of explanation became a desired attribute of a theory, and the idea that the observations must guide theoretical refinement became

---

[1]The astronomical unit (AU) corresponds to the Earth's semi-major axis, with 1 AU = 149,597,870.7 km.

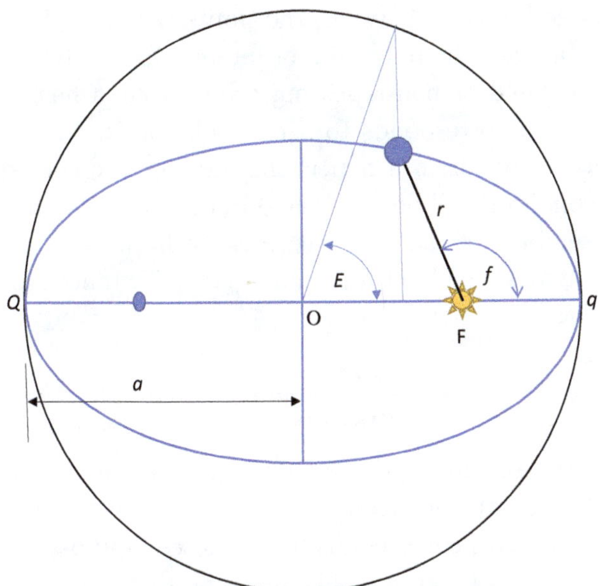

Figure 3.1.   The geometrical properties of an ellipse. The semi-major axis a is half the major axis QOq, and f is the true anomaly as measured from the perihelion point q, r is the heliocentric distance as given by Equation (3.2).

all important. With Newton, the theory of dynamics also became universal, with the laws that govern motion on the Earth (in the laboratory) being the same laws that govern the motion of the planets and the stars in deep space.

To begin the unpacking process of Kepler's three laws, take a look at Figure 3.1 which illustrates the basic features of an ellipse. The semi-major axis a is simply half the major (or longest) axis of the ellipse, and the Sun is taken as being located at one of the focal points. The two focal points, OF_1 and OF_2, which define the geometrical properties of an ellipse, are located equidistance from the center O, so that $OF_1 = OF_2 = OF$, and the eccentricity of the ellipse is defined as the ratio $e = OF/a$. If the focal points are collapsed to the center at O, with $OF = 0$, so the eccentricity is zero and the resultant figure is a circle. As the distance OF increases away from the center of the ellipse, moving towards the perimeter, so the eccentricity approaches 1. For an elliptical, closed figure, the eccentricity is bounded between the limits $0 \leq e < 1$. In general, elliptical orbits are the only ones of interest in this text, but open, that is, non-closed or non-repeating orbits result when $e \geq 1$. The $e = 1$ condition corresponds to a parabolic orbit, while the $e > 1$ condition corresponds to a hyperbolic orbit — the consequences of the latter condition will be considered in more detail in Section 7.12. Once the semi-major axis and eccentricity have been given, so an ellipse is fully described, with the heliocentric distance r being given in terms of the equation

$$r = \frac{a(1 - e^2)}{(1 + e \cos f)} = a(1 - e \cos E) \tag{3.2}$$

where f is the so-called angle of true anomaly, as measured from the perihelion point, and E is the so-called eccentric anomaly. The perihelion distance q corresponds to the closest approach distance to the Sun, when $f = 0$ degrees. The aphelion distance Q corresponds to the greatest distance between the Sun and the orbiting body, and this will occur when $f = 180$ degrees. The line drawn between the perihelion point, the Sun and the aphelion point is called the line of apsides.

While the elliptical orbit is fully defined by the parameter pair (a, e), the geometrical orientation of the orbit with respect to the plane of the Earth's orbit about the Sun — the ecliptic plane — requires that three more parameters need to be specified. Figure 3.2 illustrates the spatial orientation parameters i, ω, Ω, where i is the inclination of the orbit to the ecliptic, ω is the argument of parhelion, and Ω is the longitude of the ascending node. The ω term is the angle between the ascending node and the perihelion point. There are two nodal points to the orbit, and these correspond to the locations at which the orbit cuts through the ecliptic — they are also the points at which a collision between the orbiting body and the Earth might occur (providing, that is, at least one of the nodal point distances corresponds to 1 AU). The ascending node (N_{asc}) corresponds to the point where the orbiting body cuts through the ecliptic passing from below, while the descending node (N_{des}) is the location at which the orbiting body cuts through the ecliptic from above — the line drawn

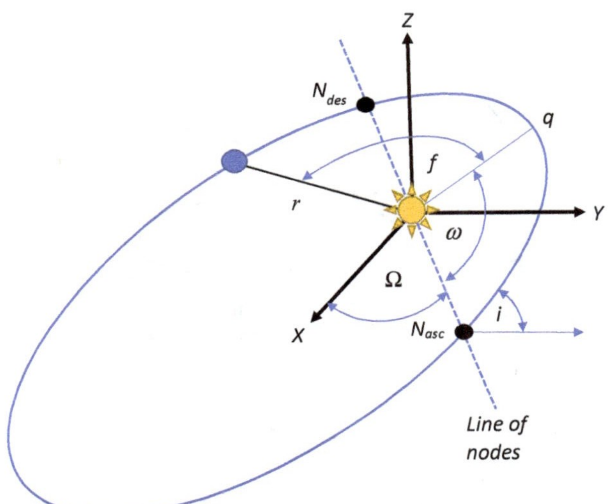

Figure 3.2. The parameters and geometry for describing the spatial orientation of an elliptical orbit. The X-axis is orientated in the direction of the first point of Aries, the Y-axis lies on the plane of the ecliptic, and the Z-axis is perpendicular to the ecliptic.

between the ascending node, the Sun, and the descending nodes is called the line of nodes. The heliocentric distances to the nodal points are determined via Equation (3.2) when the true anomaly takes on values of $-\omega$ and $180 - \omega$. The Ω parameter is measured in terms of the angle away from the first point of Aries, which is the location in the sky at which the ecliptic intercepts the celestial equator (Figure 3.3).

Having defined the geometry and orientation conditions for an elliptical orbit, the next task is to determine the location and speed of the orbiting body, at some specified time T. To do this a little more unpacking of Kepler's laws is required, and then a solution to Kepler's problem needs to be found. To begin these steps let us begin by looking at Kepler's second law.

The key aspect of Kepler's second law is that an orbiting body sweeps out equal areas in equal intervals of time. This condition simply asserts that $dA/dt = \text{constant} = h/2$. Now, consider a small triangular element, interior to the elliptical orbit, at heliocentric

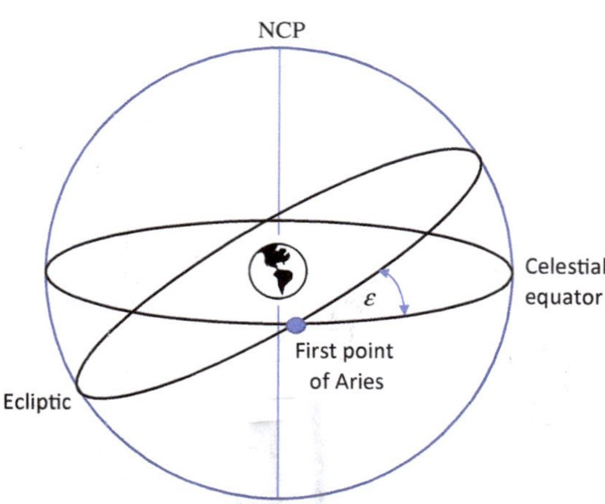

Figure 3.3.   The celestial sphere, indicating the celestial equator, the ecliptic, and the first point of Aries. The ecliptic and celestial equator are inclined by an angle 23.5 degrees (this is the so-called obliquity of the ecliptic, ε). The points at which the Earth's extended spin axes intercepts the celestial sphere correspond to the north (NCP) and south (SCP) celestial poles.

radius r, having a base length ΔS, and subtending an angle Δf at the Sun focal point. The area of the triangle element will be $\Delta A = \frac{1}{2}r\Delta S$, and since we are dealing with a small triangle, $\Delta S = r\Delta f$, which gives

$$\frac{dA}{dt} = \frac{1}{2}r^2\frac{df}{dt} = \frac{h}{2} = \text{constant} \tag{3.3}$$

where h is the angular momentum and specifically, Equation (3.3) established the result that $h = r^2\,df/dt = \text{constant}$ (this is effectively a statement of the conservation of angular momentum). To take this analysis further, we would need to look at solutions to the equation of motion for a central force: $F(r) = m(d^2r/dt^2)$, where m is the mass of the orbiting body, and $F(r) = -GMm/r^2$ is the gravitational force between the orbiting body and the Sun (having mass M), and where G is the gravitational constant. Finding solutions to the equation of motion will take us too far away from the topic in hand, and accordingly, we state the result that the angular momentum is determined as [1]

$$\left(\frac{h}{m}\right)^2 = GM\,a(1 - e^2) \tag{3.4}$$

We are now in a position to determine an expression for the total energy E (the symbol E here not to be confused with the eccentric anomaly defined earlier) associated with an orbiting body of mass m at radial distance r from the Sun, traveling with speed v. Accordingly, the total energy will be the sum of its kinetic energy K and gravitational potential energy V, such that

$$E = K + V = \frac{1}{2}mv^2 - G\frac{Mm}{r} \tag{3.5}$$

At perihelion, where the velocity of the body is V_q and where $r = q = a(1 - e)$, the angular momentum, by definition, will be $h = mqV_q$. Casting V_p in terms of the h and q, and substituting for h from Equation (3.4), the total energy is

$$E = -\frac{GMm}{2a} \tag{3.6}$$

The negative sign is important in Equation (3.6) since this indicates a bound orbit. As the orbital semi-major axis increases, however, so the orbital energy will decrease towards zero, and accordingly the orbiting object will be less tightly bound to the Sun and the solar system. If the energy were greater than zero then the orbit would no longer be closed (that is, the orbit would have the form of a parabola for $E = 0$, or a hyperbola for $E > 0$). Such open, positive energy orbits will come about under the condition that the eccentricity $e \geq 1$, and this condition can be applied in the search for interstellar material entering into the solar system (see Section 7.12).

Substituting Equation (3.6) back into (3.5) provides a formula for the velocity of the orbiting body at any given heliocentric distance r. Accordingly,

$$V^2 = GM\left(\frac{2}{r} - \frac{1}{a}\right) \tag{3.7}$$

Equation (3.7) indicates that the ratio of the velocity at perihelion to that at aphelion is

$$\frac{V_q}{V_Q} = \left(\frac{1+e}{1-e}\right)^2 \tag{3.8}$$

showing, as required by the observations and Kepler's second law, that the velocity at perihelion must be greater than the velocity at aphelion when $0 < e < 1$. For circular orbits, where $e = 0$, Equation (3.8) gives the required result that $V_q/V_Q = 1$ — that is, the velocity is constant.

With Equation (3.7) we have an expression for the speed with which an orbiting body is moving at any specific point in its orbit, but to know where an object is in its orbit at some specified time T requires the development of a few more equations. Given that an object orbits the Sun with an orbital period P, the mean anomaly M is defined so that $M = T(2\pi/P)$, where T is the time since perihelion passage. A relationship between the mean anomaly and the eccentric anomaly can be developed [2], giving

$$M = E - e\sin E \tag{3.9}$$

and this equation is to be solved for the eccentric anomaly E. With E determined, Equation (3.2) gives the heliocentric distance while Equation (3.7) gives the velocity at the specified time T. While Equation (3.9) looks relatively harmless, it is problematic in that it has no analytic solution. Accordingly, an iterative scheme (see the Appendix) is required to find the value of the eccentric anomaly E for a given value of M and the specified orbital eccentricity e.

3.2   Journey to the Earth

The orbits of several well-observed meteorite-producing bodies are shown in Figure 3.4, and the orbital data relating to these objects are shown in Table 3.1. Several details are immediately discernible from the figure and table: firstly, all the orbits have an aphelion point between Mars and Jupiter, and all of the orbits lie within a few tens of degrees to the ecliptic plane. These basic features clearly implicate

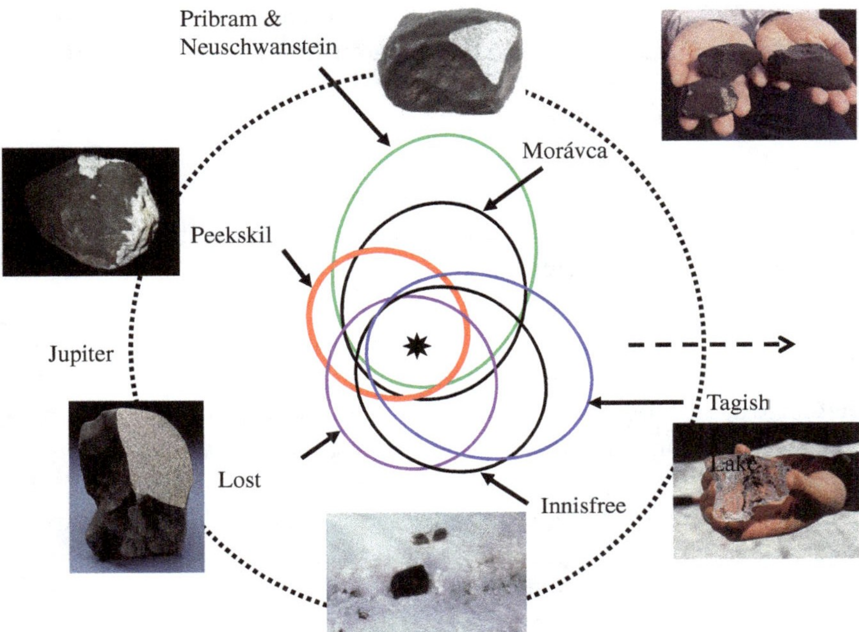

Figure 3.4. The orbits of several well-observed meteorite-producing bodies. The dashed line indicates the direction to the first point of Aries, and the orbital parameter data are given in Table 3.1. The outermost (dashed) circle corresponds to the orbit of Jupiter at 5.2 AU from the Sun.

Table 3.1. Orbital parameters deduced for eight well-observed meteorite falls. Notice the extreme similarity between the orbits for the Pribram and Neuschwanstein meteorites.

Name	Year	Type	a (AU)	e	i (deg.)	ω (deg.)	Ω (deg.)
Pribram	1959	H5	2.40	0.67	10.48	241.75	17.11
Lost City	1970	H5	1.66	0.41	12.00	161.00	283.00
Innisfree	1977	LL5-6	1.87	0.47	12.27	177.97	316.80
Peekskill	1992	H6	1.49	0.41	4.90	308.00	17.03
Tagish Lake	2000	C	2.10	0.57	1.40	222.00	297.90
Morávka	2000	H5-6	1.85	0.47	32.20	203.50	46.26
Neuschwanstein	2002	EL6	2.40	0.67	11.41	241.20	16.83
Chelyabinsk	2013	LL5	1.76	0.58	4.93	108.3	326.44

the main-belt asteroid region as the source location of meteorites — that is, meteorites are the shards spun off into space when asteroids collide.

While Figure 3.4 and Table 3.1 implicate the main-belt asteroid region as the source reservoir for the meteoroids that eventually become meteorites, they also provide us with what appears to be a mystery. Specifically, the Pribram and Neuschwanstein meteorites have essentially identical orbital parameters, and yet they fell 43 years apart and are of entirely different meteorite types. While it is expected that meteorites will fall at random times and random locations, it seems odd (even highly unlikely) that two very different meteorites, out of what is a very small total of orbits, will have the same five orbital parameters. It turns out, however, that since there is a relatively restricted parameter space that an Earth-orbit-crossing meteorite can be on, and that orbits are rapidly randomized by planetary perturbations, so the similarity of orbits for meteoroids produced in completely different asteroid collision events, as demonstrated by the Pribram and Neuschwanstein meteorite types, is much more likely than one might at first suspect (see Section 19.2 for further discussion).

The other key mystery embedded within Figure 3.4 and Table 3.1 is that, given that meteorites are derived from main-belt asteroids, how come they have such different orbital characteristics — especially with respect to their perihelion distances (at least 1 AU for an Earth collision) and their orbital eccentricities? The main-belt asteroids are confined to the region between 2.1 and 3.3 AU from the Sun and generally have near circular orbits with eccentricities $e \approx 0.1$ (Table 3.2). How then do meteoroids find themselves on orbits that can intercept the Earth? At least one possibility immediately springs to mind. Given a collision origin for the meteoroids that might eventually become meteorites, is it the case that the Earth-crossing condition automatically comes about through the ejection process?

The orbital parameters for the first 6 asteroids to be discovered are given in Table 3.2, and while it is thought that there are likely to be between 1 and 2 million asteroids larger than a few tens of meters across in the main-belt region, these initial orbits

Table 3.2. The orbital parameters for the first 6 main-belt asteroids. Column 2 corresponds to the year of discovery, while column 3 indicates the asteroid spectral type. The C and B spectral types have characteristics in common with carbonaceous meteorites; the V spectral type are the pyroxene-rich subgroup of the S, siliceous (or stony) asteroids. Asteroid 4 Vesta is specifically associated with the HED meteorites (see Chapter 13), while asteroid 6 Hebe is thought to be a major contributor of ordinary chondrite (stony) meteorites.

Name	Year	Type	a (AU)	e	i (deg.)	ω (deg.)	Ω (deg.)
1 Ceres	1801	C	2.977	0.076	10.593	72.52	80.33
2 Pallas	1802	B	2.773	0.231	34.837	310.01	173.08
3 Juno	1804	S	2.671	0.256	12.892	248.41	169.87
4 Vesta	1807	V	2.362	0.089	7.140	151.20	103.85
5 Astraea	1845	S	2.574	0.191	5.368	358.75	141.58
6 Hebe	1847	S	2.426	0.202	14.751	239.49	138.75

are representative. Inserting appropriate values in Equation (3.7) the typical orbital velocity of a mid-main-belt asteroid is of order $V_{\text{orbit}} \sim 20\,\text{km/s}$, and detailed numerical modeling of asteroid orbits indicates that when asteroids collide the typical collision speed is about $V_{\text{coll}} \sim 5\,\text{km/s} \ll V_{\text{orbit}}$. Furthermore, impact modeling indicates that while most of the collision energy goes into the heating and deformation of the two interacting asteroids, about 10% of the collisional energy goes into the kinetic energy of the fragments produced. This latter estimate indicates collisional ejection velocities of potential meteorite-producing meteoroids of $V_{\text{eject}} \sim 0.1$–$0.2\,\text{km/s}$.

That asteroids do collide, even in the modern era, has been amply illustrated in recent times through the direct observations of debris trails. Figure 3.5 shows Hubble Space Telescope images of one particular collision event that was observed in early 2010. Initially designated P/2010 A2 (and later 345P/LINEAR[2]), the collision event, which actually occurred sometime in February/March 2009, came into prominence and caught the attention of astronomers

---

[2]The LINEAR label indicates that the discovery was made by the Lincoln Near-Earth Asteroid Research survey team located at the White Sands Missile Range in New Mexico.

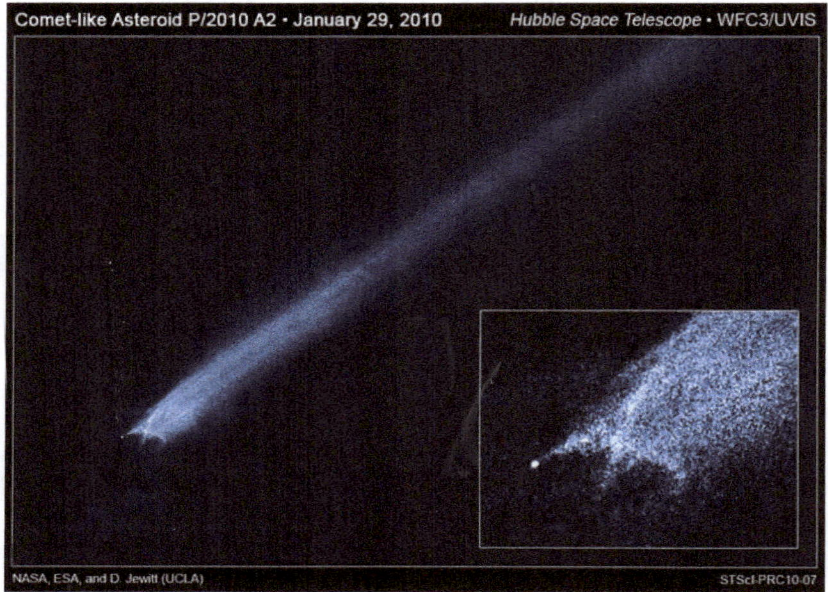

Figure 3.5. Collision and rubble tail associated with asteroid P/2010 A2 (later designated 345P/LINEAR). The impact event occurred in early 2009, but was not discovered until almost a year later. The inset shows a complex X-shaped debris trail structure, and highlights the target asteroid 345P/LINEAR (the bright dot to the lower left of the image). Image courtesy of NASA.

when the parent object rounded perihelion. The target asteroid is estimated to be some 120 m in diameter while the impactor was perhaps some 5–10 meters across. The prominent trail is composed of small dust and pebble-sized debris scattering sunlight, but no doubt, within the trail are larger meteoroids in the many-centimeter to meter size range.

Collisions between asteroids is just one mechanism by which dust grains and potential meteorite-producing meteoroids can be ejected into space. Rotational spin up is another potential mechanism. In this case it is an interaction with sunlight that, over the course of many millennia, increases the spin rate.[3] A critical rotation period,

---

[3] The specific spin-up mechanism is the so-called Yarkovsky–O'Keefe–Radzievskii–Paddack (or YORP) effect (see Section 13.3. for further details).

for which an asteroid is close to physical break up and at which material might be ejected from its surface, can be set according to the condition that the centripetal acceleration becomes equal to the surface gravitational acceleration. Accordingly, for an asteroid of mass M and radius R, spinning with a velocity V_{rot}, a surface rock of mass m will be close to ejection once

$$\frac{m V_{\text{rot}}^2}{R} = G \frac{mM}{R^2} \qquad (3.10)$$

where G is the gravitational constant. Writing now, $V_{\text{rot}} = \omega R$, where $\omega = 2\pi/P$ is the angular velocity and where P is the spin period, so the critical rotation period is

$$P_{\text{crit}} = \left(\frac{3\pi}{G\rho}\right)^{1/2} \qquad (3.11)$$

where ρ is the bulk density of the asteroid. Once the spin period $P \le P_{\text{crit}}$, loosely associated material can be spun off the surface of an asteroid. Taking a characteristic asteroid bulk density of $1500\,\text{kg/m}^3$, Equation (3.11) indicates a critical spin period of order 2.7 hours. Notice that this critical spin rate is independent of the actual size of the asteroid.

Once the rotation period of an asteroid approaches that of the critical limit, material located upon its surface can literally be spun off into space. Figure 3.6 shows a close-up segment of asteroid 101995 Bennu, where numerous, loosely grouped boulders are seen to litter its surface. The spin period of asteroid Bennu is 4.3 days, however, well below the critical spin period. An example of an asteroid that has reached the critical spin period is that of 311P/PANSTARRS[4] which was discovered on August 27th, 2013. Figure 3.7 shows the multiple debris trails associated with this particular asteroid.

Upon ejection into space a meteoroid will acquire a slightly different velocity to that of its parent asteroid. The change in velocity will be small compared to the velocity of the parent body, and it will

---

[4]PANSTARRS corresponds to the Panoramic Survey Telescope and Rapid Response System, located in Hawaii.

Figure 3.6.   Detail of the surface of asteroid 101955 Bennu, as revealed by the camera aboard the *OSIRIS REx* spacecraft. Bennu is a B spectral type (carbonaceous) asteroid, about 492 meters in diameter, and it is a potentially Earth-impacting object (see Section 12.6). The surface is littered with centimeter to meter-sized boulders. Image courtesy of NASA.

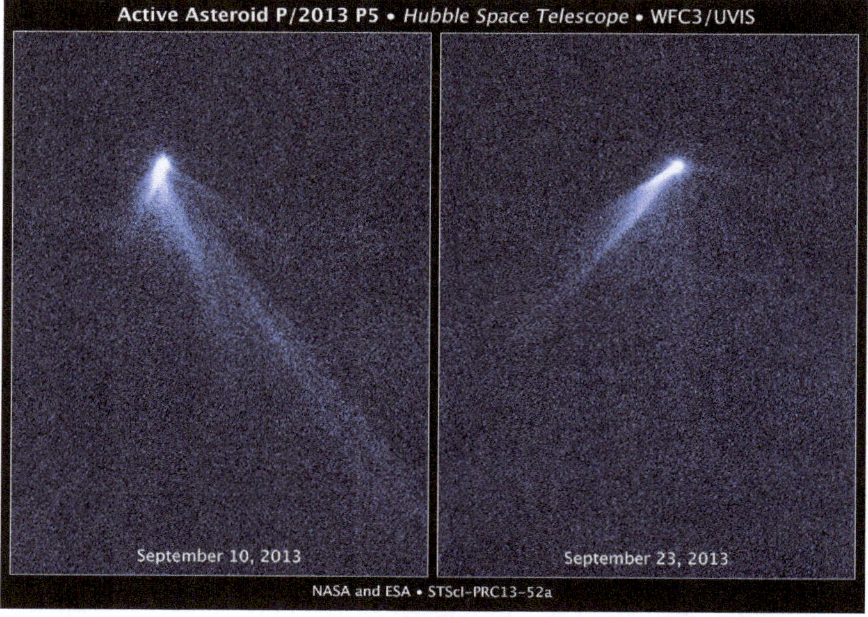

Figure 3.7.   Debris trails resulting from material being spun off the surface of asteroid 311P/PANSTARRS. The asteroid is estimated to be about 240 m across. Image courtesy of NASA.

depend upon where in the parent body's orbit the ejection takes place — recall, an asteroid will be moving more rapidly at perihelion and more slowly at aphelion. The change in velocity will also depend upon the angle of ejection compared to the parent body's direction of motion. The effects of the specific change in the energy ΔE associated with the ejection of a meteoroid can be determined via Equation (3.6) and the application of Kepler's third law (3.1). If a and P are the semi-major axis and orbital period of the parent body, then differentiating Equation (3.6) indicates that:

$$\frac{\Delta E}{E} = -\frac{\Delta a}{a} = -\frac{2\Delta P}{3P} \tag{3.12}$$

and Equation (3.12) indicates that the change in the energy of the meteoroid, as a consequence of the ejection event, must inevitably result in a change in its orbital semi-major axis and its orbital period. Since, however, the ratio $\Delta E/E$ is going to be very small, so the meteoroid's new semi-major axis and orbital period are additionally going to be small compared to that of the parent asteroid. The initial change in the semi-major axis is hardly noticeable, but since the change in the meteoroid's orbital period is cumulative, after m orbits, the asteroid and ejected meteoroid will pass perihelion (say) at different times, with the time difference being $m\Delta P$. After many orbits all of the various meteoroids ejected from the surface of an asteroid, as a result of a specific collision event, will be spread around the entire orbit of the parent asteroid.

In general, there will be some transverse component to the ejection velocity of a meteoroid, and this will result in a change in the specific orbital angular moment h. As seen in Equation (3.4), the orbital angular moment is related to the orbital eccentricity e, and accordingly it is to be expected that meteoroid ejection will not only involve a change in the semi-major axis and orbital period, but additionally a change in the orbital eccentricity — and by analogy and association the argument of perihelion ω. These orbital parament changes will additionally result in changes to the meteoroid's nodal crossing distances. An ejection velocity component perpendicular to the plane of the parent asteroid will additionally introduce a change to the plane of the meteoroid's orbit, resulting in a variation in its longitude of the ascending node Ω. Accordingly, the effect of the initial ejection event is to change all of the orbital parameters by a small amount, but importantly, it is these small initial changes, the continued effects of gravitational perturbations due to the planets (especially Jupiter and Saturn), the effects of radiation pressure and thermal re-radiation by the spinning meteoroid (see Section 13.3), that ultimately result in the meteoroid adopting, over time, a very different orbit to that of its parent body (indeed, eventually, an Earth-intercepting orbit).

3.3 Resonances

The key mechanism that drives large changes in the orbit of a meteoroid, however, is the attainment of a resonance state with either Jupiter or Saturn. The location of important resonance locations in the main-belt asteroid region are indicated by what are known as Kirkwood gaps. Named after Daniel Kirkwood, who first described them in 1866, the gaps fall at specific locations in heliocentric distance; it is called a gap as it is a small region in the semi-major axis distribution where very few asteroids are found (see Figure 3.8). Kirkwood realized that the semi-major axis locations at which gaps are observed had a very specific relationship with the orbit of Jupiter. Indeed, they were resonance locations at which the orbital period of the asteroid was some small fractional value of the orbital period of Jupiter. These are so-called mean-motion resonances.

Jupiter orbits the Sun with a period of $P_{\mathrm{Jup}} = 11.862$ years, and a resonance will occur if the orbital period of the asteroid P_{ast} is some small integer fraction of P_{Jup}. Specifically,

$$mP_{\mathrm{ast}} = nP_{\mathrm{Jup}} \qquad (3.13)$$

where m and n are small value integers. A resonance condition occurs when after m orbits of the asteroid about the Sun, Jupiter has made n orbits. A 3:1 (three to one) resonance, for example, indicates that the asteroid makes three orbits ($m = 3$) about the Sun for every one orbit of Jupiter ($n = 1$). Figure 3.8 shows the location of mean-motion resonances when $n = 1$ and $m = 3$ (at about 2.5 AU); $m = 5$, $n = 2$

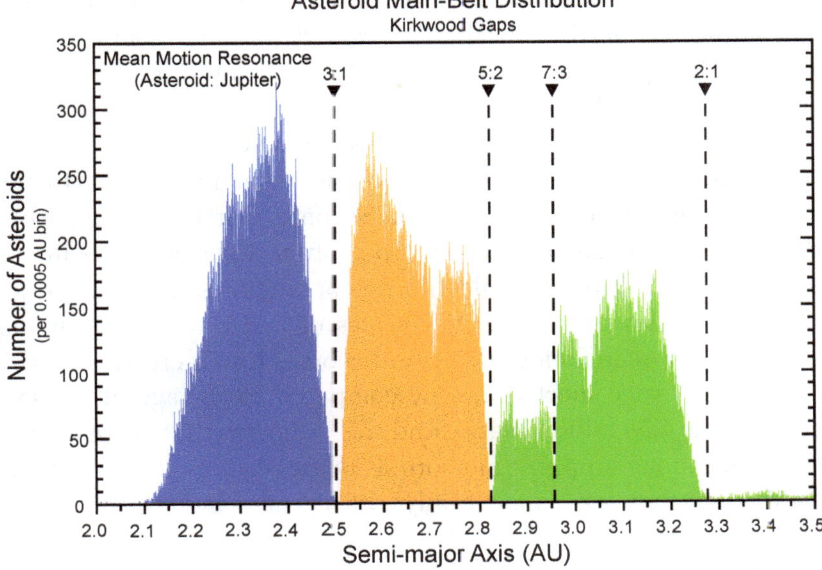

Figure 3.8. The distribution of main-belt asteroids with respect to semi-major axis. The vertical dashed lines indicate a series of strong mean-motion resonances associated with Jupiter.

(at about 2.83 AU); $m = 7$, $n = 3$ (at about 2.96 AU); and $m = 2$, $n = 1$ (at about 3.27 AU). The semi-major axis values at which the various mean-motion resonances occur are determined by Kepler's third law. Combining Equation (3.13) with (3.1) gives

$$a_{\text{resonance}} = a_{\text{Jup}} \left(\frac{n}{m}\right)^{2/3} \tag{3.14}$$

where $a_{\text{Jup}} = 5.2$ AU is the semi-major axis of Jupiter's orbit about the Sun. Accordingly, for example, the 3:1 resonance will fall at a semi-major axis value of $5.2 \times (1/3)^{2/3} = 2.499$ AU.

The key point about mean-motion resonances is that an asteroid will undergo repeat alignments with Jupiter in regular intervals and in the same relative location in its orbit. These regularly timed alignments result in small repeated gravitational tugs from Jupiter and the net effect (detailed numerical models reveal) is to gradually, on a timescale of the order of hundreds of thousands of years, increase the asteroids orbital eccentricity — and this, of course, is exactly what is required to drive the perihelion point closer inwards towards the Sun. Accordingly, the asteroid orbit first becomes Mars-crossing, once $q < 1.666$ AU, and Earth-crossing once $q < 1.017$ AU. What works for asteroids that chance to be in a mean-motion resonance works for meteoroids as well, and it is thought that the 3:1 mean-motion resonance, in particular, is responsible for delivering meteorites onto Earth-crossing orbits. Indeed, this particular Kirkwood gap is often described as being the meteorite *escape hatch*. While the repeat alignments for an object in a resonance zone occur on timescales of just a few years, it may take 10^5 to 10^6 years before the orbit becomes Earth-crossing — this, of course, is a very short timescale compared to the age of the solar system (some 4.56 billion years), and this indicates that there must be a near continuous supply of meteoroids being placed within resonance locations so that we can actually collect new meteorite falls, on Earth, at the present time. The asteroids in the main-belt region are a collisionally evolved population of objects, and the collisions are still happening to this very day (as witnessed, for example, in the case of 345P/LINEAR).

There is a second very important set of resonances that can bring meteorite-producing meteoroids (and asteroids) onto Earth-crossing orbits. These are the so-called secular resonances. Such resonances occur when the entire orbit of a meteoroid (or asteroid) precesses at the same rate as the orbital precession of either Jupiter or Saturn. These resonances come about because neither the orbits of the planets nor the asteroids (or meteoroids) are permanently fixed with respect to space and time. Accordingly, the orbits rotate about the focal point of the Sun with the direction of the line of apsides (and the line of nodes — recall Figure 3.2) gradually changing as time goes by. For Jupiter the precession rate, that is the rate of change in its argument of perihelion, is $\dot{\omega}_J = 4.3$ arc seconds[5] per year; that of Saturn is $\dot{\omega}_S = 27.77$ arc seconds per year. Accordingly, the lines of apsides for the orbits of Jupiter and Saturn rotate through

---

[5]One arc second corresponds to $1/60^{\text{th}}$ of an arc minute, and 1 arc minute corresponds to $1/60^{\text{th}}$ of one degree.

360 degrees once every 300,000 and 47,000 years respectively. If the orbit of an asteroid chances to precess at the same rate as Jupiter it is said to be in a ν_5 secular resonance, while if it precesses at the same rate as Saturn then it is said to be in a ν_6 secular resonance. There is yet another secular resonance that has been identified as being important with respect to sculpting the main-belt asteroid region. In this case the resonance relates to the precession rate of the object's ascending node Ω (recall Figure 3.2). For Jupiter the precession rate is $\dot{\Omega}_J = -25.73$ arc seconds per year (taking some 50,000 years, therefore, to rotate through 360 degrees). Once more, if the longitude of an object's ascending node precesses at the same rate as Jupiter's then it is said to be in a ν_{16} secular resonance.

There are several other secular resonances that can be identified, but it is the ν_5, ν_6, and ν_{16} resonances that appear to be particularly important in shaping the inner edge of the main-belt asteroid region and in the delivery of meteorites to the Earth. The important factor in secular resonances is that the strength of the perturbation depends both on the semi-major axis, the eccentricity, and the inclination of an object's orbit, and the ν_6 resonance appears to be particularly important in determining the upper limit to the inclination of asteroid orbits and in delivering meteoroids into the inner solar system. Figure 3.9 shows the complex structure that is mapped out by the ν_6 and ν_{16} resonance boundaries at the inner edge of the main-belt asteroid region. Also shown in Figure 3.9 is the location of asteroid 6 Hebe (recall Table 3.2), and since this particular asteroid is located very close to the ν_6 boundary, it has been identified as a major contributor to the influx of H-group ordinary chondrite meteorites on Earth.

Once a meteoroid has been liberated from its parent asteroid and chances to enter into a resonance location (secular and/or mean-motion), then detailed numerical simulations reveal that its orbital eccentricity and inclination begin to vary in a chaotic manner, with the trend, however, typically being towards a gradual increase in the orbital eccentricity. This increase in the eccentricity brings the perihelion point inwards towards the Sun and eventually, if it avoids a direct impact and/or strong gravitational interaction with Mars,

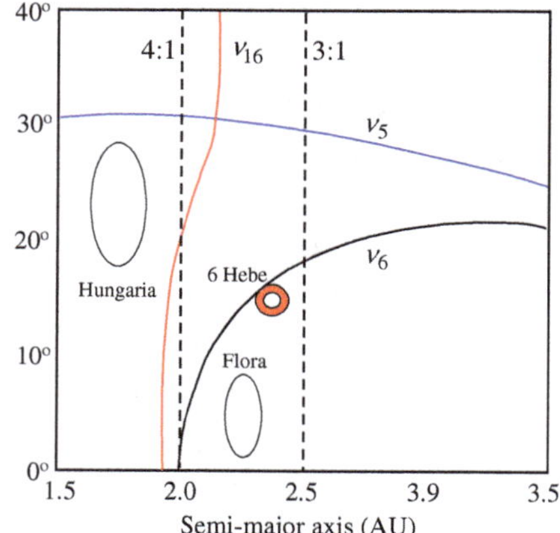

Figure 3.9.   Boundary map in the inclination–semi-major axis plane for the ν_5, ν_6, and ν_{16} secular resonances in the main-belt asteroid region; also shown are the 4:1 and 3:1 mean-motion resonance locations. The location of asteroid 6 Hebe is indicated along with the regions associated with the Hungaria and Flora asteroid families.

the meteoroid can evolve onto an orbit that intercepts that of the Earth — and a meteorite might accordingly result.

The exact path by which a particular meteorite finds its way to the Earth is sufficiently chaotic that it is impossible to trace it back to a specific parent asteroid and/or specific collision event (although see later). For all this, however, encoded within a meteorite's orbit is some indication of which resonance zone it came from. The semi-major axis values listed for the meteorites described in Table 3.1 place them as being from the inner region of the main-belt asteroid zone (the blue colored zone in Figure 3.8), situated between the 4:1 and 3:1 mean-motion resonance locations with Jupiter. Table 3.3 further indicates that for the meteorites under consideration none pass closer at perihelion to the Sun than 0.75 AU — indicating that they are potentially Mars and Earth-crossing, but that they do not interact with either Mercury or Venus. Likewise, the largest (aphelion) distance of the meteorites from the Sun is 4.02 AU indicating that they are all well interior to the perihelion distance of Jupiter, and are not under that planet's direct gravitational influence. Again, this is all indicative of meteorites being derived

Table 3.3. Physical properties and orbital characteristics of eight well-observed meteorite falls. Columns 1 and 2 give the name and meteorite type (see Table 3.1), columns 3 and 4 give the perihelion and aphelion distances, column 5 is the measured bulk density of the meteorite, column 6 is the cosmic-ray exposure age in millions of years (see Chapter 4), and column 7 is the most probable source region as derived by Granvik and Brown [3].

Name	Type	q (AU)	Q (AU)	ρ (kg/m^3)	CRE (Myr)	SR
Pribram	H5	0.79	4.02	3570	12	3:1
Lost City	H5	0.97	2.33	3570	8	ν_6
Innisfree	LL5-6	0.99	2.75	3325	26	ν_6
Peekskill	H6	0.89	2.09	3590	32	ν_6
Tagish Lake	C	0.88	3.08	1640	>3	ν_6
Morávka	H5-6	0.98	2.72	3590	7	4:1
Neuschwanstein	EL6	0.79	4.00	3500	47	3:1
Chelyabinsk	LL5	0.75	2.80	3325	1	ν_6

from the main-belt asteroid region, and specifically from the inner regions of the main belt.

In addition to the perihelion and aphelion distances Table 3.3 reveals (in column 7) the most probable source regions for the meteorites being considered. These results are taken from a study published in 2018 by Mikael Granvik (University of Helsinki, Finland) and Peter Brown (University of Western Ontario, Canada) [3]. The importance of the ν_6 secular resonance is revealed in the column data, with the 3:1 mean-motion resonances coming in as a secondary source region. The CRE ages (column 6), which are indicative of the time intervals between meteoroid launch and meteorite arrival on Earth, vary from about 1 million years (Chelyabinsk) to some 50 million years (Neuschwanstein). The CRE ages indicate the typical timescale (millions to tens of millions of years) upon which the various resonance clearing mechanisms work.

While only a small data set is available with respect to well-observed meteorite falls, it is apparent that no (or at least very few) meteorites appear to be derived from the outer main-belt region (the green zone between the 5:2 and 2:1 mean-motion resonance zones shown in Figure 3.8). The reasons for this are probably many-fold, but one reason in particular is that which relates to their composition. The outer region of the main belt is dominated by carbonaceous, C-type asteroids and these objects have particularly low bulk densities (column 5 in Table 3.3 — specifically the Tagish Lake meteorite[6]). The C-type asteroids are generally thought, as a consequence of their low density, to have weaker material strengths. Accordingly, if material ejected from a C-type asteroid is going to survive passage to the Earth, then it had better make the journey as rapidly as possible. Indeed, there appears to be a general correlation between meteorite density and CRE age, with low density meteorites typically having the shortest CRE ages, and higher density meteorites the longest CRE ages. Additionally, meteoroids composed of relatively weak carbonaceous material are less likely to survive atmospheric passage (see Chapter 6) and thereby deliver a meteorite to the ground.

Asteroid 6 Hebe has been identified as one of the parent asteroids to the H-type chondrite meteorites due to its close proximity to the ν_6 secular resonance (Figure 3.9). This result is further strengthened by its S-type (stony) spectral characteristics (see Section 12.3). It is by looking at meteorites (in the laboratory) and measuring asteroid spectral characteristics that another group of meteorites can be linked to one specific asteroid — these are the HED achondrite meteorites and asteroid 4 Vesta. The story surrounding this particular connection will be delayed until Chapter 13.

References

[1] C. Pask. *Magnificent Principia — Exploring Isaac Newton's Masterpiece.* Prometheus Books, New York (2019).
[2] M. Beech. *The Wayward Comet — A Descriptive History of Cometary Orbits, Kepler's Problem and the Cometarium.* Universal Publishers, Boca Raton (2016).
[3] M. Granvik and P. Brown. "Identification of meteorite source regions in the solar system." *Icarus*, **311**, 271–287 (2018).

---

[6]Tagish Lake is actually an exception to this rule, in that it is an ungrouped carbonaceous meteorite derived from the inner (rather than outer) main-belt region according to the analysis by Granvik and Brown [3].

Chapter 4

Rock of Ages

Those who contemplate the beauty of the Earth find reserves of strength that will endure as long as life lasts

Rachel Carson. *The Sense of Wonder*

American chemist Clair Patterson (Figure 4.1) was inspired by the beauty of the Earth, and he was the first person to successfully determine its age. His scientific journey was illustrious and carried him from early career contributions to the Manhattan Project at Chicago and Oak Ridge Laboratories, to highly public and often vitriolic battles with petrochemical companies to stop them from adding lead to gasoline. His work that resulted in the age determination of the Earth followed on from his doctoral research, which concerned lead isotope measurements, conducted at the University of Chicago. The experimental protocol for measuring trace elements of lead isotopes was exacting, and required exquisite care in avoiding contamination. In 1953, Patterson, who by that time had established an ultra-clean lab at Caltech, and coworkers were able to measure the uranium and lead isotope abundances in samples of troilite (iron sulfide: FeS) extracted from a Canyon Diablo iron meteorite. With the measurement techniques proven, Patterson then set about measuring the uranium and lead abundances in four additional meteorites,[1] and by 1956 was ready to publish his results.

---

[1]These were the Nuevo Laredo (eucrite), Forest City (H5 chondrite). Modoc (H6 chondrite), and Henbury (iron) meteorites.

Figure 4.1.   Clair Patterson at work in a Caltech lab. Image courtesy of Caltech Archives.

The age of the five meteorites, and by default the age of the Earth, was 4.55 billion years plus or minus 70 million years. What an incredible result this is, placing, as it does, the very epoch at which the solar system began to form. Not only are we made of stardust (see Chapter 5), incredibly the primordial stirrings and machinations that, after many twists and turns, resulted in the appearance of us, wise man (*Homosapiens*), just some 500,000 years ago can be set to a specific moment in the unfolding flow of the universe. So much toil is held within the range of that simple (but hard won) number[2] of 4.55 billion years. Patterson, much latter in 1994, recorded his thoughts on being the first person to know the age of the Earth — he wrote

> *True scientific discovery renders the brain incapable at such moments of shouting vigorously to the world "Look at what I've done! Now I will reap the benefits of recognition and wealth." Instead such discovery instinctively forces the brain to thunder "We did it" in a voice no one else can hear, within its sacred, but lonely, chapel of scientific thought.*

---

[2]If the unit of years is exchanged for that of dollars, and your bank account held 4.55 billion dollars, then according to the Forbes 2020 list, you would not even break into the list of top 200 wealthiest people.

Indeed, the brain *thunders* at the thought of knowing such deep results — but what exactly was it that Patterson was measuring, and how can we be sure that the number thus derived is correct? The answer lies in the instability of large atomic nuclei.

4.1 Radioactivity

Natural philosopher Henri Becquerel first described the photographic film blurring properties of uranium salt in two research papers read to the French Academy of Sciences in April and March of 1896. Mysteriously, the uranium salt could produce an effect upon the silver halide crystals coating a photographic plate even when it was shielded by layers of card. Initially it was unclear if the uranium salts were emitting some kind of electromagnetic radiation (similar, perhaps, to the newly discovered Röentgen-rays) or some kind of elementary particle. That the uranium salts must be emitting particles (rather than radiation) was first demonstrated in experiments conducted by Ernst Rutherford and Frederick Soddy in 1903, and soon thereafter it was realized that the transformation of one element into another was at play. In modern terms, radioactivity is the process by which an unstable atomic nucleus loses energy by emitting either a gamma-ray photon, an alpha particle or a beta particle. The latter two particles are now recognized as being helium atom nuclei (two protons and two neutrons) and electrons respectively. It is by systematically shedding alpha and beta particles that an unstable atomic nucleus can transform itself into a stable nucleus. In the case of the radioactive decay of uranium-238 (the 238 here indicates the atomic mass, and in this case involves 92 protons and 146 neutrons), for example, the end stable product is lead-206, being produced after 14 steps involving the emission of 8 alpha particles and 6 beta particles.

In addition to realizing that radioactivity involved the transmutation of one element into another, Rutherford additionally demonstrated that the rate of transformation followed an exponential decay law. The time required for an individual nucleus to decay is not fixed, but rather there is a characteristic lifetime for each specific kind of radioactive element. Indeed, the probability that a nucleus will decay is independent of the age of the nucleus, and accordingly after a time T, the number of atoms $N(T)$ of a given radioactive element drops exponentially from some initial value $N(0)$ in accord with a specific element-dependent characteristic decay constant λ (with units of inverse time). The decay law is expressed as:

$$N(T) = N(0) \exp(-\lambda T) \tag{4.1}$$

The smaller the value of the decay constant, so the slower the rate of decay [if $\lambda = 0$, so $N(T) = N(0) = $ constant]. The decay rate of a specific radioactive element is often expressed in terms of its half-life $T_{1/2}$, which represents the time required for half of a given sample to decay. The relationship between the decay constant and the half-life is easily found, and accordingly, from Equation (4.1), after one half-life,

$$N(T_{1/2}) = \frac{N(0)}{2} = N(0) \exp(-\lambda T_{1/2}) \tag{4.2}$$

which can be rearranged to give the result:

$$T_{1/2} = \frac{\ln 2}{\lambda} \tag{4.3}$$

The key point here is that the decay constant (or half-life) is a quantity that can be measured experimentally or calculated directly. For uranium-238, the half-life is 4.47 billion years and $\lambda = 1.55 \times 10^{-10}$ (yr^{-1}). Conservation of nuclei dictates that as the number of parent radionuclides decreases so the number of stable daughter nuclei must increase (Figure 4.2), and in terms of finding the age of a rock, this is crucial. Specifically, a unique age can be attached to any set of simultaneously made measurements for the number of parent and daughter nuclei.

4.2   Formation age

In terms of meteorites and specifically the pioneering work by Clair Patterson, the key measurement to be made is that of the formation age — that is, the determination of the time since the material in the meteorite was last molten. For this measurement, long-lived radio

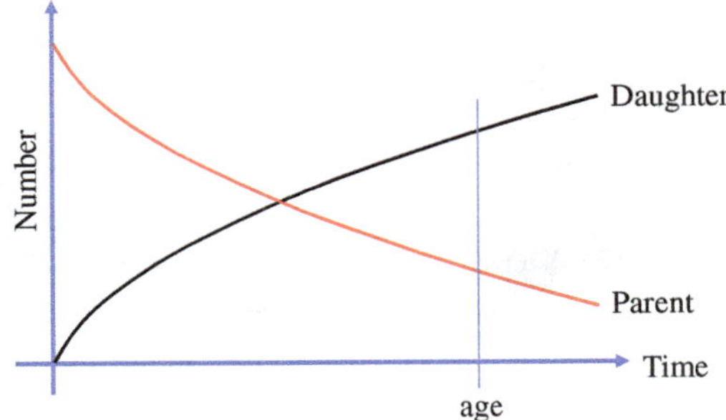

Figure 4.2.   The conservation of nuclei requires that as the number of parent radionuclide decays (red line) so the number of daughter nuclides increases (black line). Measurement of the abundances of the parent and daughter nuclides are used to determine the age.

Table 4.1.   A selection of radionuclides giving the parent and daughter configurations along with their half-lifes ($T_{1/2}$) and decay constants (λ).

Parent	Daughter	Half-life (year)	Decay Constant (λ)
Uranium-238	Lead-206	4.47×10^9	1.55×10^{-10}
Uranium-235	Lead-207	7.04×10^8	9.84×10^{-10}
Rubidium-87	Strontium-87	4.88×10^{10}	1.42×10^{-11}
Potasium-40	Argon-40	1.35×10^9	5.13×10^{-10}
Iodine-129	Xenon-129	1.57×10^7	4.41×10^{-8}
Sodium-22	Neon-22	2.6	0.267

isotopes need to be studied. Patterson specifically looked at the decay of uranium to lead, but many other parent–daughter decay chains are available, each with its own specific half-life (see Table 4.1).

The technique for finding the age of a meteorite proceeds by measuring the parent and daughter abundances, $N_P(T)$ and $N_D(T)$ respectively, where T now corresponds to the age. Because the number of nuclei is conserved, it is evident that at time T, the number of daughter nuclei can be written as $N_D(T) = N_P(0) - N_P(T)$. Now, using Equation (4.1), the unknown quantity $N(0)$ can be substituted for to yield:

$$N_D(T) = N_P(T)[\exp(\lambda T) - 1] \qquad (4.4)$$

Figure 4.3. Strontium versus rubidium data plot for the Guarena (H6) chondrite meteorite (see inset). The best-fit line to the data points indicates a slope of $m = 0.0658$, and an original strontium-87 to strontium-86 ratio of 0.6997. Data from Wasserburg *et al.*, *Earth and Planetary Science Letters*, **7**, 33–43 (1969).

Equation (4.4) indicates that in a diagram in which the values of $N_D(T)$ are plotted against $N_P(T)$ a straight line will result of slope $m = \exp(\lambda T) - 1$, and from this slope the age can be calculated as $T = [\ln(m + 1)]/\lambda$. Figure 4.3 shows a data plot for rubidium and strontium measurements derived from a selection of grains extracted from the Guarena (H6) meteorite. After loud detonations were heard, this particular meteorite fell on July 20th, 1892, in the Spanish Iberian Peninsula, delivering a total of 39 kg of material. A straight-line best fit to the data points provides a slope of $m = 0.0658$, and accordingly, from Equation (4.4), the formation age of the Guarena meteorite is 4.49 billion years. Figure 4.4 shows rubidium and strontium data for the Guarena meteorite, in comparison to those measured for a Martian meteorite (ALH77005) and a lunar rock sample (a highly vesicular olivine basalt) collected during the Apollo 15 mission. The best-fit lines to the data points indicate that the lunar rock sample is some 3.2 billion years old, while the

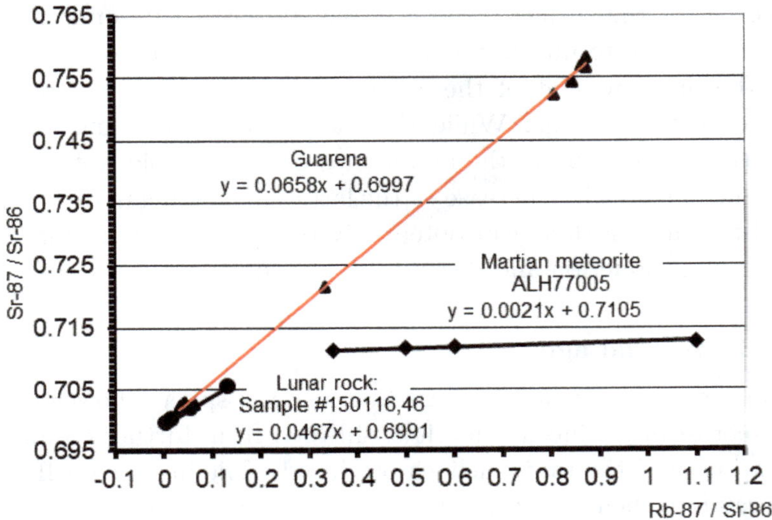

Figure 4.4. Strontium versus rubidium data plots for the Guarena (H6) chondrite meteorite, Martian meteorite ALH77005, and Moon rock sample #150116,46 returned from the Hadley Rille landing site.

Martian meteorite is merely 148 million years old. The informative (and magical) point about this latter age is that it tells us that Mars must have been volcanically active as recently as 150 million years ago.

4.3   Cosmic-ray-exposure age

While long-lived radionuclides such as uranium and strontium are used to find the crystallization (that is formation) ages of meteorites, other radionuclides are used to determine their cosmic-ray-exposure (CRE) ages. This latter determination is a measure of how long the meteorite has been in space between ejection from its parent asteroid to its arrival on Earth. All of the interstellar medium and no-less interplanetary space is pervaded by a flux of cosmic rays — largely protons, but also a smattering of other atomic nuclei and electrons, moving at relativistic speeds. These particles carry sufficient energy that they can enable the transformation of atomic nuclei situated to a depth of a meter or so within a meteoroid or the surface of an asteroid. Importantly, some of these transmutations result in the formation of unstable isotopes, and it is the measurement of these specific isotopes than can be used to infer the CRE age. While the parent body to a meteorite is in interplanetary space, the abundance of both stable isotopes and radionuclides steadily increase — the latter, depending upon the half-life of the radionuclide, can potentially reach some equilibrium level where the production and decay rates are equal.

4.4   Terrestrial age

As soon as the meteoroid enters Earth's atmosphere, however, and a meteorite is produced, then it is shielded from further cosmic-ray alteration, and the radionuclides produced while in space will begin to decay into their stable end products. This decay can be studied in order to determine the terrestrial residency time of the meteorite — literally, the time since its fall (see Figure 4.5). A commonly used radionuclide for gauging the terrestrial residency age is ^{10}Be, which has a half-life of 1.39 million years. In order to determine CRE ages the isotopes of noble gases are used, such as ^{21}Ne and ^{38}Ar. These isotopes are used since they are exclusively produced by

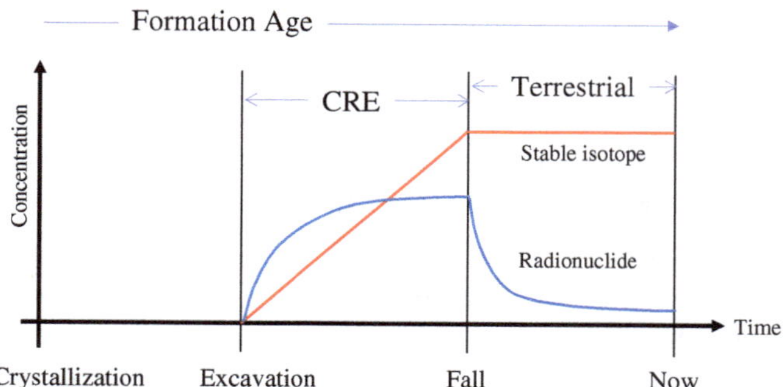

Figure 4.5.   Concentration versus time diagram for the growth of stable isotopes (red line) and radionuclides (blue line) within a meteoroid. A meteoroid is only exposed to cosmic rays once it has been excavated from the deep interior of its parent asteroid. Once on Earth and shielded by the atmosphere, the radionuclide concentration will begin to fall, while that of the stable isotopes will remain constant.

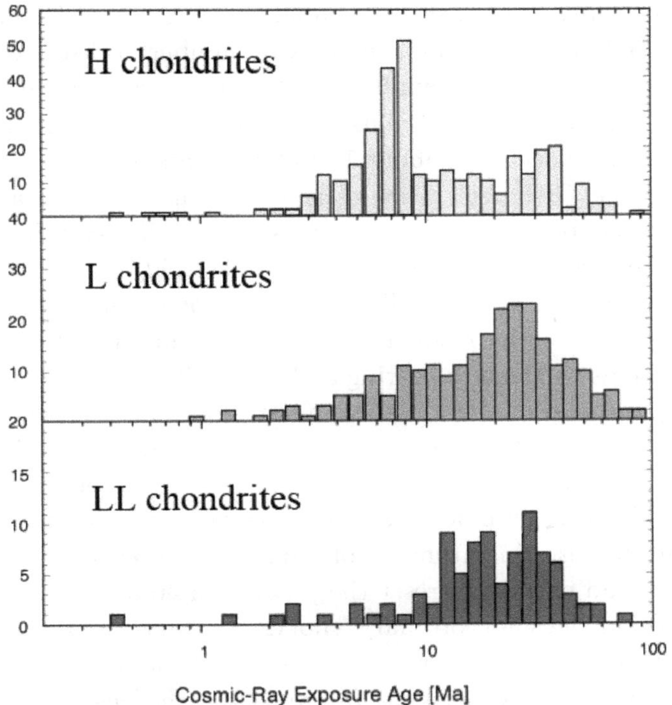

Figure 4.6. Cosmic-ray exposure ages deduced for H, L, and LL chondrite meteorites.

cosmic-ray impacts. Typical CRE ages vary from 10^5 to 10^6 years for carbonaceous chondrites, 10^7 years for ordinary chondrites, and 10^8 to 10^9 years for the iron meteorites.

Figure 4.6 shows a frequency (number versus age) plot for the deduced CRE ages of ordinary chondrites. The values show a wide range in ages, which in turn indicates a wide range in the orbital evolution times for meteoroids to find their way to an Earth-intercepting orbit and the eventual delivery of a meteorite upon the ground. The L chondrites show a reasonably smooth variation in CRE ages, with a peak being apparent between 20 and 30 million years. The H chondrites, however, show a strong peak at an age of about 8 million years, and a second weaker peak at about 35 million years. These latter peaks are suggestive of at least two distinctive break-up events, where asteroids capable of delivering H chondrites have collided. The LL chondrites show a number of distinctive CRE age peaks, suggestive of a number of collisional events between parent asteroids. The most recent collision event producing LL chondrites appears to have occurred about 15 million years ago.

While the CRE age gives a clear idea of how long a specific meteoroid has been in interplanetary space, earlier collision events, when the meteoroid was part of a larger asteroid, may be recorded. This latter record is especially evident in breccias showing shock metamorphism — the result of an asteroid upon asteroid collision. Such collisions will lead to heating and shock alteration of constituent material, and also enable the escape of trapped internal gasses. The so-called gas-retention age can be measured by looking at those radionuclides that decay into a stable daughter gas. For example, the radionuclide potassium-40 decays into the gas argon-40. Accordingly, shock effects resulting from a collision can enable the loss of the argon gas, and effectively reset the potassium-40-decay-based clock. Gas-retention ages for ordinary chondrites vary from about 4 to a few hundred million years, indicative of randomly timed impact events across the asteroid belt. Interestingly, the heavily shocked L chondrites have a common gas retention age of about 500 million years, and this indicates a catastrophic break-up event at that time, with subsequent collisions producing smaller asteroids and meteoroid fragments that go on to produce the observed spread in CRE ages.

Chapter 5

Dust to Dust

For dust thou art, and unto dust thou shalt return

Genesis 3:19

5.1 Complex elements

The beginnings of some stories start way back, and in the case of
the solar system, the story really begins with the Big Bang and the
origin of the universe itself. We need not follow every step in the nine-
billion-year time interval between the appearance of the universe
and the first stirrings in the interstellar natal cloud out of which
the solar system formed, but it was the Big Bang that produced
the key bulk ingredients responsible for our existence. Indeed, out
of the primordial nuclear synthesis, the first minutes of the universe
produced a vast reservoir of hydrogen and helium. It was, and is, from
this simple atomic stock pile, the relic matter of Big Bang creation,
that all else has been made [1]. Key to making other elements, beyond
helium, is the formation of stars. Indeed, it is through the nuclear
fusion reactions that take place within stellar interiors, and supernova
explosions, that all the other elements are forged. To reiterate the
famous words of Carl Sagan, "the cosmos is within us — we are made
star stuff" (Figure 5.1).

Only about 9.5 percent of the human body is made up of
hydrogen, while the Sun and stars, in contrast, are predominantly
hydrogen and helium — indeed, in terms of mass fraction, the Sun is
75 percent hydrogen, 24 percent helium, and just 1 percent all other
stable elements up to uranium. The Sun is not a pure hydrogen plus

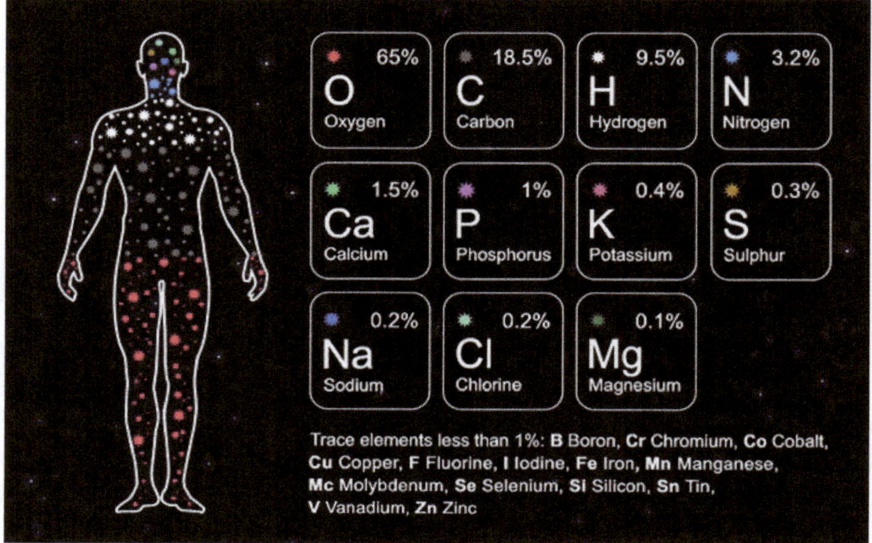

Figure 5.1.   All of the crucial atoms that allow life to exist and constitute the human body (except for the hydrogen contribution) were formed within the hot central cores of past generations of stars. Image courtesy of the Natural History Museum, London.

helium star since it is not an original, first-generation star, and this is betrayed by the fact that it contains about 1 percent mass fraction of elements other than hydrogen and helium (these are collectively called the heavy elements); this, and the fact that we know the solar system is just 4.56 billon years old (as seen in Chapter 4), much less than the 13.8-billion-year age of the universe. The one percent mass fraction of elements other than hydrogen and helium in the Sun is the result of an extensive cosmic recycling scheme in which the nuclear-processed material generated within the cores of stars is returned to the interstellar medium. Slowly, but assuredly, the stars are converting the original stock pile of hydrogen and helium into heavier atomic elements, and each stellar birth–death cycle results in the chemical enrichment of the gas clouds out of which a subsequent generation of stars will appear. The universe and specifically the

Milky Way galaxy had undergone some 9¼ billion years of recycling and evolution prior to the formation of the Sun and solar system, and humanity (with all of its genius and foibles) is composed of just a miniscule fraction of the heavy elements that constituted the original interstellar cloud.[1] Incredibly, we are dead stars made animate — the *disjecta membra* of supernovae and the terminal exhalations of red giants.[2]

5.2   Solar nebula

Star formation is driven by gravity and rotation. When these two physical effects lock in to an interstellar gas cloud then the collapse of that cloud, to produce a star, is set in motion and the collapse is rapid. An estimate of the collapse time, T_{coll}, can be gauged by considering the motion of a small mass m of gas initially located

---

[1]The mass of the solar system is of order 1.001 $M_\odot = 1.99 \times 10^{39}$ kg (the 0.001 part of this number being the total mass of the planets, comets, and asteroids). The current estimate for the human population is some 7.7 billion individuals, and the biomass of humanity (in the form of carbon atoms) is estimated as being some 60 billion kilograms. The carbon mass fraction for the Sun is 0.004, giving a total mass of carbon atoms of 7.96×10^{27} kg. This latter number is effectively a lower bound on the mass of carbon atoms in the original interstellar cloud which collapsed to produce the solar system. Humanity therefore constitutes at most a fraction of about 7.5×10^{-18} of the original carbon budget. For all our abundance on Earth, we are indeed, small fry.

[2]The matter content of the universe is arranged according to 5% ordinary matter, 27% dark matter, and 68% dark energy. In this sense, ordinary matter is that composed of atoms and includes stars, interstellar/intergalactic gas, planets, and us. The dark matter component is deduced by the dynamics of stars within galaxies, and galaxies within galaxy clusters, which all move with speeds faster than would be predicted on the basis of the observed ordinary matter. Dark matter has a gravitational influence but does not emit and/or absorb electromagnetic radiation, and so cannot be directly observed with a telescope. The dark energy component is estimated according to the rate at which the expansion of the universe is accelerating. Neither dark matter nor dark energy have a clear theoretical underpinning, and they provide present-day physicists with a profound problem that has yet to be solved.

at the edge of an interstellar cloud of mass M and radius R. Since the collapse is driven by gravity, the acceleration a experienced by the small mass of material will be given by Newton's second law, with

$$a = \frac{F_{\text{grav}}}{m} = \frac{GMm/R^2}{m} = G\frac{M}{R^2} \tag{5.1}$$

where G is the gravitational constant. Now, the acceleration term can be approximated as $a \approx V/T_{\text{coll}} = (R/T_{\text{coll}})/T_{\text{coll}}$, where V is the characteristic in-fall velocity. Upon substitution into Equation (5.1), this approximation gives

$$\frac{R}{T_{\text{coll}}^2} = G\frac{M}{R^2} \tag{5.2}$$

and finally, substituting for the bulk density $\rho = M/\frac{4\pi}{3}R^3$ the dynamic collapse time of the parcel of material to fall from the cloud edge to the cloud center becomes

$$T_{\text{coll}} = \sqrt{\frac{3}{4\pi G\rho}} \tag{5.3}$$

The characteristic density of an interstellar cloud is $\rho \approx 10^{-19}\,\text{kg/m}^3$, giving a collapse time of $T_{\text{coll}} \sim 6$ million years. Within this characteristic time the majority of the material within the interstellar cloud will have found its way to the cloud center, where a proto-star will have begun to form. Figure 5.2 shows a schematic outline of the collapse and planetary system formation process: the collapse time just derived essentially covers the time to complete steps 1 and 2, where the scale decreases from that of a large, spherical(ish) interstellar cloud (some several thousand AU across) to that of a flattened, rotating disk having a diameter of several hundred AU. The kinetic energy of the infalling gas results in the heating of the central proto-star, and conservation of angular momentum results in

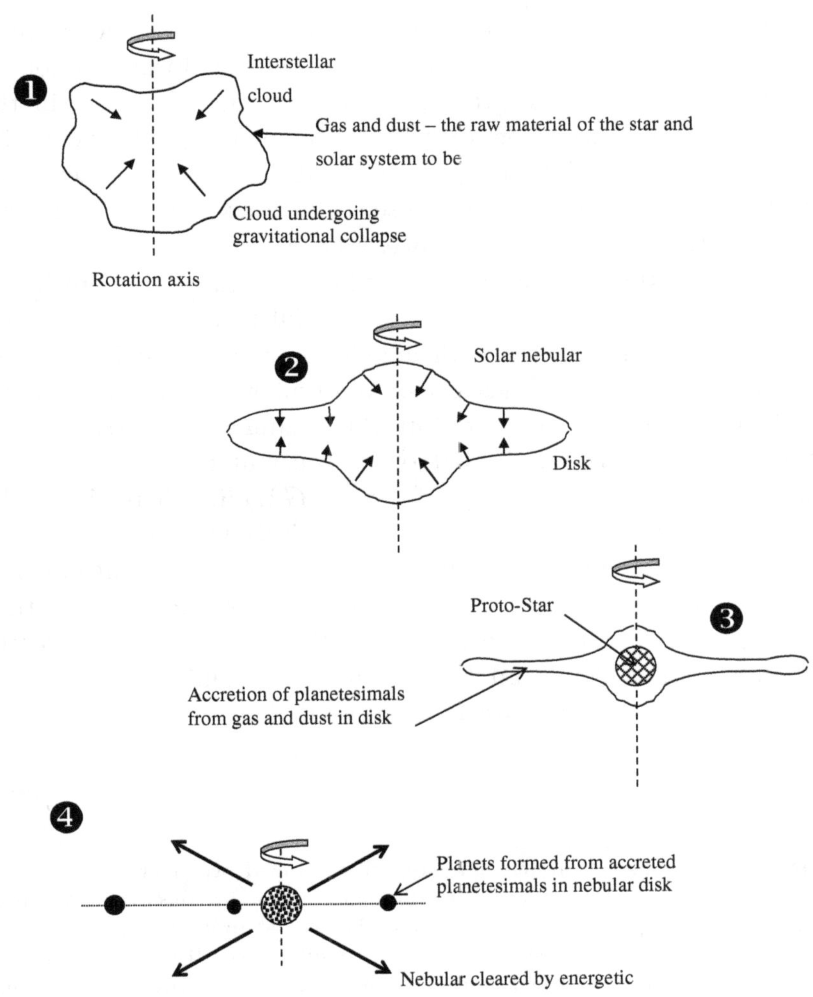

Figure 5.2. Key stages in the formation of the Sun and solar system. (1) Gravitational collapse of an interstellar cloud begins — this is the raw material for the solar nebula. (2) Gravitational collapse and rotation result in the collapsing gas cloud forming a rotating disk of gas and dust around the central mass concentration. (3) As the proto-Sun begins to grow in mass, temperature, and luminosity, planetesimals begin to form in its surrounding accretion disk. (4) Planetesimals collide, accrete, break apart, and re-form to produce the planets. After about 10 million years from the onset of cloud collapse, the Sun begins to generate energy via fusion reactions in its central core, and a strong surface wind sweeps any remaining gas and dust from the solar nebula.

material rotating more rapidly as it approaches the cloud center.[3] Indeed, the collapse and rotation result in the formation of an accretion disk about the growing proto-star, and it is within the remnants of this accretion disk that the planets will eventually form — stages 3 and 4 in Figure 5.2. After a time T_{coll} accretion onto the central proto-star will essentially cease, and it is at this stage that the planet-building processes can begin.

The approximate size of a protoplanetary (accretion) disk can be estimated on the basis that the material settles into a circular orbit about the proto-star with angular momentum being strictly conserved — technically such a disk won't allow material to actually accrete onto the central star, but the argument will give us the approximate size. For a circular orbit, the material in the disk at radius R will have a velocity of $V^2 = GM/R$, where M is the mass of the proto-star and G is the gravitational constant. If we imagine material moving in from the edge of the star-forming cloud to radius R, then the conservation of angular momentum requires that $h = V_{\text{initial}} R_{\text{cloud}} = VR$, where V_{initial} is the initial rotation velocity of the interstellar cloud and R_{cloud} is its initial radius. Substituting for the final, circular velocity V term, then gives,

$$R = \frac{(V_{\text{initial}} R_{\text{cloud}})^2}{GM} \tag{5.4}$$

---

[3]There are many complications that are being skipped over in this discussion. While angular momentum is a conserved quantity, there are mechanisms in play that enable the accretion of material onto the central proto-star to occur, but which leave the angular momentum in the surrounding disk. If this did not occur, then material would spin ever faster as it approached the central proto-star and accretion would be severely disrupted. To see this, the conservation of angular momentum requires that the initial angular momentum (per unit kilogram) of the rotating gas cloud must equal the angular momentum (again per unit kilogram) at the surface of the proto-star: $h = V_{\text{initial}} R_{\text{cloud}} = VR$. The numbers now reveal the outcome; given an initial rotation velocity of say 0.1 km/s, a cloud size of 0.01 parsec ($= 3 \times 10^{11}$ km), and a proto-star size comparable to that of the Sun $R = 7 \times 10^5$ km, so $V = 4.4 \times 10^4$ km/s which is of order 15% the speed of light. That accretion disks are able to accommodate the inflow of material, but leave the angular momentum in the disk, is demonstrated within the solar system with the angular momentum of the planets being over ten times larger than the spin angular momentum of the Sun.

Figure 5.3. The protoplanetary disk about the young Sun-like star HL Tauri. The disk is some 200 AU across, and prominent gaps in the disk (where planets are thought to be forming) occur at about 15, 30, and 70 AU. The system is estimated to be no more than 100,000 years old. Image courtesy of ALMA.

Typical values for the initial star-forming interstellar cloud size and rotation velocity are 0.01 parsecs ($\sim$2000 AU) and 100–200 m/s, and accordingly for a 1 solar mass star, the outer radius of the accretion disk will be at about $R \sim 100$ to 150 AU. This scale is exactly what is observed for the solar system out to the Kuiper Belt region, and for protoplanetary disks about newly forming stars (see Figure 5.3).

Before planets can form within a protoplanetary disk, the basic chemical composition of the building material has first to be established, and this is set according to a temperature sequence (Figure 5.4). Since the central proto-star will have a surface temperature of many thousands of kelvins, the inner-most regions of the disk will be far too hot for any solid material to form — essentially, the disk temperature is higher than the melting point of any solid material, be it rock or metal, that might find itself in the inner region. Moving outward into the disk, however, the temperature will begin

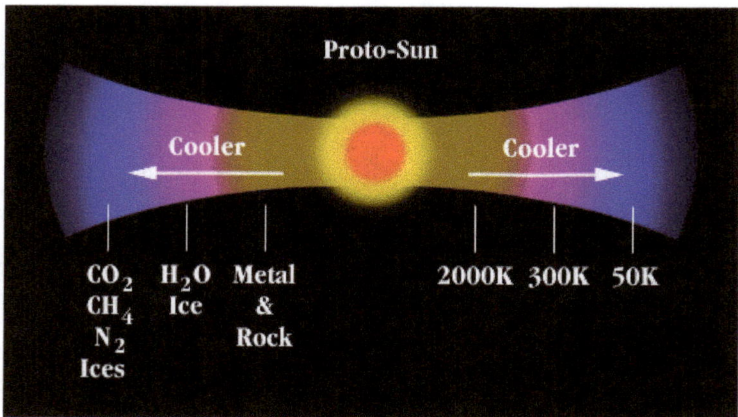

Figure 5.4. The temperature structure of a proto-stellar disk. The highest temperatures are realized at the inner edge of the disk, closest to the central proto-star. High-melting-point material will form in the inner region of the disk, at temperatures below 1500 kelvins, while ices can only form in the outer disk once the temperature has dropped below 273 kelvins.

to drop, and this will result in the development of specific domains where metals, with high melting points, will form interior to the region where silicate material, with lower melting points, will form, and so on with carbonaceous compounds and ices forming in the outermost regions of the disk.

The condensation sequence that comes about, due to the steady temperature decrease within the disk with radius, dictates that the basic planet-building material in the inner regions of the disk will be composed of metals and silicates, while the solid planet-building material in the outer disk will be predominantly ices of one form or another. The radial distance beyond which water ice will remain in its solid form is called the ice line (technically, of course, an ice circle about the proto-star), and in the case of our solar system detailed numerical calculations indicate that the ice line will fall at a distance of about 3 AU from the proto-Sun. This distance is significant, since it falls in that region which separates the terrestrial planets from the Jovian planets. Interior to the ice line the planets are predominantly composed of metals and silicate material. Beyond the ice line the planets are large and mostly composed of hydrogen and helium, and they are rich in their number of accompanying icy moons. This planetary dichotomy is related to the basic building units involved in their construction. Interior to the ice line the predominant building blocks will be asteroid-like planetesimals; beyond the ice line the predominant building blocks will be comet-like planetesimals.

5.3   Planetesimals

Planetesimals are the first formed, solid structures that begin to coalesce and then grow by accreting material from the proto-stellar disk. The essential aphorism that describes the growth of planets is "from little acorns to might oak trees grow." Indeed, planet formation proceeds from the bottom up. First small grains begin to condense and coalesce in the disk mid-plane (where the density of material is at its highest), these small grains interact via direct accretion (which is basically a hit-and-stick process) to produce pebble-sized structures, and these structures grow by accretion to produce boulder-sized structures, and so on. Once the planetesimals have grown to a size of a few tens of kilometers across, the accretion cross-section grows dramatically and larger-sized bodies will start to rapidly form. In the direct accretion process, the cross-section area for accretion σ is simply the area presented by a body to the on-coming flow of material (Figure 5.5).

With reference to Figure 5.5, imagine that we are following the growth of the large seed particle as it moves through a sea of

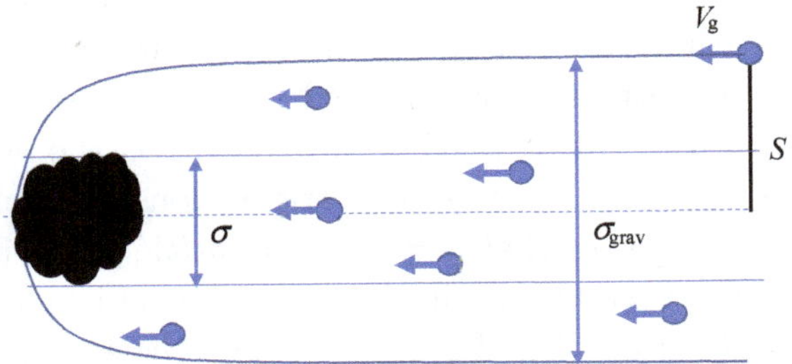

Figure 5.5.   The hit-and-stick accretion scenario has a cross-section of interaction σ, but when gravity becomes important the cross-section of interaction increases to $\sigma_{grav} > \sigma$.

smaller grains. If the number of grains is n_g per cubic meter and the characteristic grain velocity is V_g, then the number N of grains encountered by the seed particle in time ΔT will correspond to the volume of a cylinder of cross-section area σ and length $\Delta T V_g$ multiplied by the number of grains per unit volume, giving $N = \sigma (V_g \Delta T) n_g$. So, after time ΔT, the mass of the seed particle will have grown by the amount $\Delta M = N M_0$, where M_0 is the grain mass (here, all grains are assumed to have the same mass). Substituting for N accordingly reveals the accretion rate as

$$\frac{\Delta M}{\Delta T} = M_0 \pi R^2 V_g n_g \tag{5.5}$$

where we have substituted for $\sigma = \pi R^2$, where R is the radius of the seed particle. Accordingly, Equation (5.5) indicates that when direct, hit-and-stick, accretion dominates then the mass of the seed particle increases as its radius squared. The bigger the particle is, so the more rapidly does its mass increases. This result is characteristic of the accretion process in general, with the biggest particles growing at the expense of the smaller ones — the appropriate aphorism is "the rich get richer at the expense of the poor."

When gravity becomes important, the accretion rate changes to a cross-section area $\sigma_{\text{grav}} > \sigma$. To determine the mass growth rate under the gravity-dominant condition, we will need to apply the conservation laws of momentum and energy. If we follow the trajectory of a test particle that approaches the seed grain with a velocity V_g at a displacement distance of S from the mid-plane, then the initial angular momentum is $h = mSV_g$, where m is the mass of the test grain. When the test grain is accreted onto the seed particle, the angular momentum will be mRV_{hit}, where V_{hit} is the velocity with which the test particle hits the seed particle. Conservation of angular momentum now requires that $h = mSV_g = mRV_{\text{hit}}$. Turning now to the energy of the system, the initial potential and kinetic energies of the test particle are $\frac{1}{2}mV_g^2 - GMm/D$, where D is the initial separation of the seed and test particle, and G is the gravitational constant. When the test particle hits the seed particle, the sum of its kinetic and potential energies will be $\frac{1}{2}mV_{\text{hit}}^2 - GMm/R$.

These initial and final energy terms, by the conservation of energy, must be equal and by adopting the approximation that $D \gg R$ and that the initial potential energy is very small, we have $V_g^2 = V_{\text{hit}}^2 - 2GM/R$. Substituting for V_{hit} in the conservation of angular momentum equation now gives

$$S^2 = R^2 \left[1 + \left(\frac{V_{\text{esc}}}{V_g} \right)^2 \right] \tag{5.6}$$

where V_{esc} is the escape velocity of the seed particle: $V_{\text{esc}} = \sqrt{(2GM/R)}$. Looking at the escape velocity in terms of the radius R and density δ of the seed particle, $V_{\text{esc}} = R\sqrt{(8\pi G\delta/3)}$. Written in this manner and returning to Equation (5.6), we see that S^2 varies as a quantity depending on the radius of the seed particle to the fourth power: R^4. Having found an expression for S, so, in time ΔT the mass of the seed particle will increase by $\Delta M = NM_0$, where, as before, $N = \sigma(V_g \Delta T)n_g$, but now $\sigma = \pi S^2$. Accordingly, the accretion rate of the seed particle, when gravity is important, will vary as R^4, as opposed to R^2 in the case of hit-and-stick accretion.

In this manner, once planetesimals reach a critical size where gravity becomes important, so the accretion rate picks up dramatically, and the bigger seed particles rapidly sweep up the smaller particles to produce even larger planetesimals, and eventually, once a size several hundreds to a thousand or so kilometers across is achieved, proto-planetary embryos are produced. The planets themselves grow by the accretion of these large planetary embryos. At every step in the planetary formation process, chance is the key factor; some planetesimals will be bigger than others, and these will grow faster, but they might later be destroyed or accreted by an even larger planetesimal. The process plays out over hundreds of thousands of years, and only stops once a few dominant masses (the future planets) have chanced to form in regions of the disk that are sufficiently far apart that their mutual gravitational interactions are small and where their orbits are relatively stable (but see below).

The picture of planetesimal and eventually planet growth by collisions and accretions is certainly correct in outline, but there are still many details that require further study. Most recently,

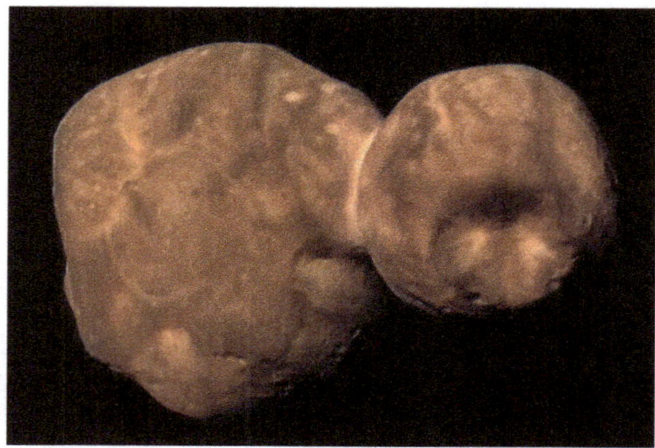

Figure 5.6. Kuiper Belt object 2014 MU65 = 486958 Arrokoth. This leftover icy planetesimal is some 36 km in total length. Image courtesy of NASA.

for example, observations gathered by the *New Horizons* spacecraft as it flew past Kuiper Belt object 486958 Arrokoth (Figure 5.6) have been used to reinterpret the early stages of planetesimal growth. The spacecraft imagery revealed that Arrokoth is a bi-lobed structure, composed of two disks that must have joined together at low velocity. Not only this, the observations also revealed that Arrokoth is remarkably uniform in its composition. These two observations have resulted in the argument that Arrokoth did not grow by simple, small-to-large hierarchical process. Rather, Allan Stern (Southwest Research Institute in Boulder, Colorado) and coworkers argue that the two lobes condensed directly out of a spinning cloud of icy particles in the primordial solar nebula. As with the solar nebula itself, but on a much smaller scale, the spin of the collapsing particle cloud produces the flattened disk-shaped lobes; and if two collapsing particle clouds are close enough, they will begin to slowly orbit each other until a gentle surface contact occurs, thus producing a bi-lobed structure. One of the problems this scenario of particle-cloud collapse solves is that concerning how to make small, pebble-sized particles stick together given that they will be moving and colliding with relatively high speeds. Once structures like Arrokoth have formed from collapsing particle clouds, however, then the accretion process can proceed apace and begin to build up larger and larger structures.

An additional issue with respect to the early phases of planetesimal formation that has yet to be solved is when, where, and by what mechanism(s) did chondrules form. Since chondrules (recall Section 1.2) are the main component of the abundant chondrite meteorites, the question is, did the chondrules form independently within the protoplanetary disk, before being incorporated into asteroid-like planetesimals, or did they form at the same time as, and in cahoots with, the planetesimals? Radionuclide ages indicate that the very first, that is oldest to us, objects to form in the solar nebula are the calcium-aluminum inclusions (CAIs) — see Chapter 8. These irregular-shaped granules form in the high temperature environment near to the inner edge of the protoplanetary disk, and have ages that date, with very little variation, to 4.567 billion years. Chondrules, on the other hand, appear to have mostly formed 1 to 3 million years after the CAIs. As of the present there are many models capable of explaining chondrule formation, one researcher poignantly using the phrase "a distressingly long list of possibilities" [1], and it is not clear if they all contribute to the chondrule population, or if one method dominates over the others.

Chondrules are the solidified droplets of once molten silicate material, and their crystallization structures indicate that they must have cooled rapidly, within a timeframe of just minutes to hours. Whatever the process that resulted in the production of chondrules, it must have been extreme, in order to raise the progenitor material to temperatures of at least 1600 K, and transient, so as to allow for quick cooling. Among the processes that have been suggested for chondrule formation are, that they are the cooled droplets ejected when two near-molten asteroids[4] collide; that they are molten spray droplets produced as a jetting effect when two stony (non-differentiated) asteroids collide; that they are the result of the Sun going through an FU Orionis outburst triggered by abrupt mass transfer from the protoplanetary disk; that they are due to lightning discharge events within the protoplanetary disk; and last, but by no means

---

[4]This process invokes the idea of radiometric heating and asteroid differentiation — see Chapter 13.

least, that they are the result of shock waves generated by fast moving planetesimals. This latter model is predicated on the rapid formation of Jupiter since it is this planet that is required to excite the orbital eccentricity of the planetesimals. It also invokes the idea of a concomitant origin for the rocky planetesimals and chondrules. Other issues with chondrule formation that have yet to be resolved are, why are they so chemically diverse, and why are they so small, being characteristically limited to no more than a few millimetres in size. That chondrules formed early on within the solar system seems clear, but their origins are still largely a mystery.

5.4   Rock or gas

The accretion process plays out at different rates according to location in the disk. This is a reflection of the decreasing density of material within the solar nebula moving outwards from the Sun. The density decrease, however, is also composition dependent. Interior to 3 AU (the location of the ice line) the only stable, solid materials are silicates and metals. Beyond 3 AU, however, ices can form and remain in their solid phase, producing an overall density increase. Eventually, however, the amount of material from which to form planets, whether from gas, rock, metals, or ice, drops towards zero. This steady decrease in available building material sets the timescale for planet growth according to heliocentric distance, and (as ever) these reveal problems with the simple accretion model. Numerical studies indicate that Earth and inner-solar-system planets take several hundred million years to acquire their final masses. In contrast, however, Jupiter, in spite of its high mass, forms much more rapidly, perhaps within a few tens of thousands of years. This rapid formation time is due to the quick development of a large central core, which can then accrete gas directly from the disk. The core grows rapidly because of the efficient accretion of icy planetesimals. Once the core is above several Earth masses then it can begin to accrete gas directly from the nebula disk.

The ability of a gas giant planet to hold on to its gas envelope is dictated by its mass and radial distance from the Sun. It is the mass of the planet that sets the escape velocity V_{esc}, while it is the radial distance from the Sun that sets the temperature and hence characteristic thermal velocity V_{therm} of the envelope molecules. Provided $V_{therm} < V_{esc}$ then the gas in the envelope will not be able to escape from the planet. There is, however, a tail-waging-the-dog effect here in that the thermal velocity is based upon statistics (the Maxwellian distribution) and there are always some molecules in the envelope that will have $V_{therm} \gg V_{esc}$, and this means that some molecules will slowly leak into space. Detailed numerical calculations indicate that provided $20V_{therm} < V_{esc}$ then a planet should be able to retain its atmosphere for at least the age of the solar system — a timespan of some 4.5 billion of years.

The typical thermal velocity for molecules of mass μm_H, where m_H is the mass of the hydrogen atom and μ is the mean molecular weight, in a gas at temperature T is

$$\frac{1}{2}\mu m_H V_{therm}^2 = kT \tag{5.7}$$

where k is the Boltzmann constant. The retention condition that $20V_{therm} < V_{esc}$ can now be reworked by substituting for the escape velocity $V_{esc} = \sqrt{2GM/R}$, where M and R are the mass and radius of the planet. Accordingly, the requirement for atmospheric retention over the age of the solar system is

$$M(\text{kg}) > 5.4 \times 10^{24}(T/\mu)/\sqrt{\rho} \tag{5.8}$$

where ρ is the bulk density of the planet.

The temperature of the atmosphere is controlled by the distance of the planet from the Sun and the greenhouse heating effect that results in accordance with the atmospheric composition. The blackbody temperature of a planet is derived upon the basis of a balance between the energy flux received at a distance D away from the Sun, and the energy radiated back into space at the planet's surface. Accordingly,

$$\frac{L}{4\pi D^2}(\pi R^2)(1 - A) = (4\pi R^2)\varepsilon\sigma T^4 \tag{5.9}$$

where L is the Sun's luminosity, R is the planet's radius, A is the albedo (accounting for the fraction of energy that actually gets through the atmosphere to heat the ground), ε is the emissivity (which accounts for the efficiency of reradiation of heat), and σ is the Stefan–Boltzmann constant. The factor of 4 on the righthand side of Equation (5.9) is set according to the rapid-spin limit in which it is assumed that energy is effectively reradiated from the entire surface of the planet. The slow-rotation limit would reduce the factor of 4 to 2, with the idea being that only half the planet's surface — the night side — is efficiently reradiating energy back into space.

Reworking Equation (5.9) to express the temperature as a function of distance from the Sun, we have:

$$T(\text{K}) = \frac{279}{\sqrt{D}} \left(\frac{1 - A}{\epsilon}\right)^{1/4} \tag{5.10}$$

where D is now expressed in astronomical units. From Equation (5.10) it is seen that the characteristic temperature decreases as the inverse square root of distance from the Sun. Taking an albedo of $A = 1/3$ and an emissivity of $\varepsilon = 0.9$, at $D = 1$ AU from the Sun the blackbody temperature is $T(\text{K}) = 259$. The actual characteristic temperature of Earth's atmosphere is $288\,\text{K}$, indicating some 29 degrees of greenhouse heating. Furthermore, taking a characteristic bulk density of the terrestrial planets to be $4800\,\text{kg/m}^3$, so Equation (5.8) indicates that the Earth would need to have a mass of $M(\text{kg}) \approx 2.2 \times 10^{25}$ in order to hold on to a hydrogen ($\mu = 1$) atmosphere — that is the Earth would need to be at least 4 times more massive than it actually is. At 5.2 AU from the Sun the characteristic temperature is of order $110\,\text{K}$, taking a characteristic density of a Jovian planet to be $1200\,\text{kg/m}^3$, so a hydrogen atmosphere can be maintained provided $M(\text{kg}) > 1.7 \times 10^{25}$, that is at least 3 times more massive than the Earth.

So-called super-Earths have masses up to about 8 to 10 times the mass of the Earth, and these objects mark the upper limit for terrestrial planets. That is, they have a solid rocky mantle that sits on top of a metallic core, and they will additionally have an atmosphere, and hydrogen will form part of that atmosphere. Once the planet mass becomes larger than about 10 Earth masses, however, then run-away accretion of gas from the disk will set in. In this way, a gas giant planet will form, with the mass of the envelope (predominantly composed of hydrogen and helium) dominating over that of the core. Gas giants, in principle, will keep acquiring mass for as long as there is gas within the protoplanetary disk for them to sweep up. While Jupiter and Saturn are expected to grow rapidly, because of their relatively close location to the ice line, Uranus and Neptune will take longer to form because they are further out in the disk where there is less material for them to sweep up. Numerical studies suggest that, at their presently observed distances from the Sun, it will take at least many tens to perhaps a few hundred million years for the outermost planets to form. This result, however, is problematic since it appears that protoplanetary disks do not survive for much more than 5 to 10 million years.

The lifetime of a protoplanetary disk is dictated by the parent star, and specifically the tantrums that low-mass stars undergo as they approach the main sequence. The main sequence is a feature in the Hertzsprung–Russell diagram where the vast majority of stars reside when plotting luminosity against temperature,[5] and it corresponds to the phase where a star is generating internal energy via the conversion of 4 protons into one helium atom nucleus. Prior to the initiation of the hydrogen fusion reactions within its core, a proto-star generates energy through gravitational contraction. While in the transition stage from being a proto-star to a *bona fide* star, a low-mass Sun-like star undergoes what is known as a T Tauri phase.

---

[5]Constructed independently by Ejnar Hertzsprung (c. 1911) and Henry Norris Russell (1913), the HR diagram is a plot of stellar temperature versus luminosity. When stellar data is arrayed in this diagram, the vast majority of stars plot on a diagonal locus (the main sequence) that runs from the very hot, high luminosity stars to low temperature, low luminosity stars. Above the main sequence are the large, low temperature, high luminosity red giants; and below the main sequence are the small, low luminosity white dwarfs. Stars on the main sequence generate internal energy via hydrogen fusion reactions, while those in the red giant region via helium fusion reactions. White dwarfs are on a cooling sequence and no longer generate energy within their interiors.

Named after the prototype variable star T in the constellation of Taurus, this phase sees the star develop a strong outflow of gas from its surface. It is this strong wind and associated magnetic field that pick up the dust and gas resident in the disk and carry it away into interstellar space. In essence the T Tauri wind sweeps the larder clean, and in short order destroys the disk and thereby stops the growth of gas giant planets. Observations of young star clusters indicate that the T Tauri phase typically limits protoplanetary disk lifetimes to about 5 million years.

That the solar nebula lasted for a little less than the canonical 5 billion years is evidenced by information teased from angrite meteorites. Indeed, the message lies within the reconstruction of the Sun's magnetic field strength. This latter information can be gauged by carefully measuring the remnant magnetization of meteorites in the laboratory. Huapei Wang (MIT) and coworkers have developed a highly sensitive technique for studying the magnetic properties of meteorites and have combined this data with formation age estimates (based upon uranium to lead composition measurements — recall Chapter 4). Magnetic measurements related to the Semarkona meteorite (an LL3 chondrite that fell in India on October 26th, 1940) indicate that it formed in an environment when the Sun's magnetic field strength was some 5 to 50 microtesla, about 2 million years after the solar system formed (set according to the CAI crystallization age at 4.567 billion years ago). When the team repeated the measurements on four angrite meteorites, which (recall Section 1.2) are examples of basaltic achondrites, they found that these bodies formed in an environment where the Sun's magnetic field strength was near zero (or at least smaller than 0.6 microtesla). The D'Orbigny angrite (which was found in Argentina in 1979), for example, formed in an environment where the magnetic field strength was smaller than 0.3-microtesla, 4 million years after solar system formation. The combined meteorite measurements enabled Wang and coworkers to establish that the Sun's magnetic field strength fell from 5–50 microtesla at 1 to 3 million years after the formation of the solar system to smaller than 0.6 microtesla at 3.8 million years. The relevance of the dramatic decline in the Sun's deduced magnetic field strength is that this is thought to indicate its arrival on the main sequence and the end of its T Tauri phase. In other words, by the time that the solar system was about 4 million years old, its surrounding disk of gas and dust had been cleared away, and the accretion growth of the gas giants must have ended. The paleomagnetic data from meteorites, therefore, implies that Jupiter, Saturn, Uranus, and Neptune must have acquired their (present-day) observed mass within 4 million years of the formation of the solar system.

5.5 Numerical simulations

The problem of the long, 10 to 100 million-year, growth timescales canonically derived for the outermost gas giants prompts the question, did they actually form where we see them today, or did they form closer in towards the Sun where the disk density is higher and the growth time smaller. And, the answer to this question is a clear no we do not see them where they formed, but where and when, and under exactly what conditions they formed remain open issues yet to be decided upon (if ever). It now seems clear that Uranus and Neptune must have formed much closer in towards the Sun, in a higher density part of the solar nebula, than their present orbital radii indicate. Not only this, it is also additionally clear that Jupiter and Saturn must have also undergone some significant migration prior to finding their currently observed positions. These two ideas, close-in formation and orbital migration, are encapsulated within the so-called Nice Model and the Grand Tack model.

The Nice Model, so-called as it was largely developed at the Observatoire de la Côte d'Azur, in Nice, France, was first described in a series of three articles published in 2005. The model was a collaborative work of Rodeny Gomes, Alessandro Morbidelli, and Kleomenis Tsigans (all then researchers at the Observatoire), and Hal Levison (Southwest Research Institute). The aim of the papers [2] was straightforward and required intensive numerical simulations to be made. The starting primordial disk was taken to contain the four gas giants, all of which were moving on circular orbits, at distances between 5.5 and 17 AU from the Sun. Surrounding the planets was a dense disk, containing some 35 Earth masses of rocky and icy planetesimals, extending out to about 35 AU. The model then numerically followed the orbital evolution of the planets as they interacted with the planetesimals. The calculations revealed that the evolution effectively started from the outside in, with the outermost planet interacting with and scattering the planetesimals located in the inner edge of the disk. These interactions resulted in planetesimals being scattered inwards, to interact with the other planets, and the outward movement of Uranus, Neptune, and Saturn. After several hundred million years, Saturn moved into a 1:2 mean-motion resonance with Jupiter (recall Chapter 3), and this started to increase the orbital eccentricity of Saturn, which then destabilized the orbits of Uranus and Neptune. At this stage very rapid changes began to take place. The outermost planets moved outwards into the disk, essentially acquiring their presently observed orbits, and in the process, they additionally disrupted the planetesimal disk, causing a massive influx of material into the inner solar system.[6]

One simulation from the Nice Model is shown in Figure 5.7. In this case, the computer simulation followed the orbital evolution of the outer four planets and 3500 planetesimals. Three curves are shown for each planet, corresponding to their perihelion, semi-major axis, and aphelion distance as a function of time. The entry of Saturn into a 1:2 resonance with Jupiter occurs 600 million years into the simulation, and this produces a dramatic change in orbital spacing, resulting (in this specific simulation) with the orbits of the two outermost ice giants exchanging position.

The Nice Model provides a scenario under which the accretion and growth timescales for Uranus and Neptune are solved, since they form much closer in towards the Sun where the disk density is higher and the accretion time more rapid. It also provides for an explanation of the Late Heavy Bombardment (LHB) in terms of the dramatic scattering of disk planetesimals when Neptune rapidly

---

[6]An additional feature of the Nice model is that it allows for a capture explanation of the Trojan asteroids that are in a 1:1 mean-motion resonance with Jupiter.

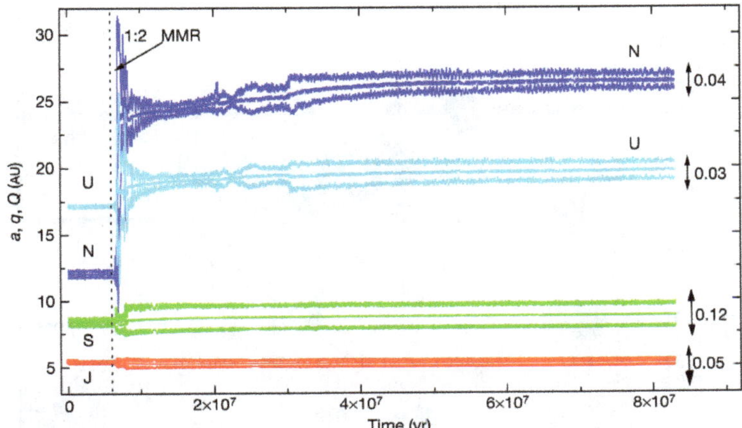

Figure 5.7.   Simulation sequence from the Nice Model. Each set of color-coded curves shows the perihelion (q), semi-major axis (a) and aphelion distance (Q) of the planets from Jupiter to Neptune. The numbers to the far right in the diagram correspond to the maximum orbital eccentricity experienced by each planet during the final 2 million years of the simulation. Note, in this simulation the initial position of Neptune is interior to the orbit of Uranus. Image courtesy of Wikipedia.

acquires its final orbital displacement. It also allows for the final planetary orbits to be nearly circular. For all these positive results, however, the Nice Model, in many ways, opens up more questions than it solves. Specifically, the starting conditions are all fine-tuned so as to achieve the delay of 600 million years before the onset of the LHB. There is no theory or observations to constrain the starting positions of the planets, and/or the size and mass of the primordial planetesimal disk. That the model can be made to work is its main selling point.

The Nice Model cannot be the full story of planet growth and migration within the solar system. Indeed, current theory and observations require that prior to its starting phase, the newly formed (and still forming) planets must have already undergone some significant migration within the solar nebula. The orbital migration that takes place within the first half-million years of the solar system is described according to the so-called Grand Tack Model (Figure 5.8). This is, once again, a model based upon extensive

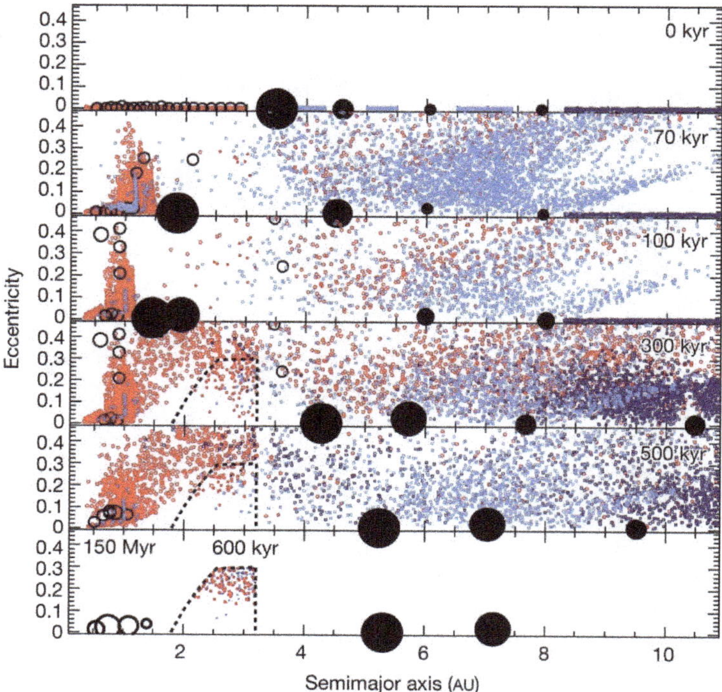

Figure 5.8. A series of images showing the position versus eccentricity of planets, planetesimals, and planetary embryos in a numerical simulation of the Grand Tack Model. The simulation follows the orbital evolution of 4830 planetesimals [3]. The gas giant planets are shown as black circles, stony and icy planetesimals are shown as red and blue dots respectively, and outer disk (Kuiper-Belt-like) planetesimals are shown as black dots. Open circles indicate planetary embryos, and the dashed region between 2 and 3 AU corresponds to the asteroid zone. The first panel shows the starting configuration, with subsequent panels showing the state of evolution after 70, 100, 300, and 500 thousand years. The last (bottom) panel shows the configuration after 150 million years — at this time the embryos that will eventually become the terrestrial planets have formed in near circular (eccentricity ~ 0) orbits. Also (although not specifically modeled) at this stage, the collision with the proto-Earth responsible for the Moon's formation will have occurred. Image from Walsh *et al.* [3].

numerical calculations, but in this case, the evolution is invoked according to observations related to exoplanetary systems. One of the inescapable facts that emerged from the first exoplanet detections was that planetary migration must be much more important than had previously been allowed for. Indeed, the first exoplanet 51 Pegasi b is what has become known as a hot Jupiter — that is, it is a $1/2$ Jupiter mass gas giant planet that orbits its parent, Sun-like star, at a distance of just 0.5 AU with an orbital period of just 101.5 hours. From the very outset, if we indeed know anything at all about planet formation, it is clear that this gas giant cannot have formed where we see it now. It must have formed beyond the 3 AU ice line of proto-51 Pegasi, and then moved inwards to its currently observed position. This dramatic orbital migration is explained by planet–disk interactions that were, prior to the last 25 years, ignored (largely because available computing power could not handle such complex calculations). In the Grand Tack Model, planet–disk interactions are taken into account, and it is the orbital migration of Jupiter and Saturn specifically that turns out to be important — the most important part of the simulation being the mechanism that stops the migration of Jupiter all the way inwards, to be consumed by the proto-Sun. Once again, it is the attainment of a resonance condition that saves the day.

Planet migration comes about because of interactions with the material in the disk. The Grand Tack Model (Figure 5.8) follows this migration, and indicates that some 70,000 years after its formation, Jupiter will have migrated some 2 AU inwards to an orbital radius of about 1.5 AU. In this manner it will deplete the supply of planetesimals in the inner solar system — these, recall, will be the planetesimals out of which the terrestrial planets must eventually form. At this stage Saturn will not have migrated inwards very much, but in the ensuing 30,000 years it begins to grow in mass and move inwards, so that some 100,000 years into the simulation, it enters into a 3:2 mean-motion resonance with Jupiter. The numerical simulations then reveal that due to the development of this resonance and continued disk interactions, Jupiter and Saturn now move outwards — this is the *tack* part of the model. It is the evolution of this resonance that specifically stops Jupiter from becoming a hot Jupiter (or a solar meal). The outward motion of the resonance-locked Jupiter–Saturn pair continues for another 4 to 5 hundred-thousand years, after which time they, along with Uranus and Neptune, acquire orbital locations consistent with the starting conditions required of the Nice Model.

One result of the Grand Tack Model is that it truncates the inner edge of the planetesimal disk at about 1 AU, and thereby reduces the planetesimal mass available to form Mars — this solves the so-called stunted Mars problem, whereby its observed mass is much smaller than standard (without the *tack*) accretion models would otherwise predict. Also, since Jupiter and Saturn cross the zone where we currently find the main-belt asteroids twice, this region is depleted of planetesimals, and this explains its low total mass (and why no additional planet is found there[7]). The model also indicates that the region between Mars and Jupiter should contain planetesimals scattered from inside and outside of Jupiter's orbit, thus accounting for the primitive asteroids being located towards the outer edge of the present-day main-belt asteroid zone (see Section 12.1). Once again, however, the Grand Tack Model presents almost as many new problems as the old ones it attempts to solve, but, for our purposes, it is one of the stages through which the solar system passes prior to the onset of chondrule formation. Likewise, chondrule formation will have completely ceased prior to the onset of the Nice Model instability.

With the completion of the Nice Model machinations, the solar system is deemed to have acquired its presently observed form, with a series of terrestrial and Jovian planets, separated by a low-mass asteroid zone, all surrounded by an extended Kuiper Belt of primitive, icy planetesimals (Figure 5.9). The outermost solar system structure, the Oort cloud, is additionally composed of planetesimals scattered outward during the Grand Tack migration and the Nice Model instability, with additional long-term sculpting by the random passage of nearby stars.

---

[7]The combined mass of all main-belt asteroids amounts to just 3.2% of the mass of Earth's Moon.

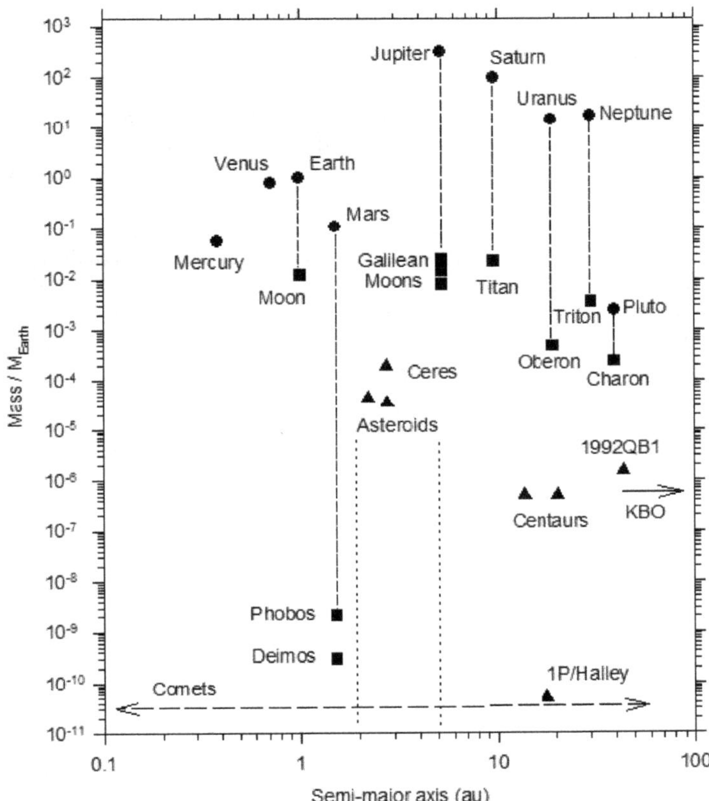

Figure 5.9. Characteristic mass versus semi-major axis for various families of objects in the solar system. The mass is expressed relative to that of the Earth. The upper righthand corner of the diagram contains the gas giant planets; the mid-upper left of the diagram contains the terrestrial planets. The lower half and middle of the diagram show the main-belt asteroid region — Ceres being the most massive asteroid. In the mid- to lower right region of the diagram are the Kuiper Belt objects (1992 QB1 being the first such object discovered) and the Centaurs (which are objects located on unstable orbits between Uranus and Saturn). The horizontal dashed line at the bottom of the diagram shows the range of semi-major axis for cometary nuclei. Vertical long-dashed lines link planets to their moons.

References

[1] M. McSween, Jr. *Meteorites and their Parent Planets.* Cambridge University Press (1999).

[2] The first paper, in a series of three, on the Nice Model is: R. Gomes, H. Levison, K. Tsiganis, and A. Morbidelli, "Origin of the cataclysmic Late Heavy Bombardment period of the terrestrial planets." *Nature*, **435**, 466–469 (2005). This was followed by, K. Tsiganis, R. Gomes, A. Morbidelli, and H. Levison, "Origin of the orbital architecture of the giant planets of the Solar System." *Nature*, **435**, 459–461 (2005), and, A. Morbidelli, H. Levison, K. Tsiganis, and R. Gomes, "Chaotic capture of Jupiter's Trojan asteroids in the early Solar System." *Nature*, **435**, 462–465 (2005). For a recent review of solar system formation see: A. Morbidelli and S. Raymond, "Challenges in Planet Formation." *JGR-planets*, **121**, 1962–1980 (2016).

[3] K. Walsh, A. Morbidelli, S. Raymond, D. O'Brien, and A. Mandell. "A low mass for Mars from Jupiter's early gas-driven migration." *Nature*, **475**, 206–209 (2011). See also, S. Raymond and A. Morbidelli, "The Grand Tack Model: A critical review." https://arxiv.org/abs/1409.6340 (2014).

Chapter 6

Atmospheric Interactions

*I think I'll dismember the world and then I'll dance
in the wreckage.*

Neil Gaiman, *Preludes and
Nocturnes* (1989)

The entry of a meteoroid into the Earth's atmosphere is a dramatic, even brutal affair. It is dramatic in terms of the physical processes that are going to occur, and dramatic in terms of what an observer will see and possibly hear. The chain of events that are enacted, once the meteoroid is in the atmosphere, result in a staggeringly violent deceleration, a large loss of material mass, and the production of light along an ionization trail. In this section a minimal set of equations will be developed to describe how a meteoroid interacts with the atmosphere. These equations are predicated upon two fundamental precepts of physics — the conservation of momentum and the conservation of energy.

6.1 Deceleration

For simplicity let us begin by assuming that the meteoroid is a solid sphere and that it remains spherical at all times as it descends through the atmosphere. The meteoroid will have some density ρ_{met}, with this density typically being of order $3000\,\text{kg/m}^3$ in the case of a stony meteoroid, and $9000\,\text{kg/m}^3$ in the case of an iron meteoroid. The meteoroid mass m, at all times during its descent, will correspond to its specific radius r and density, with $m = 4\pi\rho_{\text{met}}r^3/3$.

129

To determine an equation for the deceleration, that is the change in the meteoroid's velocity, we can use the conservation of momentum, which argues that the product of mass multiplied by velocity should be conserved. Let us consider the situation in which a meteoroid of mass m undergoes a small change in velocity δV over a time interval δt. In this same time interval, the amount of atmospheric gas intercepted by the meteoroid will be $S\rho_{\text{atm}}(V\delta t)$, where S is the cross-section area of the meteoroid (with $S = \pi r^2$) and ρ_{atm} is the atmospheric density (which recall will vary according to atmospheric height — that is, it will increase as the meteoroid descends lower into the atmosphere and closer to the Earth's surface). The momentum of the oncoming airflow is accordingly: $S\rho_{\text{atm}}(V\delta t)V$.

Our two terms can now be equated (under the banner of the conservation of momentum) through the introduction of the drag coefficient Γ. This latter term accounts for the proportion of the momentum of the oncoming airflow that actually goes into the deceleration of the meteoroid. The value of Γ will vary according to the shape of the meteoroid and the characteristics of the airflow regime — that is, is the airflow in a free-flow, transitional, or turbulent state. In principle Γ must be larger than zero and can be greater than unity. For a sphere and under typical meteoroid deceleration conditions, it is generally assumed that Γ is a constant and has a value of $1/2$. Accordingly,

$$m\delta V = -\Gamma S\rho_{\text{atm}}V^2\delta t \tag{6.1}$$

where the minus sign indicates that the velocity will be decreasing. Equation (6.1) provides the first (differential) equation for the deceleration experienced by the meteoroid

$$m\frac{dV}{dt} = -\Gamma S\rho_{\text{atm}}V^2 \tag{6.2}$$

Substitution for the mass in terms of the meteoroid radius and density could be made in Equation (6.2) at this stage, but the mathematical details and rearrangements need not concern us here.

In general Equation (6.2) has no simple solution, and numerical techniques are needed to determine the change in velocity as a function of height in the atmosphere. In the highly over-simplified situation, however, where the mass of the meteoroid as well as the atmospheric density remain constant with $\rho_{\text{atm}} = \rho_0$, say, then Equation (6.2) can be directly integrated to yield the equation:

$$V(t) = V_\infty \exp(-t/k), \text{ with } k = \Gamma S \rho_0 / m \tag{6.3}$$

where V_∞ is the initial velocity and k is a constant. Accordingly in the simplified case of no mass loss from the meteoroid and a constant atmospheric density, the velocity decreases exponentially with time t — that is, until the meteoroid hits the ground of course, at which instant the meteoroid becomes a meteorite and the velocity becomes zero.

6.2 Mass loss

As a meteoroid descends through the atmosphere, it collides with atmospheric molecules and energy is accordingly exchanged between the meteoroid and atmosphere. The net result of the myriad collisions with atmospheric molecules is to heat up the meteoroid's surface, and eventually the temperature will exceed the melting point of its constituent material. At this time vigorous ablation sets in and the meteoroid begins to lose material from its surface, and accordingly its total mass will decrease. The small amount of material lost by the meteoroid δm in the short time interval δt can be expressed in terms of the conservation of energy. Accordingly, given a fraction Λ of the kinetic energy of the impinging gas molecules being transferred to the meteoroid, then in time δt, the energy transferred will be

$$E_{\text{gas}} = \Lambda \frac{1}{2} (S \rho_{\text{atm}} V \delta t) V^2 \tag{6.4}$$

where $0 < \Lambda < 1$ is the so-called heat transfer coefficient and the bracketed term is the mass of the impinging molecules. This energy, upon being imparted to the meteoroid, goes into the melting and vaporizing of its surface layers. The amount of energy required to melt and vaporize a quantity δm of meteoroid material will be

$E_{\mathrm{vap}} = \zeta \delta m$, where ζ is the sum of the terms accounting for the energy of melting and vaporization for the specific composition of the meteoroid. Setting $E_{\mathrm{vap}} = E_{\mathrm{gas}}$, the (differential) equation describing meteoroid mass loss revealed, and:

$$\frac{dm}{dt} = -\sigma \Gamma S \rho_{\mathrm{atm}} V^3, \quad \text{where } \sigma = \Lambda/(2\varsigma\Gamma) \tag{6.5}$$

The negative sign in equation comes about because the meteoroid is losing mass, and the σ term, known as the ablation coefficient, has been introduced. This latter parameter is generally a small number, and will vary according to the composition of the meteoroid (through the ζ term), the shape of the meteoroid (through the Γ term), and the energy transfer conditions (through the Λ term). The ablation coefficient typically has a value in the range 10^{-7} to $10^{-8}\,\mathrm{s}^2/\mathrm{m}^2$.

Dividing Equation (6.5) by Equation (6.2) provides a useful expression for the variation of the meteoroid mass with velocity. Namely,

$$\frac{1}{m}\frac{dm}{dV} = \sigma V \tag{6.6}$$

and this equation can be directly integrated to yield

$$m(t) = m_\infty \exp\{[V^2(t) - V_\infty^2](\sigma/2)\} \tag{6.7}$$

where m_∞ and V_∞ are the meteoroid's initial mass and velocity at the top of the Earth's atmosphere. Given that the velocity will decrease with time, as the meteoroid descends through the atmosphere, Equation (6.7) reveals that the meteoroid mass will also decrease — since the exponential term is becoming increasingly negative. With respect to the ablation mass loss condition it is generally assumed that the meteoroid velocity must be greater than some limiting value V_{dark}, $\sim 2\,\mathrm{km/s}$. Once $V(t) < V_{\mathrm{dark}}$, then mass loss via ablation stops, and as discussed below the light production mechanism will turn off — at this stage the now constant mass meteoroid has entered into its dark flight regime.

Equation (6.7) indicates that there is a strong selection effect against meteoroids having a high initial velocity, V_∞, from producing meteorites on the ground. Taking an ablation coefficient of

$\sigma = 5 \times 10^{-8}\,\mathrm{s^2\,m^{-2}}$, for example, then a 90% reduction in the initial mass [that is, $m(t) = 0.1 m_\infty$] will occur once $V_\infty > 10\,\mathrm{km/s}$. It is this velocity condition that largely explains why meteorites are not, in general, associated with meteor showers derived from comets. Although potentially meteorite-producing bodies most likely reside in cometary meteoroid streams, the characteristically high initial velocities (of order 30 to 60 km/s) associated with stream encounters mitigates against them surviving the effects of ablation, and delivering any appreciable amounts of material (beyond possibly minuscule dust grains) to the ground.

6.3 Light production

Equations (6.2) and (6.5) provide a set of equations that describe the variation of a meteoroid's mass and velocity as a function of time, and the next equation to be developed is that which describes the meteoroid's brightness — or more correctly the radiation intensity I. The light produced as a meteoroid descends through the atmosphere (giving it the appearance of a meteor) is derived through emission. That is, the ablated atoms from the surface of the meteoroid collide with and scatter from atmospheric molecules and this results in the atoms being placed in an excited state. It is the subsequent de-excitation of these atoms, through the emission of photons, that produces the light that is observed. Contrary to popular belief, the light from a meteor is neither a flame (that is the result of chemical burning), nor is it light emitted from the meteoroid's incandescent surface. The radiation intensity I is taken to vary in proportion to the kinetic energy of the mass δm evaporated in time δt, and accordingly,

$$I = \tau \left(-\frac{dm}{dt} \right) \frac{V^2}{2} \tag{6.8}$$

where the τ term is called the coefficient of radiative efficiency. This term is very poorly known, since it needs to take into account the total amount of light emitted by all of the different types of atoms, in their various excited states, that are involved in the emission process. The radiative efficiency is also likely to be dependent upon the velocity, and typically it is written in the form $\tau = \tau_0 V^n$, where τ_0 is a constant, V is the velocity, and where n is a constant usually taken to be of order one. In very general terms the coefficient of radiative efficiency is typically found to be of order 0.1 to 0.01 percent.

6.4   Electron line density

The atoms vaporized from the surface of meteoroid can also become collisionally ionized after ejection and this results in a trail of electrons being laid down along the path of the meteoroid. If the ionization probability is β (the probability that an ejected meteoroid atom will produce an electron–ion pair), then the number of electrons produced per second, and placed within the ion trail behind the meteoroid, will be $(\beta/\mu)(-dm/dt)$, where μ is the average mass of a meteoroid atom. Radio astronomers typically work in terms of the number of electrons per meter along the trail. Accordingly, the electron line density q (in units of electrons per meter) will be

$$q = \frac{1}{V} \left(\frac{\beta}{\mu} \right) \left(-\frac{dm}{dt} \right) \tag{6.9}$$

The relevance of the electron line density will be discussed in Section 7.17 later, but for the moment it suffices to say that it is the electron trail that enables meteor trails to be studied via back (as in radar) and forward (radio wave reflection) scatter techniques. Such systems, for example, can be used to probe high altitude winds (via trail distortion effects), allow radio communications, as well as determine meteoroid structure and meteoroid ablation characteristics.

Large quantities of research time, experimentation, and computation have been directed towards understanding, and predicting, the values for Γ, L, τ, β, and σ under various meteoroid–atmosphere encounter conditions, but for all the calculations to be discussed in this text, typical or average values will be assumed. In general, however, they are complex functions of meteoroid shape, velocity, and airflow regime.

In a plane-parallel atmosphere, where no account is taken for the curvature of the Earth's surface, the atmospheric path of a meteoroid is determined according to its initial zenith angle Z. This latter term is literally the angle that the meteoroid trajectory makes with respect to the vertical when it is at the top of the atmosphere, with $Z = 0$ indicating a vertical downward path. It will be shown in Section 7.15 that the most likely value for the zenith angle is $Z = 45$ degrees. In general, the variation in a meteoroid's atmospheric height is described by the equation

$$\frac{dh}{dt} = -V \cos(Z) \qquad (6.10)$$

which is to be solved according to some set starting height $h = h_{\text{start}}$ (say, for example, 200 km altitude). The negative sign in Equation (6.10) indicates that the altitude decreases as a function of time, with $h = 0$ indicating the attainment of ground contact.

6.5   Atmospheric density

The final relationship required to solve for the equations of meteoroid mass loss, deceleration, light production, and the electron line density is that accounting for the variation of atmospheric density ρ_{atm} with height h. The most straightforward equation for this variation is that derived for a constant temperature atmosphere, and in this case the density decreases exponentially with increasing height above the Earth's surface

$$\rho_{\text{atm}} = \rho_0 \exp(-h/H) \qquad (6.11)$$

where $\rho_0 = 1.225 \, \text{kg/m}^3$ is the atmospheric density at sea level and $H = 8.5 \, \text{km}$ is the so-called atmospheric scale height.

6.6   Impact cratering

Some indication of a meteorite's ground-impact speed can be gauged by considering the amount of atmospheric gas that resides within the column through which it must pass to reach the ground. The total mass of atmospheric gas within a cylindrical column, of $1 \, \text{m}^2$ cross-section area, that cuts from the top of the atmosphere to the ground is given by the quantity: $M_{atmos} = P_S/(g \cos Z)$, where P_S is the pressure at sea level, g is the surface gravity, and Z is the angle that the cylindrical (flight path) column makes to the vertical. The rule-of-thumb argument is that a meteoroid of constant mass M_{met} can be considered as being "stopped," that is reduced from its cosmic encounter velocity to sub-sonic speeds, once it encounters more than $\eta > 10$ times its own mass in the form of atmospheric gas. With this constraint in place, the high-ground-impact condition (under the assumption of no mass loss through ablation) becomes

$$(\pi R^2)M_{atmos} > \eta M_{met} \tag{6.12}$$

where R is the meteoroid radius. Condition (6.12) accordingly sets a minimum size limit for crater production (that is a meteoroid reaches the Earth's surface with its cosmic speed largely intact — see Section 7.5) as

$$R > \frac{3}{4\eta}\frac{P_S}{\varrho_{met}\, g \cos Z} \tag{6.13}$$

where ρ_{met} is the meteoroid density. Inserting for $P_S = 10^5\,\mathrm{Pa}$, $g = 9.8\,\mathrm{m/s^2}$, and setting $Z = 45$ degrees, so $R > 0.31$ meters for stony meteoroids with $\rho_{met} = 3500\,\mathrm{kg/m^3}$, and $R > 0.12$ meters for iron meteoroids with $\rho_{met} = 9000\,\mathrm{kg/m^3}$.

These *illustrative* numbers shown in Table 6.1 indicate that the larger the initial size (and hence mass) the more likely it is that the velocity will remain high as the meteoroid descends through the

Table 6.1. Critical meteoroid diameters (in meters) for impact cratering on various planetary bodies. Three meteoroid compositions are adopted, ice, rock and iron, with the associated densities being 1000, 3500 and 9000 kg/m³ respectively.

	g (m/s²)	P_S (Pa)	Ice	Rock	Iron
Venus	8.9	9.2×10^6	219.28	62.65	24.36
Earth	9.8	1.0×10^5	2.16	0.62	0.24
Mars	3.7	636	0.04	0.01	0.004
Titan	1.4	1.5×10^5	22.73	6.49	2.53

Figure 6.1. A deep plunge pit (here being excavated) produced by the fall of one of the Campo Del Cielo meteorites. Image courtesy of https://www.dogonews. com/2016/9/22.

atmosphere. Eventually, the size of the meteoroid will be such that it will essentially pass through the atmosphere, barely noticing that it is there, in fact, and impact the ground with its cosmic (that is initial) velocity hardly changed. Such events will result in the meteoroids either burying themselves within deep plunge pits (Figure 6.1), or in the formation of an impact crater (Figure 6.2). In the latter case it is the kinetic energy of the meteoroid $K = \frac{1}{2}mV^2$ that results in ground-material excavation — see Section 7.5 for a description of the impact cratering process. The atmospheric filtering effect is clearly revealed in Table 6.1 with Venus and Titan being particularly prominent in that only large impactors are likely to penetrate the atmosphere to reach the ground. Indeed, spacecraft observations indicate that Venus and Titan show very few impact craters, and then only large ones, upon the respective surfaces.

Figure 6.2. The Carancas crater. This crater, some 14 meters across, was produced by a 1-m diameter stony meteoroid (Carancas meteorite). The impact velocity was relatively high in this particular case since the fall location, close to Lake Titicaca, is in the high Andean mountains. Image courtesy: http://meteoritegallery.com/carancas-h4-5.

6.7   Brightness

Meteoroids with initial sizes appreciably smaller than the limits set by Equation (6.13) and indicated in Table 6.1, will typically be near-fully vaporized as they pass through the Earth's atmosphere. A typical shooting star, for example, seen at any hour on any night (and technically day) of the year, has a diameter of perhaps just a few millimeters, and is fully vaporized by the time it has descended to an altitude of about 80 to 90 km. The more rarely observed, bright fireball, or bolide (see Chapter 19), which can dramatically light up the night sky, just like day-light, for a few seconds, might have a size of just a few centimetres across — and it too will be fully destroyed in the Earth's atmosphere. There is, accordingly, a sweet spot in the mass–velocity domain that allows for the existence of meteorites — too small and too fast and a meteoroid is destroyed in the atmosphere; too large and it is destroyed in an impact crater.

The characteristic brightness of a meteor, as it descends through the atmosphere, can be determined via Equation (6.8). Specifically, combining Equations (6.8) and (6.5) indicates that:

$$I = K_1 \rho_{\text{atm}} m^{2/3} V^5 \tag{6.14}$$

where K_1 accounts for all the constant terms. The mass enters as the $2/3^{\text{rd}}$ power in Equation (6.14) by substitution for the cross-section area S term in Equation (6.5) — this term varies as the radius squared, which in turn varies as the mass to the $1/3^{\text{rd}}$ power, and accordingly, $S \sim m^{2/3}$. If I_{max} is the maximum brightness, then Equation (6.14) can be rewritten as

$$\frac{I}{I_{\text{max}}} = \frac{\rho_{\text{atm}}}{\rho_{\text{max}}} \left(\frac{m}{m_{\text{max}}} \right)^{2/3} \tag{6.15}$$

where ρ_{max} and m_{max} are the atmospheric density and meteoroid mass at the moment of maximum brightness. Now, differentiating Equation (6.14) under the condition of constant velocity (i.e., taking $V = $ constant so $dV/dt = 0$), and setting $dI/dt = 0$, the mass of the meteoroid at maximum brightness I_{max} can be found:

$$\frac{1}{I}\frac{dI}{dt} = 0 = \frac{1}{\rho_{\text{atm}}}\frac{d\rho_{\text{atm}}}{dt} + \frac{2}{3m}\frac{dm}{dt} \tag{6.16}$$

Substituting from Equations (6.11) and (6.14) when $I = I_{\text{max}}$, we recover the result that at the time of maximum brightness the meteoroid mass will be:

$$m_{\text{max}} = \frac{4HI_{\text{max}}}{3\tau V^3 \cos(Z)} \tag{6.17}$$

Integrating Equation (6.5) under the condition of constant velocity, a little algebra and substitutions from Equations (6.11) and (6.17), reveals that the meteoroid mass at maximum brightness is related to its initial mass m_0 by the reduction ratio of 8/27, namely

$$M_{\text{max}} = \frac{8}{27} m_\infty \tag{6.18}$$

that is, the initial mass m_∞ is reduced by a factor a little under $1/3^{\text{rd}}$. Equation (6.18) is derived on the basis that the atmospheric density $\rho(h)$ at the starting height h_0, where $I(h_0) = 0$, is also zero. Integrating Equation (6.5) — again assuming a constant velocity — it can be shown that the mass at any specific height in the atmosphere is to be expressed via the relationship

$$m = \frac{m_\infty}{27}(3 - \rho_{\mathrm{atm}})^3 \tag{6.19}$$

From Equation (6.19) it follows that the variation in brightness can be expressed as

$$\frac{I}{I_{\max}} = \frac{1}{4}X(3 - X)^2 \tag{6.20}$$

where $X = \rho_{\mathrm{atm}}/\rho_{\max}$. The author along with graduate student Megan Hargrove [1] have shown that the allowed solutions to the cubic Equation (6.20), given that $0 \leq I/I_{\max} \leq 1$, are such that the light curve must always be late peaked, with the ratio $t_{\mathrm{rise}}/t_{\mathrm{fall}} > 1$, where the rise and fall times correspond to the time to reach maximum brightness, and then decay away from that maximum back to zero.

While the constant velocity assumption for a meteoroid holds approximately true before the moment of maximum brightness, it does not hold true thereafter, and accordingly Equations (6.2), (6.5), (6.11), and (6.14) must be solved for via a numerical integration scheme. Such a numerical solution to the equations of meteoroid ablation is shown in Figure 6.3. In this example simulation the initial mass and velocity were taken as $1000\,\mathrm{kg}$ and $15\,\mathrm{km/s}$ respectively. The height range of greatest variation in the mass and velocity is located between $50 < h(\mathrm{km}) < 10$.

Observed meteor and fireball light curves are typically more complicated than that revealed by simple numerical simulations, showing flares and domains of near constant brightness (Figure 6.4), and sometimes periodic flickering (see Figure 6.7 below). In order to model these latter effects very specific model-dependent assumptions need to made — such as variations in the shape of the meteoroid, spin of the meteoroid, change in the ablation coefficient, or a change in the mode of mass loss and ablation. In general, the equations of ablation are solved with respect to observational constraints which

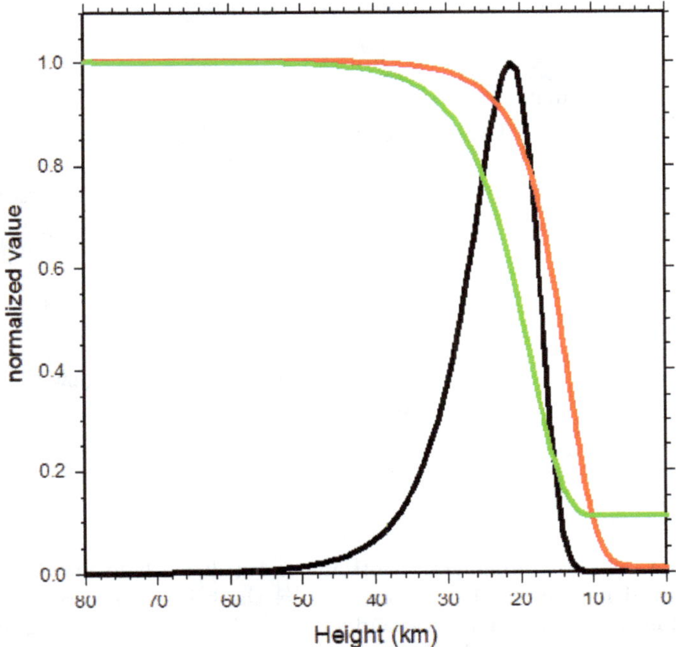

Figure 6.3.   The characteristic brightness (black line), mass (green line), and velocity (red line) variation of a meteoroid as it moves through the Earth's atmosphere from a starting height of 150-km altitude and zenith angle $Z = 45°$. Each curve is scaled according to the maximum value; the initial mass and velocity were taken to be 1000 kg and 15 km/s respectively, and the density adopted is that of a stony meteorite with $\rho_{met} = 3000\,kg/m^3$ — giving an initial diameter of 0.86 meters across. The ablation coefficient was taken as $\sigma = 5 \times 10^{-8}\,s^2/m^2$. In this specific simulation, dark flight began at an altitude of 9.2 km, and a meteorite of mass 108 kg (indicating that 90% of the original mass was ablated away) impacted the ground at a speed of 115 m/s.

provide information on the actual variation of brightness, height, and velocity along a fireball's path as it moves through the atmosphere. In essence the physical characteristics of mass loss and ablation parameters are reverse-engineered from the observations. Breaking the meteor/fireball path into small segments, and knowing the height and velocity range in these segments, the ablation equation can be solved for to determine the mass loss appropriate to the brightness change. Effectively, the brightness variation is governed by solving for

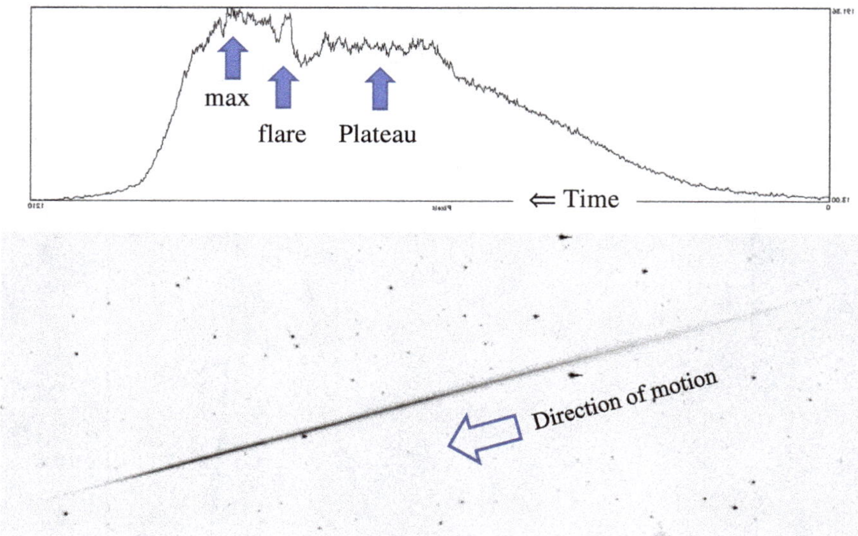

Figure 6.4.   Photographic trail (bottom) and deduced light curve (top) for a bright Perseid meteor. In this particular case, the brightness variation shows a plateau of near constant brightness and a distinct flare before reaching maximum brightness, but otherwise the light curve is late-peaked with a slow rise to maximum and a rapid decline thereafter.

Equation (6.8), and if the velocity is derived from the observations relating to the meteor, so the mass loss rate dm/dt can be estimated. If, for example, the meteor shows a brightness plateau of near constant magnitude, so dm/dt must also be constant (assuming that the velocity is constant over the trail segment being considered), and accordingly, the mass variation will be of a form $m(t) = m(0) - Kt$, where K is a constant related to the ablation coefficient and t is the time interval that spans the specific path segment being analyzed. Flare events can be modeled analytically, over a restricted height range, by assuming a small amount of material, of mass Δm, is ejected from the main body, and following its ablation under the approximation that the density varies in a linearly manner [see e.g. Reference 1] — this approximation can be used to model symmetric brightness humps superimposed upon the main-body light curve.

6.8   Fragmentation

Up to this point in the discussion it has been assumed that as a meteoroid descends through the atmosphere it remains as a single, coherent body. While this appears to be true in many cases, it is certainly not always true. Indeed, the existence of strewn fields, containing hundreds and even thousands of meteorites, attest to the fact that atmospheric fragmentation is an important physical effect. In terms of the atmospheric interaction, the important quantity to consider is the so-called ram pressure, P_{ram}, which results from the on-coming air molecules being divert around the descending meteoroid. This quantity varies according to the atmospheric density — which in turn varies with atmospheric height, recall Equation (6.11) — and the square of the meteoroid's velocity, with

$$P_{\text{ram}} = \frac{1}{2}\Gamma \varrho_{\text{atm}} V^2 \tag{6.21}$$

where, as before, Γ is the drag coefficient.

The condition for meteoroid fragmentation is determined by setting the ram pressure equal to the compressive (or mechanical) strength σ_{mec} appropriate to the meteoroid composition. This gives the fragmentation condition of $P_{\text{ram}} = \sigma_{\text{mec}}$. The mechanical strength is a property that can be measured directly in the laboratory, or it can be estimated from fireball observations. The maximum mechanical strength of rock and iron are typically given as $\sigma_{\text{mec}}(\text{iron}) \approx 500\,\text{MPa}$ and $\sigma_{\text{mec}}(\text{rock}) \approx 100\,\text{MPa}$, but these are maximal values that do not generally apply to observed meteoroids. The main reason for this is that the parent bodies to meteorites are invariably (for so it seems) pre-fractured and this results in the reduction of the mechanical strength. In the case of the Peekskill fireball (Figure 6.5) and meteorite dropping event of October 9th, 1992, fragmentation occurred at an altitude of 41.5 km when the meteoroid was travelling at some 14 km/s, and the atmospheric density would have been of order $0.004\,\text{kg/m}^3$. These conditions indicate a material strength of some 0.8 MPa [2] for the Peekskill meteoroid. Measurements can,

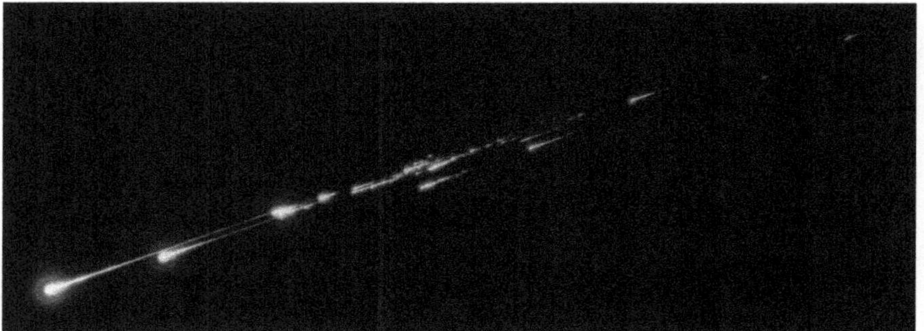

Figure 6.5. Multiple fragments produced during the atmospheric flight of the Peekskill meteorite on October 9th, 1992. Only one of the fragments has so far been found, and this largely because it struck a parked car. Image courtesy of Sara Eichmiller (Altoona Mirror).

of course, be made of σ_{mec} in the laboratory but these generally tell us about the mechanical strength on the smaller size scale, and these measurements need not necessarily relate to the mechanical strength of the initial, much larger, parent meteoroid.

In general, however, it is usually assumed that larger objects fragment more easily, that is have lower values of σ_{mec}, than smaller objects. This condition is often expressed according to Weibull's statistical strength theory, which attempts to describe the distribution of structural defects, with the probability of larger structural defects increasing in step with the volume of the object concerned — basically, big objects can contain larger defect zones and longer cracks than smaller objects, and this reduces its mechanical strength. The ultimate strength of a meteoroid is accordingly written

$$\sigma_{\mathrm{mec}} = \sigma_0 \left(\frac{m_\infty}{m} \right)^\alpha \tag{6.22}$$

where σ_0 and m_∞ correspond to the laboratory measured values for the mechanical strength and mass of some known meteorite sample, and α is a constant. Unfortunately, there is little agreement on what the value of the power α should be, although a value between 0.25 and 0.5 is typically assumed. Many tables have been published for the measured values of σ_0 and m_∞ relating to various meteorite types, with Table 6.2 below indicating a selection of results. Clearly, from

Table 6.2. Representative, laboratory measured, values for the constants σ_0 and m_0 appearing in Equation (6.22). The entry for MacAlpne Hills 88118 (an Antarctic meteorite find) is noteworthy in that the study by Kimberley and Rames [3], from which the data is derived, indicates that the mechanical strength is dependent upon how rapidly the load is applied, and that perhaps contrary to expectation the mechanical strength is lower when the load is applied slowly (as in the *stat* case) than when it is applied more rapidly (as in the *dyn* case) — the reason for this apparent effect is not currently understood.

Meteorite	Composition	m_∞ (kg)	σ_0 (MPa)
- - - - - - - - - -	ice	0.5	0.5
Allende	CV3	0.5	40
Tamdakht	H5	0.5	124
Seminole	H4	4.4×10^{-4}	174
Tsarev	L5	0.5	370
MAC 88118	L5	4.0×10^{-4}	50 (stat)
			200 (dyn)
Sikhote-Alin	IIAB	1.0	410

the table, there is much variation in the mechanical properties of meteorites, not just between different petrological types, but within a given petrographic group as well.

One of the key observational outcomes of fragmentation, apart from producing multiple components and a strewn field, is to drastically change the shape of a meteor's light curve. The variation of brightness with time will be very different from the classical, non-fragmenting light curve (recall Figure 6.4). Generally, when fragmentation occurs, many smaller-mass particles are produced, and as these rapidly ablate a sudden and short-duration brightness flare is produced. Such flares were recorded during the fall of the Buzzard Coulee meteorite on November 20$^\text{th}$, 2008 — recall Figure 1.14.

6.9   End result

In terms of modeling meteoroid entry, there appears (as hinted at by the Table 6.2 entries) no general way of predicting what a given meteoroid will do, *a priori*, as it descends through the atmosphere.

In reality, the fragmentation history needs to be determined from the specific observations pertaining to each meteorite producing fireball. Equations describing the lateral spread and differential deceleration of the various components following a fragmentation event have been developed, but they will not be studied here, although the reader is referred to the Earth Impact Effects webpage[1] for more details (also see Section 7.5).

Combining Equations (6.21) and (6.22) with the meteoroid ablation equations (6.2), (6.5), (6.11), and (6.14), one can map out the domain of possible end results for a specific meteoroid entering into the Earth's atmosphere; the end results being (i) complete vaporization with no meteorite(s) being produced, (ii) a single meteorite being produced, or (iii) multiple meteorites, in a strewn field, being produced as a result of atmospheric fragmentation. Such a study has been made [4] for the fall of the Mazapil meteorite on November 27th, 1885. In this particular case an iron meteorite with a mass of 4.656 kg was observed to fall, but no information with respect to its initial velocity and parent body mass is available. Figure 6.6 shows the end state domains in the initial mass versus initial velocity plane for iron meteorites. If the initial mass is too big, then fragmentation is favored, whereas if the velocity is too high then complete ablation is favored. The Mazapil meteorite is particularly interesting since it fell at the same time of a spectacular outburst of the Andromedid meteor shower. This particular storm event was caused by the Earth sweeping through the material produced during the break up of comet 3D/Biela — the parent comet to the Andromedids (recall Section 2.4). If, and there is no compelling reason for it to be the case, the meteorite was somehow related to the Andromedid meteoroid stream, then its initial velocity would be about 15 km/s, indicating an initial meteoroid mass of about 10^5 kg, corresponding to an initial diameter of 2.9 meters.

The ability of a meteoroid to penetrate the atmosphere is often described by the so-called PE criterion (or parameter). First introduced by Zedenek Ceplecha (Ondrejov Observatory, Czech Republic)

---

[1]https://impact.ese.ic.ac.uk/ImpactEarth/ImpactEffects/.

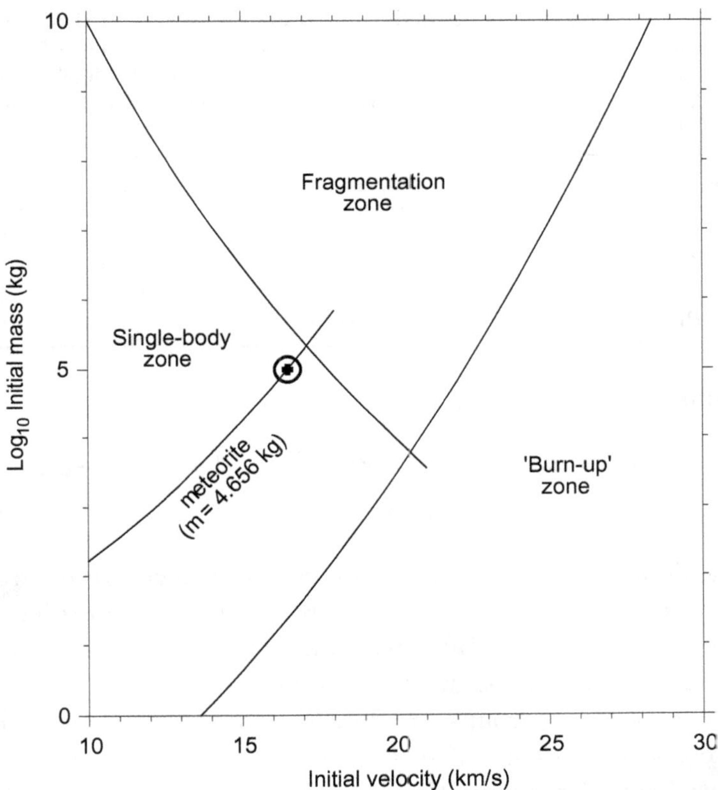

Figure 6.6. Mazapil meteoroid end state diagram. The various zones indicate, for a given combination of initial mass and velocity, whether a single meteorite will be produced, a meteorite strewn field will form as a result of atmospheric fragmentation, or whether the meteoroid will be fully ablated in the atmosphere leaving no collectable meteorite(s) upon the ground — the 'burn-up zone.' The arc labelled 'meteorite' indicates the range in initial mass and velocity in which a single meteorite (with a mass of 4.565 kg) reaches the ground; the dotted-circle indicates the initial mass of the parent body if the Mazapil meteorite is a member of the Andromedid stream with an initial velocity of $V = 16$ km/s.

and Richard McCrosky (Smithsonian Astrophysical Observatory), the PE criterion is parameterized according to the initial meteoroid mass M_∞, the initial velocity V_∞, the atmospheric density at the meteor end height ρ_{end}, and the zenith angle Z measured away from the vertical, and calibrated against the available observations.

Table 6.3. PE criterion values and deduced fireball types and their associated physical characteristics. Column 1 provides the range of PE values, while column 2 gives the associated fireball type. Column 3 indicates the likely meteoroid composition (asteroid or cometary in origin), while columns 4 and 5 indicate the typical bulk density and ablation coefficient associated with each taxonomic group.

Criterion	Type	Composition	Density (kg/m^3)	σ (s^2/km^2)
$-4.60 < \mathrm{PE}$	I	Stony	3700	0.02
$-5.25 < \mathrm{PE} < -4.60$	II	Carbonaceous	2000	0.04
$-5.70 < \mathrm{PE} < -5.25$	IIIa	Cometary	750	0.10
$\mathrm{PE} < -5.70$	IIIb	Weak cometary	300	0.21

Accordingly,

$$\mathrm{PE} = \log(\rho_{\mathrm{end}}) - 0.42 \times \log(M_\infty) + 1.49 \times \log(V_\infty)$$
$$- 1.29 \times \log(\cos Z) \qquad (6.23)$$

The detailed analysis of many fireball trajectories resulted in the introduction of four main meteoroid groups, gauged according to the range of PE values (see Table 6.3), and against each group a set of characteristic physical parameters can be assigned.

The taxonomic groups are intended as a guide and indicate the likely source origins of meteoroids. Meteorites are likely to be associated with Type I and II groups, which have a high bulk density and low ablation index, and are derived from asteroid parent bodies. Types IIIa and b are cometary in origin, have a low associated density and a large ablation coefficient, and as such fireballs are not likely to produce meteorites on the ground. In general, it is taken that a meteorite may well have reached the ground if the luminous end height (that is the start of the dark flight regime) is lower than 35-km altitude, and the terminal velocity is less than 10 km/s. Rather than using rule of thumb and/or parameter-fit formula (such as the PE criterion), in order to gauge the likelihood of a given fireball presaging a meteorite dropping event, recent attempts have been made to cast the condition in terms of purely physical parameters [5]. Specifically, the condition is based upon the ballistic parameter

α and the mass-loss parameter β, where:

$$\alpha = \frac{\Gamma}{2}\,\frac{\rho_0 H}{\cos Z}\,\frac{S}{m_\infty} \tag{6.24}$$

with the various symbols having their earlier defined meanings,[2] and

$$\beta = (1 - \mu)\frac{V_\infty^2}{2}\sigma \tag{6.25}$$

where μ is a shape factor defined according to the relationship between the cross-section area S of the meteoroid and its mass m, with

$$\frac{S}{S_\infty} = \left(\frac{m}{m_\infty}\right)^\mu \tag{6.26}$$

Equation (6.26) accounts for the manner in which ablation affects the shape of the meteoroid during atmospheric flight. If the shape remains the same, that is it is self-similar (always spherical, for example), then $\mu = 2/3$; if the ablation is such that it proceeds from the end of a constantly orientated cylinder then $\mu = 0$; if the meteoroid is wedge-shaped and the ablation proceeds to preserve the apex angle, then $\mu = 1/2$. The α-β combination will vary according to the adopted values of the ablation coefficient σ and the shape factor μ, the other terms in Equations (6.24) and (6.25) being deduced from the observations. In very broad terms, however, it appears that if the simultaneously $\beta < 3$ and $\alpha \cos Z < 35$, then a meteorite is likely to have reached the ground. As with all such conditions, no matter how parameterized, the delivery of a meteorite to the ground and the actually finding of that meteorite are two separate things, and the latter remains largely a question of good fortune.

---

[2]Note that these α and β terms are not to be confused with the parameters used in the mechanical strength formula — Equation (6.22) — and the electron line density formula — Equation (6.9).

6.10   Shape and flight dynamics

Rotation of an irregularly shaped meteoroid will also modulate the light curve, the brightness going up and down according to the presentation of the large and then small cross-section area to the on-coming airflow — the fireballs associated with the Innisfree and Peekskill meteorites, for example, showed a prominent periodic flickering effect [6]. Smaller-mass meteoroids can also show distinct flickering effects as they descend through the atmosphere, and this is thought to be an indication of spin and non-constant cross-section profiles. Both the Geminid and Taurid meteor showers are noted for delivering meteors that show a distinct flickering effect, and Figure 6.7 shows the light curve for a bright Geminid fireball recorded on December 13th, 2002. Analysis of the video data for this particular fireball indicated an initial meteoroid mass of 0.4 kg (giving an initial diameter of about 10 cm) and a flickering period (and hence spin period) of 0.2 seconds [7]. The video record lasted some 2.5 seconds

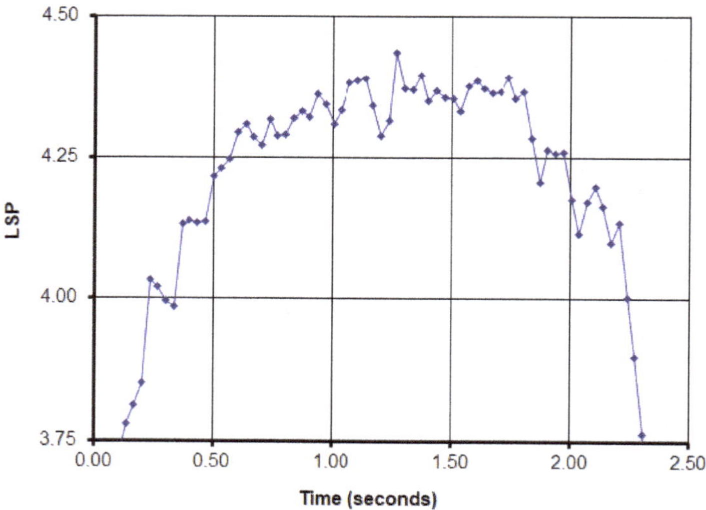

Figure 6.7.   The light curve of a periodically flickering Geminid fireball recorded by a video camera system on December 13th, 2002. The y-axis corresponds to the brightness and is calculated according to the log sum pixel (LSP) count derived from each video frame (there are 30 frames per second) of the video recording [7].

Figure 6.8. The oriented profile of the Karakol (LL6) meteorite. This was a witnessed fall, and observers reported hearing a distinctive whistling sound before it struck the ground. The meteorite was described as being warm to the touch and sulfurous smelling when first found. Just a single 2.7-kg mass was recovered from the fall location. Meteorite image courtesy of the Vernadsky Institute, Moscow.

in duration and with a peek brightness of magnitude -9 the fireball was nearly as bright as the full Moon.

While meteoroid shape and spin can result in the formation of flickering meteor trails, it can also result in the production of oriented meteorites. Such meteorites are relatively rare, but they are aesthetically pleasing and much sought after by collectors. Figure 6.8 shows the oriented profile of the Karakol, a Kazakhstan meteorite which fell on May 9th, 1840. The profile is exactly like that of a projectile, with an apex angle of about 90 degrees. This profile is most likely brought about by a spin-effect stabilizing, that is orientating, the meteoroid as it descends through the atmosphere. In this way the ablation will preferentially ablate the front surface only rather than the entire surface, as would be the case if the meteoroid tumbled through the atmosphere.

In addition to spin stabilization, some meteoroids might simply have, by chance, the right profile to enable oriented flight. This latter possibility has been investigated by Khunsa Amin (University of New York) and coworkers [8]. Conducting experiments in a fluid-flow tank, the motion of different cone-shaped models was observed, and Amin *et al.* found that there was a restricted range of cone apex angles for which orientated flight was possible (Figure 6.9) — for apex

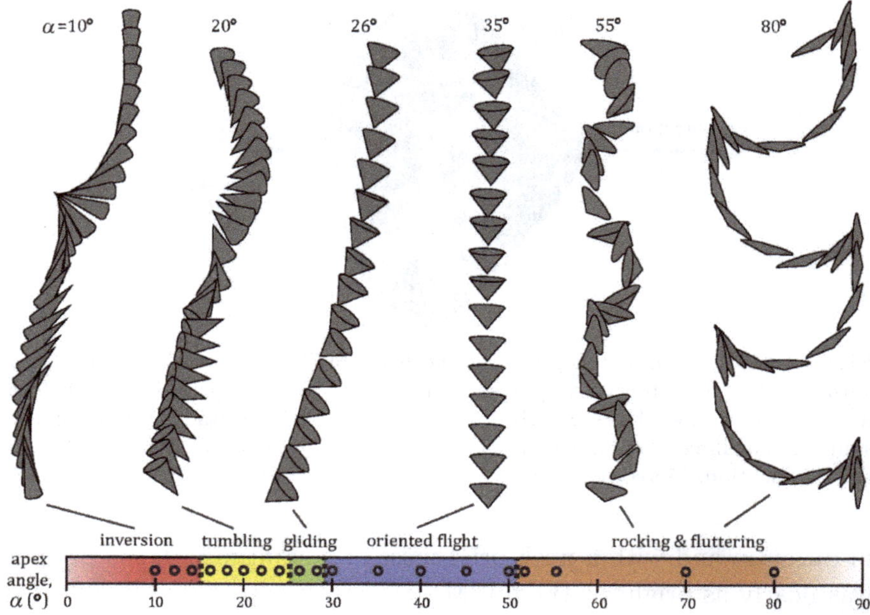

Figure 6.9. Free-fall dynamics of conical bodies of varying apex half-angle α (i.e., $\alpha = 45°$ corresponds to a cone angle of $90°$). Only in the range between $30 < \alpha(\text{deg.}) < 50$ is the flight found to be stable and orientated. Image courtesy of K. Amin *et al.* [8].

angles outside of the oriented stability range, the motion observed was either that of a tumbling or a rocking/fluttering motion. While these experimental results are compelling, there is still much work that needs to be done with respect to their application to meteoroid flight dynamics. A meteoroid, for example, is travelling at supersonic speeds through the atmosphere, and undergoes mass-loss and re-shaping due to surface ablation.

Meteoroid shape and spin effects may additionally be at play with respect to another mysterious phenomenon of meteoroid flight. Reported only very rarely, but consistently over the past several hundred years, some visually observed meteors appear to follow non-linear paths — that is, they appear to move along a spiral or a curving path (Figure 6.10) rather than that of a straight line. The physics

 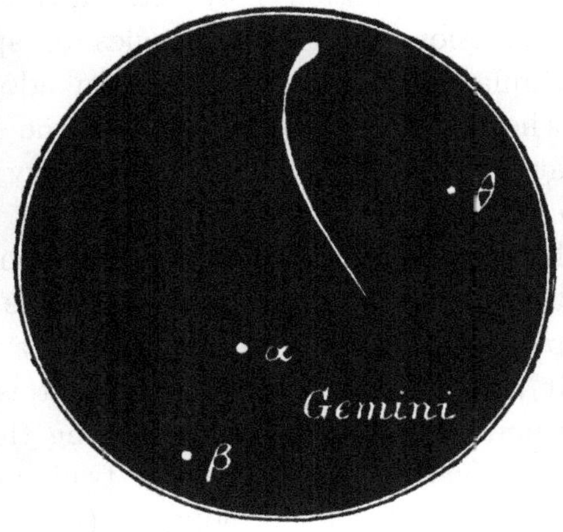

Figure 6.10. Non-linear meteor trails. To the left is a reproduction of a woodcut depicting a fireball, observed in France in October of 1852, moving along a helical path. To the right is a drawing of a meteor observed by the great amateur astronomer William Denning on December 25th, 1886; the meteor appeared to follow a curving path.

controlling such motion remains obscure but may be related to some form of the Magnus effect,[3] as occasionally observed in the flight of guided missiles and in the trajectories adopted by rifle bullets. The Magnus effect is present whenever a spinning object moves through a gas or fluid medium, and it produces a deflection (that is lift) force perpendicular to the direction of motion. It is this lift force that results in the trajectory becoming curve, with the amount of deflection being related to the speed, spin rate, and density of the medium. It is this same effect, for example, that is exploited by spin bowlers when imparting topspin and/or backspin to a cricket ball. While the Magnus effect offers a physical mechanism by which curved flight paths might be produced, it is far from clear that non-linear meteor trails are actually a real phenomenon.

---

[3] Although noted and described by Isaac Newton in 1672, after watching the flight dynamics of a ball during a tennis match, the phenomenon was first discussed in detail by the German physicist Heinrich Magnus in 1852.

Spiral motion may alternatively be caused by a torque-free precession effect that applies to spinning asymmetric bodies — similar, in fact, to the motion adopted by an American football when the quarterback induces some spin about the ball's longer axis at the time of release. The mystery here, however, is that no such curving or spiralling path has ever been recorded instrumentally.[4] It may well be that we are dealing with a psychological rather than a physical phenomenon in this case, with the eye and human perception conspiring to produce an optical illusion. Additionally, it may well be a trail effect that is being seen — that is, rather than a physical effect perturbing the path of the meteoroid body, the non-linearity is really a trail phenomenon associated with the ablated material left along the (linear) path of a meteoroid. It will no doubt require continued data collection, laboratory experimentation, numerical analysis, and the study of human night-time perception to fully resolve the question of non-linear meteor trails.

General reading

1. A detailed mathematical text (not for the faint of heart) is that by V. A. Bronshten, *Physics of Meteoric Phenomenon*. D. Reidel Publishing Company, Dordrecht (1983).
2. A gentle introduction is provided by G. S. Hawkins, *The Physics and Astronomy of Meteors, Comets and Meteorites*. McGraw-Hill Book Company, New York (1964).
3. A thorough review of modern research topics is presented in *Meteoroids: Sources of Meteors on Earth and Beyond*. G. Ryabova, D. Asher and M. Campbell-Brown (Eds.), Cambridge University Press (2019).

---

[4]Perhaps one should say that no distinctly curved or spiral paths have been recorded. Photographs reportedly showing spiral motion have been published, but it is typically the case that the motion is more readily explained in terms of camera shake than actual variations in the meteoroid's flight path.

References

[1] M. Beech and M. Hargrove. "On classical meteor light curves and utilitarian model atmospheres." *Il Nuovo Cimento* C, **26**, 77–85 (2005).

[2] P. Brown, Z. Ceplecha, R. L. Hawkes, G. Wetherill, M. Beech, and K. Mossman. "The orbit and atmospheric trajectory of the Peekskill Meteorite from video records." *Nature*, **367**, 624–626 (1994).

[3] J. Kimberley and K. Ramesh. "The dynamic strength of an ordinary chondrite." *Meteoritics and Planetary Science*, **46**, 1653–1669 (2011).

[4] M. Beech. "The Mazapil meteorite: From paradigm to periphery." *Meteoritics and Planetary Science*, **37**, 649–660 (2002).

[5] M. Moreno-Ibanez, M. Gritsevich, J. Trigo-Rodriguez, and A. Silber. "Physically based alternative to the PE criterion for meteoroids." *Monthly Notices of the Royal Astronomical Society*, **494**, 316–324 (2020).

[6] M. Beech and P. Brown. "Fireball flickering; the case for indirect measurement of meteoroid rotation rates." *Planetary and Space Science*, **48**, 925–932 (2000).

[7] M. Beech, A. Illingworth, and I. S. Murray. "Analysis of a 'flickering' Geminid fireball." *Meteoritics and Planetary Science*, **38**, 11–7 (2003).

[8] K. Amin *et al.* "The role of shape-dependent flight stability in the origin of oriented meteorites." *Proceedings of the National Academy of Sciences*, **116**, 16180–16185 (2019).

[9] M. Beech. "Non-linear meteor trails." *Earth, Moon and Planets*, **42**, 185–199 (1985). See also, Y. Shigeno, M. Toda, and M. Kobayashi, "A spiral meteor train." *WGM, The Journal of IMO*, **26**(5), 220–225 (1998).

American artist Charles Willson Peale draws back the curtain to his cabinets of curiosities — in this case, a collection of natural history flora and fauna. To the lower left is a turkey set atop a box of taxidermy tools, to the lower right are mastodon bones. Painting by Peale, 1882

Chapter 7

Gems from the Cabinet

*The feeling of awed wonder that science can give is one of the
highest experiences of which the human psyche is capable.*

Richard Dawking

There is nothing more controversial than a list — everyone makes
them, and everyone disagrees about their content. For all this,
however, it would seem that lists abound: from the mundane
shopping list, to top news stories, to advertising, to music charts, to
TV programs and internet recommendations. Problematically, there
are many different ways of making lists, arranging and selecting
their content — this is where the controversy generally lies: who
made the list, what criteria did the composer use to generate the
content, who gains (or loses) by being on the list, what does the
list actually tell us. Here, however, the ordering will be controversial
only in the sense of the provisional nature of the content — we do
not know where, for example, the largest meteorite on the Earth is,
or how much it might actually weigh. What we do know, of course,
is that of the meteorites actually found to date there is a biggest
in terms of physical mass, there is an oldest in terms of terrestrial
residence age, and there is a strewn field with the greatest number
of meteorites collected from within it, and so-on. All such records,
however, change as time and discovery proceed, and some of the
meteorites described on the oddities list may only appear odd to
the author, but be perfectly natural objects (in some fashion) to
somebody else. From the cabinet of meteorite curiosities, however,
the following lists seem mostly reasonable, but some no doubt are

157

doomed to have a short life — the list concerning impact craters, for example, will certainly change as time progresses. Other lists will possibly retain their ordering for a long time yet to come.

7.1   The 5 most massive iron meteorites

The Hoba or (unofficially) Grootfontein meteorite is the most massive single-piece meteorite known, leading by a factor of 2 times that of the other top contenders (Table 7.1). The meteorite was discovered in 1920 and has an unusual slab-like appearance with side dimensions of $2.7 \times 2.7 \times 0.9$ meters. It has been estimated that the Hoba meteorite fell some 80,000 years ago, but what is most remarkable about the fall, given the meteorite's great mass, is that there is no associated crater (see Section 11.1 later). The meteorite was declared a state monument by the Namibian Government in 1955, and it is now a major tourist attraction (Figure 7.1).

Ahnigito, one of the Cape York meteorites, squeaks in as the second most massive meteorite (recall Figure 1.3), beating out by just 80 kg the recently discovered Gancedo meteorite fragment that was excavated from the Campo del Cielo (literally, *the field of stars*) strewn field in Argentina. El Chaco, also from the Campo del Cielo strewn field, comes in as the 3rd most massive known meteorite.

The Aletai (formerly the Armanty and also Altay) meteorite, *The Silver Camel*, was found in 1898, and it is the largest meteorite so-far identified in China. It has recently been suggested that this meteorite is a member of the world's largest strewn field with a long axis extending over of some 500 km (see below).

Table 7.1.   The 5 most massive iron meteorites.

Meteorite	Location	Discovery	Type	Mass (tonnes)
Hoba	Namibia	1920	Ataxite IVB	60
Ahnighito	Greenland	1894	Octahedrite IIIAB	30.88
Gancedo	Argentina	2016	Octahedrite IAB	30.80
El Chaco	Argentina	1969	Octahedrite IAB	28.84
Aletai	China	1898	Octahedrite IIIE	28.00

Figure 7.1. The Hoba meteorite, Grootfontein, Namibia, seated within its decorative surround. (Credit: Wikimedia commons.)

Given the large, factor of 2, gap between the masses of the Hoba and Ahnighito meteorites, it would seem highly likely that there are more meteorites in the mass range between 30 to 60 tonnes that have yet to be found.

There is no clear upper mass limit for a single meteorite. Clearly meteorites can be as massive as the Hoba meteorite, but how much more massive depends upon many physical, generally unknown, circumstances. The meteorite, for example, has to survive atmospheric passage without fragmentation, and it has to hit the ground with a sufficiently small speed that it does not shatter and/or produce a crater. Survival is enhanced by having a slow atmospheric entry velocity (that is not much larger than the Earth's escape velocity of 11.2 km/s), and up to a point by the entry angle into the atmosphere. A vertical plunge minimizes the amount of atmosphere that a meteorite has to pass through, but then it will suffer less deceleration and retain more of its initial velocity upon impact. A very shallow angle of entry will result in a long atmospheric path, and a greater decrease in a meteorite's initial entry velocity, and these effects favor a less energetic ground impact. This is all to the good for a large meteorite surviving, but the longer the atmospheric path so the more mass that is ablated, which, of course, reduces the surviving mass.

Furthermore, the size of the final meteorite on the ground will critically depend upon the internal structure of the parent meteoroid — is it solid through-and-through, or does it contain many cracks and fracture plains. If the initial meteoroid is highly fractured, then fragmentation is highly likely and this again works against the survival of a large single mass. If we take maximal compressive strength values for iron and rock meteoroids (recall Chapter 6) and assume that no fragmentation takes place, then for meteoroids entering the Earth's atmosphere at 45 degrees to the horizon, with an initial velocity of $12\,\text{km/s}$ (and an ablation coefficient of $5 \times 10^{-8}\,\text{s}^2\text{m}^{-2}$), so a single iron meteorite of mass ~ 1000 tonnes (diameter ~ 6 meters) at the top of the atmosphere might conceivably reach the ground with a mass of ~ 500 tonnes (diameter $\sim 5\,\text{m}$). For a stone meteorite this ground condition is reduced to a mass of 100 tonnes (diameter of $\sim 4\,\text{m}$). These maximal numbers are 10 to 100 times larger than the most massive iron and stone meteorites that have so-far been found, and this suggests that much larger meteorites (than the current recordholders) might conceivably exist somewhere on Earth — the problem, of course, is finding them.

7.2   The 5 most massive stony meteorites

Table 7.2 shows the most massive stony meteorites. The H5 Jilin meteorites were observed to fall on March 8th, 1976 over Kirin Province, China. The accompanying fireball was widely observed and numerous fragments were produced providing a total ground mass of

Table 7.2.   The 5 most massive stony meteorites.

Meteorite	Location	Discovery	Type	Mass (kg)
Jilin	China	1976	H5	1770
Norton County	USA	1948	Aubrite	1073
Long Island	USA	1891	L6	564
Paragould	USA	1930	LL5	371
Bjorbole	Finland	1898	LL4	330

Figure 7.2.   The main mass of the Jilin meteorite being inspected by researchers shortly after its fall to Earth.

at least 4 tonnes. The largest mass was collected from a 6-m deep plunge pit (Figure 7.2).

The Norton County, Kansas meteorite fell on February 18th, 1948. The associated fireball was widely observed and numerous sonic booms were reported. Many hundreds of small meteorite fragments have been collected, and the main mass is currently on display at the Institute of Meteoritics at the University of New Mexico in Albuquerque. As an aubrite this particular meteorite is distinctive in not showing the usual brown-black fusion crust, but is almost cream colored.

The main mass of the Long Island, Kansas meteorite was found in 4 separate pieces in 1891 — the meteorite presumably having broken apart upon ground impact. Many additional smaller meteorites have been found in the surrounding 3 by 11 km strewn field. The main mass is on display at the Field Museum of Natural History in Chicago.

The Paragould, Arkansas meteorite fell on February 17th, 1930. Accompanied by a bright fireball, many fragments, to a total mass of about 400 kg, have been collected from the region surrounding the main mass — the main mass burying itself in a 2-m deep plunge pit. The main mass is currently on display in the Mullins Library at the University of Arkansas, Fayetteville.

The Bjurbole meteorite could easily be classified as a meteorite that should never have survived. Accompanied by a dramatic detonating fireball that was witnessed from many regions around the Baltic Sea, on March 12th, 1899, the main mass punched a 5-m diameter crater through 3-m thick sea ice, and was apparently lost. A few years after the fall, however, divers were able to recover many meteorite fragments from below the impact location — these fragments were buried and preserved in sediment many meters deep under the sea water. The main mass was most probably more massive than the collected fragments indicate. Not only is the recovery of any fragments from the Bjurbole fall remarkable, but the meteorite itself is noteworthy for being (in comparison to other chondrites) highly friable and chondrule rich.

7.3   The 5 most massive stony-iron meteorites

Table 7.3 shows the most massive stony-iron meteorites. The Brenham meteorite is thought to have fallen within the last few thousand years, with the first fragments from the strewn field and associated Haviland crater (see Section 7.9) being located in 1882. The largest Brenham fragment so-far found is that discovered in 2005 (see Figure 7.3).

Table 7.3. The 5 most massive stony-iron meteorites.

Meteorite	Location	Discovery	Type	Mass (tonnes)
Brenham	USA	1882	Pallasite	4.30
Vaca Muerta	Chile	1861	Mesosiderite	3.83
Huckitta	Australia	1924	Pallasite	2.30
Fukang	China	2000	Pallasite	1.00
Imilac	Chile	1822	Pallasite	0.92

Figure 7.3. A 635-kg Brenham meteorite found by Steve Arnold (right) on Allen Binford's (left) farm in Haviland, Kansas in 2005 (photo: Marc Benavidez/AP).

The first fragments of the Vaca Muerta (literally *Dead Cow*) meteorite were found in 1861, although it is thought to have fallen (in a remote part of the Atacama Desert) some 3500 years ago.

The Huckitta meteorite was found in 1937 in the pastures surrounding the Huckitta (now Arapunya) Cattle Station. This meteorite has been paired with the so-called Alice Springs meteorite (found in 1924). The main mass of some 1.41 tonnes is on display at the South Australian Museum in Adelaide.

The Fukang meteorite was found in 2000. Bonhams auction house offered the main mass for sale in 2008, although with an expected price of two million dollars the mass remained unsold.

The Imilac pallasite fragments, like the Vaca Muerta meteorites, were found in the Atacama Desert in a strewn field at least 8-km long. Falling in pre-recorded history, and in spite of the dry dessert conditions, the Imilac pallasites are often highly weathered with the smaller fragments typically having no olivine inclusions — such specimens are often sold under the name of metal skeletons.

7.4   The 5 oldest authenticated meteorites

In Table 7.4 the idea is that the fall or find was historically recorded and that the fragments of the meteorite are extant to this day. The Nogata meteorite was observed to fall on May 19th, 861 and is still on display at the Shinto shrine in Nogata, Japan, where it was taken to after being recovered.

The Tamentit meteorite was found in Algeria in 1864 and is currently on display at the Vulcania Park in France.

Table 7.4. The 5 oldest authenticated meteorites.

Meteorite	Location	Discovery	Type	Mass (kg)
Nogata	Japan	861	L6	0.47
Tamentit	Algeria	1864	Octahedrite IIIAB	510
Elbogen	Czeckosolovakia	1400	Octahedrite IID	107
Ensisheim	France	1492	LL6	127
Campo del Cielo	Argentina	1576	Octahedrite IAB	100,000

The Elbogen (also Loket Iron) meteorite fell in the village of Loket sometime about the year 1400. There are pieces of this meteorite in many collections around the world, and it was the first meteorite to be studied by Alois von Widmanstatten, in 1808, leading to his description of the Widmanstatten lines used to characterise iron meteorites (recall Section 1.2).

The Ensisheim meteorite was observed to fall on November 7th, 1492 (the same year, of course, in which Columbus set sail for the new world). The meteorite, long on display in the 6th century building of the Musée de la Régence, landed in pasture land and was recovered from a 1-m deep plunge pit (Figure 7.4).

The Campo del Cielo meteorites were first described in 1576 (but they were known long before this to the indigenous peoples). It is

Figure 7.4.   The fall of the Ensisheim meteorite. Notice the horse rider (witness) to the lower right of the image and a startled animal hiding (towards right of center); birds are also shown falling out the sky (middle to upper right). The image also attempts to show a time sequence with the meteorite first being shown in the sky (with striations to indicate motion) and then on the ground. Image from the Vatican Library (MS Chigi G.11.36, folio, 199v).

thought that the meteorites, and their associated crater field, fell and formed some 4–5 thousand years ago. There are at least 26 craters, the largest being about 100 metres in diameter, in a fall zone spread over an area some 3 km by 19 km. The associated strewn field for smaller meteorites is at least 60 km in length, and massive meteorites are still being recovered from this fall location to this very day (recall the 2016 recovery of the Gancedo fragment).

7.5   The oldest un-authenticated meteorites

Here I will temporarily break with the top 5 ranking process — after all, rules (and lists) are made to be broken. The reason for this change is that there are many, many accounts of meteorite falls in the ancient literature. Some of these accounts may well be true but many (the majority?) are probably not true, but it is certainly the case that the meteorites related to these very early accounts have not survived to the present day (the Nogata meteorite from 861, recall, holding the current age-plus-extant record). In his extensive list (compiled from 38 historical listings and catalogues) Kaare Lund Rasmussen records 251 accounts that have been historically attributed as describing meteorite falls and finds — 30 of these accounts are set prior to 1 AD. Indeed, Rasmussen has argued that there are some 20 epochs of enhanced meteorite fall activity (varying from $\sim$1 to $\sim$100 years in length) that are evident in the 2550-year time interval between 800 BC to 1750. The veracity of such statistics is very much open to debate, but Rasmussen suggests that the times of enhancement could be the result of the Earth encountering distinct meteorite showers (see Section 19.2).

The oldest confirmed human use of meteoritic material is that of the iron beads found in Gerzeh, northern Egypt. These nine beads, found in a grave site excavated in 1910, have been assigned to a manufacturing date circa 3200 BC, but it is not known if they were produced from an iron meteorite that had recently fallen or from a find.

Ernst Chladni in his Catalog of Meteoric Stones (published in the *Philosophical Magazine* for January 1826) suggests that a stony meteorite fell in Crete circa 1478 BC, and that a massive iron fell on Mount Ida, Crete, circa 1403 BC. Chladni also suggests "fiery stones" fell during the Second Punic War circa 205 BC, and he lists falls in Constantinople (modern Istanbul, Turkey) in 459, 472, and 652, with the last event exciting considerable terror. Chladni also posits a possible meteorite fall in Bagdad in 929.

Amongst the extensive possibilities presented by ancient Greek authors, the meteorite fall at Aegospotami in 465 BC stands out. This particular event was supposedly predicted by Anaxagoras and was claimed to be a fragment of the Sun fallen to the ground. While the prediction was certainly fortuitous (given that it was actually made and that a fall actually took place), various attempts throughout history have been made to find the meteorite. Anaxagoras is better remembered to this day for his suggestion that the Earth actually orbits the Sun, rather than the other way around — an argument, in this case, that pre-dated Copernicus by over 2000 years.

The earliest meteorite fall to be recorded in China appears to be that of December 24th, 645 BC, when "five stones fell in Sung [state]." A total of 102 possible meteorite falls and finds are listed by Rasmussen for China between 645 BC and 1600 AD. Some of the Chinese accounts are particular interesting in that they suggest human injuries and structural damage (see Chapter 18 later). One event dated circa 1340 is described as damaging crops, destroying buildings, and killing numerous people.

The reality and/or accuracy of these very early meteorite fall accounts, from no matter which culture, are all highly questionable, and very little trust can be placed in their interpretation. No doubt some of the accounts do describe genuine meteorite falls, but others are more than likely based upon the misunderstanding of natural phenomena, cultural myths, down-right lies, and uncertain eyewitness accounts.

7.6 The first official meteorite (fall or find) by country

In terms of first authenticated meteorite falls of a specific compositional type, and for which samples still survive, the records are shown in Table 7.5.

Table 7.5.   First official record of each meteorite type.

Meteorite Type	Fall Location	Fall Date
H chondrite	Nogata, Japan	May 19th, 861
L chondrite	Minamino, Japan	September 27th, 1632
LL chondrite	Ensisheim, France	November 16th, 1492
Carbonaceous chondrite	Alais, France	March 15th, 1806
Iron (IID)	Hraschina, Croatia	May 26th, 1751
Pallasite	Mineo, Sicily	May 3rd, 1826
Mars	Chassigny, France	October 3rd, 1815
Lunar	- - - - - - -	Still waiting

Once again breaking with the top 5 ranking process, Table 7.6 below shows the dates and locations for the earliest known meteorites that are extant to this day, found in a selection of countries from around the world. The data for this table have been extracted from various sources, but mostly the Meteoritical Bulletin Database (July 2019): https://www.lpi.usra.edu/meteor/metbull.php. The first column in the table lists the various countries in alphabetical order, and the location refers to the first meteorite that has an authenticated composition, fall or find date, and is still extant. The last column in the table provides a total meteorite fall and find count for each country.

Table 7.6 illustrates the basic fact that meteorites can, and indeed do, fall anywhere on Earth. Figure 7.5 shows a plot of all the officially recognized meteorite falls and finds that occurred between 2500 BC and 2012 (data from the Meteoritical Bulletin Database). The total number of meteorites from any given country, however, shows no clear correlation with respect to the country's area and/or population density (see Chapter 24 later). What matters most in terms of the total number of meteorites collected appears to be the terrain. Brazil, for example, has a very large surface area and a population density of some 25 people per square kilometer, but it has a landscape largely dominated by rivers, lakes, and Amazon forest. Accordingly recovering any meteorites (falls or finds) is a difficult challenge (see the regions devoid of data points in Figure 7.5), and consequently its total meteorite count of 77 is relatively small. Oman, in contrast, with

Table 7.6. A selection of first find/fall dates for various countries. The data is largely taken from Meteoritical Bulletin Database.

Country	Location	Year	Type	Fall/Find	Total
Afghanistan	Kandahar	1959	L6	Fall	1
Australia	Barratta	1845	L4	Find	703
Belgium	St. Denis Westrem	1855	L6	Fall	5
Brazil	Bendego	1784	IC	Find	77
Canada	Mardoc	1854	IIIAB	Find	65
China	Longchang	1761	IVA	Find	345
Croatia	Hraschina	1751	IID	Fall (1)	5
Denmark	Mern	1878	L6	Fall	8
Egypt	Nakhla	1911	Martian (2)	Fall	78
England	Wold Cottage	1795	L6	Fall	13
France	Ensisheim	1492	LL6	Fall	76
Germany	Steinbach	1742	IVA	Find	53
Honduras	Rosario	1896	IAB	Fall	1
Hungary	Kaba	1857	CV3	Fall	9
India	Jalandhar	1621	Iron (3)	Fall	143
Italy	Vago	1668	H6	Fall	41
Japan	Nogata	861	L6	Fall	56
Kenya	Duruma	1853	L6	Fall	8
Lithuania	Jodzie	1877	Howardite	Fall	4
Mexico	Morito	1600	IIIAB	Find	111
Netherlands	Uden	1840	LL7	Fall	6
New Zealand	Wairarapa Valley	1863	H6	Find	9
Oman	Tarfa	1954	L6	Find	3943 (4)
Philippines	Pampanga	1859	L5	Fall	5
Poland	Schellin	1715	L	Fall	22
Russia	Doroninsk	1805	H5-7	Fall	145
Scotland	High Possil	1804	L6	Fall	4
Spain	Sena	1773	H4	Fall	31
Switzerland	Rafruti	1886	Iron	Find	11
Turkey	Adalia	1883	Eucrite	Find	16
Uganda	Maziba	1942	L6	Fall	6
USA	Weston	1807	H4	Fall	1824
Venezuela	Muenatauray	1960	IIAB	Find	3
Wales	Pontlyfni	1931	Winonaite (5)	Fall	2
Zambia	Monze	1950	L6	Fall	1

(Continued)

Table 7.6.   (*Continued*)

Notes:

(1) This was the first recorded fall of an iron meteorite.

(2) This is the defining meteorite for the Martian Nakhla group (see the myths table below).

(3) This 2-kg meteorite (mixed with additional iron) was apparently forged into two sabres, a knife and a dagger, but these weapons are all now all lost.

(4) The large number of meteorites from Oman is the result of intensive and systematic exploration of several large, desert strewn fields. The total number given in the table relates to officially recognized specimens, but many thousands more have been collected and sold on the open market.

(5) This is the only Winonaite meteorite to be collected as a fall — the Winona meteorite itself being found and identified in 1928. To date only 35 meteorites have received a Winonaite classification.

Figure 7.5.   A plot of all known meteorite fall and find locations (excluding Antarctica). The data is taken from the Meteoritical Bulletin Database. The large circles indicate falls and finds totaling more than 40 kg, while the smallest circles indicate falls and finds with masses smaller than 10 kg. Image produced by and courtesy of Ramon Martinez.

a similar population density to Brazil, but an area that is some 27 times smaller, has a total meteorite count that is 51 times larger. This result is a direct consequence of Oman having a dry, desert landscape in which meteorites are preserved (that is weather slowly), and in which meteorites can be recovered in a relatively straightforward manner.

The high Arctic regions (northern Canada, Siberia, Greenland) have additionally yielded few meteorite falls and/or finds, even though in principle these are locations that will preserve meteorites (cold, desert conditions). The main reason for this dearth of collected material is the low population density, and the extreme weather conditions that prevail in these areas — it is not because meteorites do not fall in these high northern latitudes. Indeed, in contrast to the Arctic regions, Antarctica has yielded many tens of thousands of meteorites, this being a direct consequence of numerous dedicated surveys and searches being conducted over many years and because of a special geological meteorite clustering mechanism.

7.7 The 5 most extensive meteorite showers (since 1800)

Table 7.7 is probably one of the most contentious since we have no idea how many fragments there actually are in any given fall and/or strewn field, and likewise the question is also predicated upon some (in articulated) minimum mass. This, of course, does not stop people from making various estimates to total fragment numbers. Perhaps the best studied strewn field in recent time is that of the Buzzard Coulee meteorite (fall: November 20th, 2008) in Saskatchewan, Canada. This meteorite certainly produced a well populated strewn field some 7 km by 3 km in size, and it has been estimated that perhaps some 10,000 fragments larger than 1 gram fell to the ground.

Table 7.7. The 5 most extensive meteorite showers (since 1800)

Meteorite	Location	Discovery	Type	Mass (kg)	Fragments
Pultusk	Poland	1868	H5	8860	100,000
Holbrook	USA	1912	L6	220	15,000
Sikhote-Alin	Russia	1947	IAB	70,000	15,000
Allende	Mexico	1969	CV3	2000	5000
L'Aigle	France	1803	L6	37	3000

The extensive strewn field associated with the fall of the Holbrook meteorite continues to yield small fragments to this very day. The Pultusk meteorite fall and strewn field is composed of many thousands of small fragments which weigh in at just a few grams in mass, and these are often sold under the descriptive label of *Pultusk Peas* — the largest fragment to be recovered from the 8-km long strewn field has a mass of some 9 kilograms.

The Sikhote-Alin fall was one of the most spectacular of recent times. An estimated 26 tonnes of material have been collected from the Sikhote-Alin strewn field, and many meteorites were recovered from small crater-like pits. Being a daytime event the fireball that accompanied the fall was well observed, and an associated smoke cloud reportedly hung in the air for many hours. One eyewitness, Pyotr Ivanovich Medvedev, painted the scene of the fireball's trail (Figure 7.6) — this image was later used on a 40-kopek, tenth anniversary stamp of the event.

Conspicuous by its absence from our table for the most extensive strewn fields is the Chelyabinsk meteorite (which fell on February 15th, 2013). Many tens of thousands of small, pea-sized fragments were likely laid down over the 70-km long strewn field[1] produced in this particular event, but no specific ground-based or theoretical estimates for the total number of fragments (above a given mass) have been published to date. The parent body had an estimated mass of 10 tonnes (diameter $\sim$17 m), and perhaps of order 1 tonne of meteoritic material resides in the strewn field. The largest single meteorite recovered (from a fragment that punched a hole through the frozen ice on lake Cherbarkul and was later collected by divers) had a mass of 645 kg. If just 10% of the remaining meteorites in the Chelyabinsk strewn field have a mass of say 1 gram, then there should be of order another 35,000 Chelyabinsk "peas" to find. This sounds impressive, but given a strewn field area of some 1500 km^2, this translates (assuming that they are uniformly distributed, which

---

[1]The Chelyabinsk meteorite strewn field is particularly complex, and appears to be composed of several distinct and overlapping strewn fields, with each strewn field being related to a specific fragmentation (fireball outburst) event.

Figure 7.6. P. I. Medvedev's painting of the fireball and smoke trail associated with the fall of the Sikhote-Alin meteorite. Original size 40.5 × 55 cm.

is an unlikely assumption) to about 23 "peas" per square km — that's about one pea-sized meteorite per region of $210\,\text{m}^2$ (or 1 fragment in a region equal to the area to 10 American football fields).

In general, it is to be expected that there will be many more smaller meteorite fragments in a strewn field than larger ones, but it is also the case that most of the total mass on the ground will reside in the larger meteorites. The number of fragments dN in the mass bin m to $m + dm$, is usually expressed as a power law in the mass m, with

$$dN = K\, m^{-S} dm \qquad (7.1)$$

where K is a constant and where experimental data (based upon the catastrophic disruption of rock fragments) suggest that $1.8 < s < 1.9$. The larger the value of the exponent s, so the greater the amount of material that resides in the larger mass fragments. In principle, if a thorough search of a strewn field is conducted, then a diagram of the number of meteorite fragments plotted against meteorite mass can be used to infer the value of the exponent s, and the total number of fragments N_T within a strewn field above a given mass can be determined.

$$N_T = \int_{Min}^{Max} dN = \frac{K}{1-s}(Max^{1-s} - Min^{1-s}) \qquad (7.2)$$

where the integral in Equation (7.2) is taken over the mass range *Min* to *Max*. The problem with applying Equation (7.2), of course, is that in general we do not have any *a priori* knowledge of what K, s, and *Min* and *Max* should be for a specific fall. The exponent s and constant K are typically determined from the collected number of smaller mass fragments in a strewn field, and the mass estimates *Min* and *Max* are gauged according to the total estimated mass M_T of meteorites within the strewn field — this can be gauged from observations of the pre-fall fireball (if available). The light curve data, for example, can be used to determine the initial meteoroid mass, and then we can assume that typically no more than 10 percent of the initial meteoroid mass ends up as meteorites in the strewn field. There can additionally be only one meteorite with a mass *Max*, and this object may or may not be found in the strewn field, and the practical (that is, in terms of being able to find such an object) lower limit for *Min* is of order one gram. Clearly, the fall location (is it muddy, forested, lake-covered, or snow-covered), and the time and personnel available for searching, will strongly influence how good the total fragment count for a given strewn field is going to be.

By way of example, the Bruderheim, Alberta strewn field (Figure 7.7) is relatively small, being some $5.6 \times 3.6\,\mathrm{km}$ in size, and

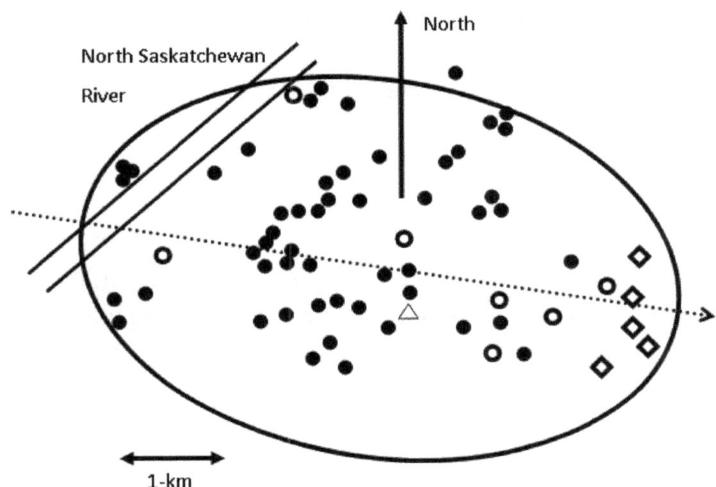

Figure 7.7.   Outline of the Bruderheim meteorite strewn field. The compact fall ellipse indicates a steep angle of fall for the meteorites. The solid circles indicate the locations where meteorites having a mass smaller than 4 kg were found. The open circles indicate where meteorites with masses between 4 and 24 kg were found, while the open diamonds indicate the location of meteorites with masses greater than 24 kg. The dotted line indicates the direction of travel. The open triangle indicates the triangulation point of latitude 53° 54′ N, longitude 102° 53′ W.

a thorough search and investigation by Robert Folinsbee (late of the University of Alberta) and coworkers has yielded 6 meteorite fragments with masses between 4.56 and 12.59 kg. The largest meteorite found within the strewn field had a mass of 31.36 kg, and over 500 meteorites with a mass less than 0.1 kg have been collected. Using Equation (7.2), assuming a constant power law exponent of $s = 1.85$, and that the 6 fragments between 4.56 and 12.59 kg represent a complete sample, indicates that $K = 32$. Now, taking *Max* to be 31.36 kg and setting (say) *Min* = 0.01 kg, Equation (7.2) further indicates that $N_T \approx 2000$ meteorites with masses between 10 grams and 31.36 kg could reside within the Bruderheim strewn field. This number, however, is based upon an assumed value for the exponent s and under the assumption (which certainly need not be true) that the exponent remains constant over all mass ranges.

Folinsbee and coworkers collected some 700 fragments from the Bruderheim strewn field.

The total mass M_T of meteoritic material within a strewn field is given by the integral

$$M_T = \int_{Min}^{Max} m\, dN = \frac{K}{2 - s}(Max^{2-s} - Min^{2-s}) \qquad (7.3)$$

And, in the case of the Bruderheim fall this gives $M_T = 250\,\text{kg}$. Now the total collected mass from the strewn field weighs in at some $300\,\text{kg}$, so we are in the right ball park with the calculations, but clearly some additional work needs to be done in refining the value used for the power law exponent. If we choose $s = 1.8$, then $K = 14.5$, $N_T = 720$, and $M_T = 115\,\text{kg}$. If, however, we choose $s = 1.9$, then $K = 35$, $N_T = 2473$, and $M_T = 270\,\text{kg}$. We see here that there must be a trade off in the final reductions. The closer the exponent s is to 2 so the greater the total number of fragments N_T in the strewn field, and the greater the total mass of meteorites M_T. With the simple theory outlined above (i.e., assuming a fixed exponent s in all mass ranges), it may well be possible to *fix* one of the numbers, N_T or M_T, but not both at the same time.

7.8   The 5 most extensive (in mass) meteorite strewn fields

In Table 7.8 the ordering is in terms of total mass recovered, and it is very similar in compilation to the previous table concerning the top 5 falls producing the greatest estimated number of fragments.

Table 7.8.   The 5 most extensive (in mass) meteorite strewn fields.

Meteorite	Location	Discovery	Type	Mass (tonnes)
Campo del Cielo	Argentina	1576	Octahedrite IAB	100
Sikhote-Alin	Russia	1947	Octahedrite IAB	70
Gibeon	Namibia	1836	Octahedrite IVA	26
Nantan	China	1958	Octahedrite IAB	9.5
Pultusk	Poland	1868	H5	8.9

The new members of the list, however, are the Campo del Cielo strewn field, Argentina, the Gibeon meteorite strewn field in Namibia, and the Nantan meteorite strewn field in Guangxi region, China.

There is no known fall date for the Gibeon meteorite, although it is typically given as some 100 million years ago. The first meteorite fragments were found in 1836 and the resultant strewn field is thought to be some 275-km long. An export ban on Gibeon meteorites was set in place by the Namibian Government in 2004, but it continues to trade as one of the most common and least expensive iron meteorites (Figure 7.8).

The Nantan meteorite is generally taken to have fallen in 1516, but the first fragments from this fall were not known or identified until 1958. Many hundreds of fragments have now been collected from

Figure 7.8. The Gibeon meteorite display in Post Street Mall, Windhoek, Namibia.

a 28-km long strewn field, with the largest found fragment having a mass of some two tonnes.

7.9   The 5 most recently formed meteorite impact craters

Table 7.9 is arranged according to the descending order of the estimated age (in all but one case) of the crater formation event. There is a large gap in the ages associated with the first two entries in this list, Carancas and Sikhote-Alin, and the remainder which all are estimated to be at least a thousand years old. Placing Haviland before Whitecourt is perhaps premature, since the Haviland age is uncertain, while that of the Whitecourt crater is based upon the carbon dating of material (fire ash) found within its interior. No doubt there are some younger craters with associated meteorites having ages <1000 years that have yet to be discovered.

The fall of the Carancas meteorite caused much local concern when nearby many villagers became ill. It was later found that this was due to mild arsenic poisoning — the arsenic, however, being released from the local ground water disturbed by the impact (recall Figure 6.2) rather than being a component of the meteorite.

The Haviland crater is associated with the Brenham, Pallasite, strewn field, while the Dalgaranga crater has yielded rare mesosiderite specimens.

Table 7.9.   The 5 most recently formed meteorite impact craters.

Meteorite	Location	Crater Diameter (m)	Age	Meteorite Type
Carancas	Peru	13	September 15th, 2007	H5 chondrite
Sikhote-Alin	Russia	26	February 12th, 1947	Octahedrite IIAB
Haviland	USA	11	1000	Pallasite
Whitecourt	Canada	36	1130	Octahedrite IIIAB
Dalgaranga	Australia	21	3000	Mesosiderite

The largest crater with a known meteorite strewn field is that of the Barringer [also Meteor] crater in Arizona. This crater, which has a diameter of 1.186 km is estimated to be some 50,000 years old, and is associated with the Canyon Diablo (Octahedrite IAB) meteorite strewn field.

7.10 The 5 most unexpected meteorite finds

Here the top 5 list is developed in terms of the improbability of both survival and discovery. At number one in our list is a small, 2.5-millimetre-sized fragment of a fossil meteorite found and described by Frank Kyte (University of California) in 1998. The remarkable property of this find is that it was recovered from a core sample of deep-ocean sediment collected from a location in the mid-Pacific Ocean. Not only is such a chance find amazing enough (the core pipe is only 6.4 cm across), but the fragment was collected from the Cretaceous/Paleogene boundary layer — the boundary layer corresponding to the great dinosaur die-off some 65 million years ago (see Chapter 11). The small fragment, identified as a chondrite, stony meteorite, is mostly oxidized, but a detailed analysis revealed a distinctive extraterrestrial composition. Kyte argues that the meteorite fragment is part of the impacting body that produced the Chicxulub impact crater. Located off what is now the Yucatan peninsula, this massive 150-km diameter crater is the Earth-scar left over from the impact of a 15-km diameter asteroid impactor — an impact that oversaw the demise of the dinosaurs and the rise of the mammals (and ultimately us).

Second on our list of improbable finds is a lunar rock (sample 14321) returned to Earth following the Apollo 14 moon landing. In this case it is not so much the fact that a lunar breccia was returned (which is marvelous enough), but that buried within the sample was a fragment of ancient Earth. Effectively, the returned Moon rock brought home a terrestrial meteorite (see Chapter 15). The analysis of sample 14321 by J. J. Bellucci (Swedish Museum of Natural History) and coworkers in 2019 revealed that it contained a small clast of terrestrial material that was excavated from Earth an estimated 3.9 billion years ago.

Third on our list of unlikely finds is that of a 25-cm sized fossilized meteorite fragment within a core sample, some 766 meters deep, in the impact-melt rocks associated with the 70-km diameter, 145-million-year-old Morokweng crater in South Africa. Not only is the probability of sampling such a meteorite extremely low, given the recovery depth, drill size, and crater diameter, but it is also one of the last places where a meteorite might be expected to survive. As discussed in Section 7.5 the production and excavation of a crater, especially a very large crater like Morokweng, entail the release of tremendous amounts of energy and this in turn is expected to fully vaporize the impacting body. Not only therefore is the survival of any impactor material surprising, but an analysis by W. D. Maier (University of Pretoria) and coworkers in 2006 revealed that the meteorite was so well preserved that it could be characterize as an LL chondrite.

Fourth on our list of exceptional finds is that relating to the discovery of what is essentially a fossilized strewn field in Sweden. In this case meteorite discoveries were made in a limestone quarry — that is, they are fossilized meteorites embedded within what is now the solidified sediments of an ancient seabed (Figure 7.9).

Not only did the Thorsberg quarry meteorites fall into an ancient seabed, some 480 million years ago, but they survived the long process of mineralization (just like the associated nautilid fossils, the original material has been replaced by calcite and clay minerals), and then they chanced to be found in a 3.2-meter-thick layer of limestone, over a region having a surface area of just 0.02 square-kilometers, being quarried to make flooring tiles. The first of the fossilized meteorites from the Thorsberg quarry was found in a waste pile, having been discarded by the quarry workers as a blemished and defective slab. The first fossilized meteorite to be recognized, the so-called Brunflo meteorite, also from a Swedish limestone quarry, was found in the early 1950s, but was not identified as being extraterrestrial until 1979. Of the nearly 62,000 meteorites currently cataloged, just over 100 are classified as fossils.

Fifth on our list of most unexpected finds is that of the Winona meteorite. In this case the meteorite was discovered in a burial cist or

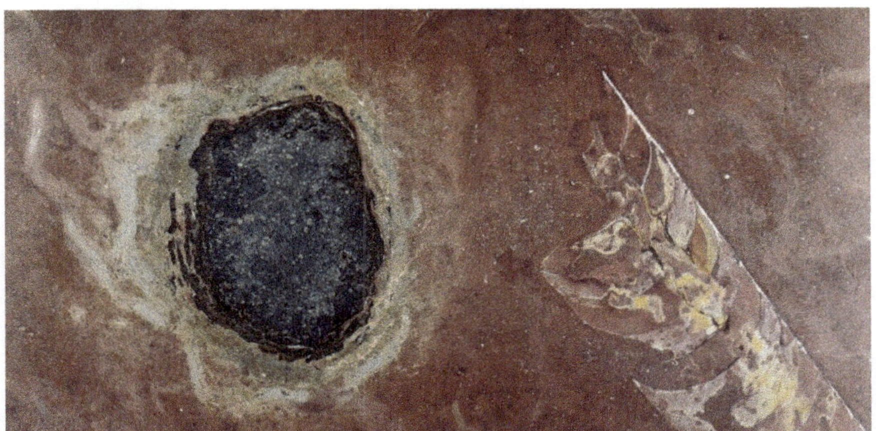

Figure 7.9. Fossilized chondrite meteorite (to the left) and a nautiloid shell (to the right) from the Thorsberg quarry in Sweden. The meteorite fell during the Ordovician period some 480 million years ago. The meteorite is some 7×8 cm in size and a careful examination of the image reveals numerous small spots which correspond to preserved chondrules. Image credit B. Schmitz.

stone casket during an archeological excavation in Winona, Arizona in 1928. The meteorite cist was found about 25 cm below ground during the excavation of a Sinagua, Native American village in what is now northern Arizona. The village was apparently occupied from the mid-12th to late 13th century, although evidence for the Sinagua culture stretches much further back in time. The fact that the meteorite was carefully stored and buried is suggestive that it was considered a sacred object — this might be especially so if the meteorite had been observed as fall. The Winona find represents the type specimen for the winonaite meteorites, and with a mass of 24 kg, it is by far the largest object in its rare class — just 35 specimens are catalogued. The winonaite meteorites are described as being primitive achondrites, with compositional characteristics falling between those of the H and E chondrites, although unlike the chondrites the winonaites show clear evidence of undergoing partial melting and thermal alteration (recall Section 1.2). And, finally, to add a twist to this entry and continuing the theme of the previous entry, Birger Schmitz (Lund University) and coworkers described in

2014 what appears to be a fossilized winonaite meteorite found in the upper (younger) layers of the Thorsberg quarry. This particular specimen (called by Schmitz *et al. the mystery object*) was found on June 26th, 2011, and of the more than one hundred fossilized meteorites that have been found in the Thorsberg quarry, it is the only one that is not an L chondrite. Schmitz and coworkers argue, in fact, that this particular fossil fragment may have originally formed part of the impactor that was responsible for the break up of the L chondrite parent asteroid some 470 million years ago (see Chapter 4).

7.11   The 5 most controversial meteorites

That something might be controversial is always a tricky label to justify — what is controversial for one person might be plain obvious for the next. There are a number of good candidate meteorites that have split the research community, however, and this list (as with all the others) will inevitably change with time and with the collection of more data, and no doubt with the collection of more controversial objects.

Number one in our list of controversial objects is a meteorite recovered in Antarctica. It is not the Antarctica connection that makes this object controversial, nor is it controversial that specimen ALH84001 (with the ALH standing for the Allan Hills region of Antarctica where it was found) is a meteorite from Mars (see Chapter 14). Rather, it is that this meteorite may contain microfossils of Martian nanobacteria (Figure 7.10) that is controversial.

The story of this particular meteorite is amazing in its own right. Found during the Antarctic Search for Meteorites (ANSMET) field season in 1984, the 1.93-kg meteorite was immediately recognized as being somewhat 'odd,' but it was not recognized as being from planet Mars until 1994. Subsequent to its Martian identity being made, it additionally turned out to be distinct with respect to the other known Martian meteorites — indeed, it is the oldest known Martian meteorite sample with a crystallization age of 4.09 billion years. This formation age is particularly relevant to the story, since at that time Mars would have supported liquid water on its surface. Detailed

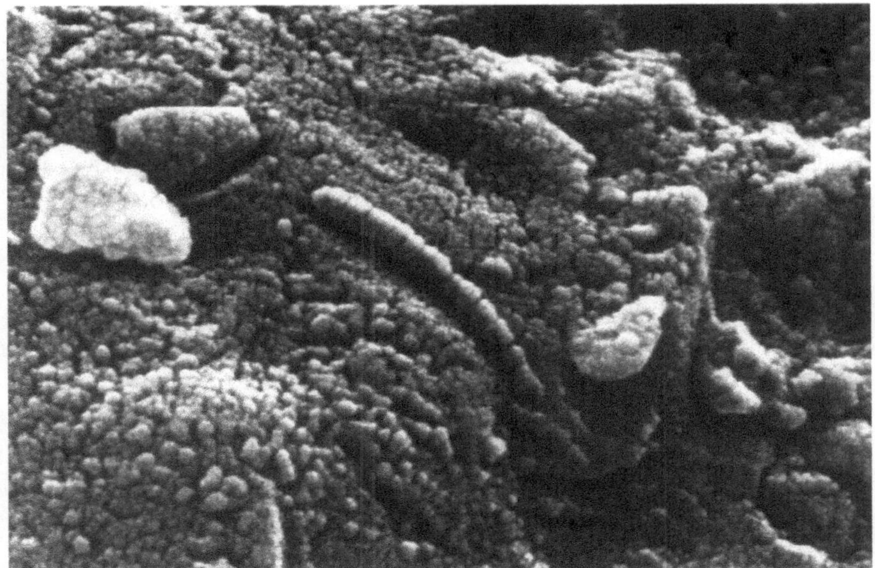

Figure 7.10. A scanning-electron-microscope image showing what may be a fossilized nanobacteria [the segmented, worm-like structure at image center] found within the ALH84001 meteorite. Image credit: Wikimedia commons.

analysis of component radionuclides (recall Chapter 4) indicates that it was launched from the Martian surface 17 million years ago, and later, after a convoluted gravitational dance, found the Earth to land in Antarctic about 13,000 years ago.

In 1996 the remarkable story of ALH84001 became even more incredible. It is at this time that David McKay (NASA Johnson Space Center) and a team of 8 coworkers published a paper in the August 16th issue of the journal *Science*. It was a paper heralded by news briefings and even a television broadcast by US President, Bill Clinton. The research paper had the relatively tame title of *Search for Past Life on Mars*, and the introductory section indicates that, "our objective was to look for signs of past (fossil) life within the pore space or secondary minerals of this [ALH84001] Martian meteorite." The great media *to-do*, however, was concerned with the claim that evidence of "relic biogenic activity" had indeed been found. The researchers made their claim based upon a series of observations concerning the discovery of carbonate globules and abundant polycyclic aromatic hydrocarbons (PAHs). PAHs are large, complex organic [that is carbon-based] molecules. Additionally, the researchers identified grains of magnetite and iron sulfides. Now, it is the case that all of the features identified by McKay and coworkers could have been produced by inorganic processes, but this, they argued, was unlikely and accordingly concluded that the features they had observed were "evidence for primitive life on early Mars": the carbonate globules having features resembling terrestrial microorganisms (Figure 7.10), and the magnetite and iron sulfides being the residue of reduction reactions known to be important in terrestrial microbial systems.

The research paper by McKay *et al.* was bold, imaginative, and well argued, but the final claim that the observations were best interpreted in terms of being signatures due to ancient Martian microbes was too much for many other scientists to accept. Ever since 1996, when the paper by McKay *et al.* first appeared, the debate has raged. Numerous research groups have since argued that in spite of the careful analysis by McKay *et al.* all the signatures that they highlighted are, in fact, inorganic in origin. To this day, the debate continues and no one side, organic versus inorganic, has yielded to the other's arguments. Perhaps, in terms of Scottish jurisprudence, all that can be said at present is "case not proven."

McKay and coworkers added a further twist to the debate concerning ancient microbial alteration of Martian meteorites in 2006, when a study of the Nakhla meteorite revealed what are interpreted as narrow dendritic channels putatively produced by Martian microbes. These micro-channels, it is suggested, are the result of corrosion from organic acids (so-called bio-alteration structures), and are similar in appearance to microbial structures found in terrestrial rocks. In 2014 a second Nakhla-type Martian meteorite, Yamato 000593 (collected in Antarctica by Japanese researchers in 2000), was found to contain similar dendritic micro-channels (Figure 7.11). Adding further zest to the Martian life debate, analyses of Y000593 indicates that it is much younger than ALH84001, having formed just 1.3 billion years ago, from basaltic rock extruded in a Martian volcano

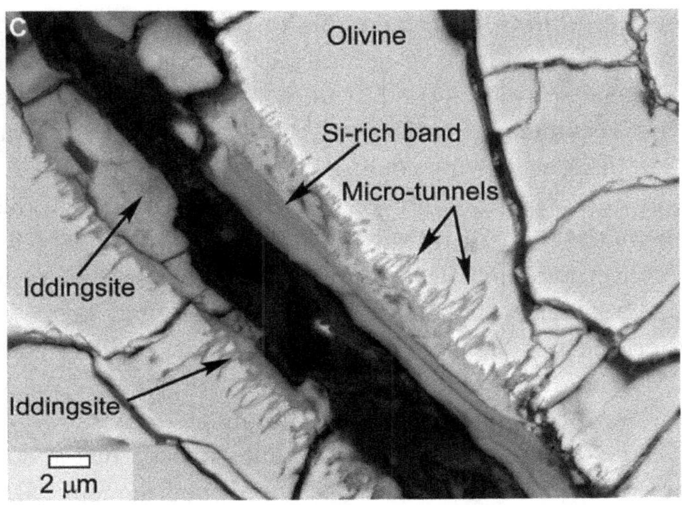

Figure 7.11. Scanning-electron-microscope image of the surface of Y000593, a Martian Nakhla-type meteorite, showing dendritic micro-tunnels. Image courtesy of NASA.

lava flow. This meteorite was launched on its journey to Earth some 12 million years ago and landed in Antarctica some 50,000 years before the present. Given the relatively recent crystallization age of Y000593, the micro-channels, on the basis they are bio-alteration structures, suggest that life existed on Mars as recently as the middle to late-Miocene on Earth — a time when Mastodons were first beginning to walk in the ancient forests of North America.

Yet one more turn in the convoluted story of ALH84001 occurred in May 2020, when Mizuho Koike (Japanese Aerospace Exploration Agency) and coworkers published a study in the journal *Nature Communications*. This particular analysis looked at the nitrogen-bearing organic compounds in ALH84001 and argued that the careful contamination-control techniques employed indicated that the organic material detected must be of Martian origin, and that it dates back to a time 4 billion years ago (to the Noachian period on Mars). Because the carbonate minerals detected typically precipitate from groundwater, the finding by Koike *et al.* indicates that a wet and organic-rich period must have existed on early Mars, and that such conditions would have been favorable to the evolution of life. While the argument concerning the actual identification of microfossils in ALH84001 has yet to be settled, it does appear, in light of the recent study, that this meteorite does tell us that early Mars was most definitely a place where life could have evolved.

Second in our list of controversial meteorites is NWA 7325. The 35 fragments that constitute the NWA 7325 meteorite were found by German collector Stefan Ralew in a Moroccan marketplace in April 2012. The fragments, totalling 345 g in mass, had been found in the southern part of Morocco earlier in the year, and were on sale as possible Martian meteorites. Ralew was not convinced of the Martian characteristics, but purchased the fragments anyway — partly because of the pleasing green-colored, almost frosty, fusion crust that they sported (Figure 7.12). Having acquired the fragments, Ralew set about having them analyzed in detail. Anthony Irving (University of Washington in Seattle) and coworkers soon identified the meteorite as being an ungrouped achondrite, containing magnesium-rich mafic

Figure 7.12.   A slice through the NWA 7325 meteorite. This may be one of the rarest meteorites on Earth. (Credit: Wikimedia commons)

silicates, highly calcic plagioclase, and chromium-bearing sulfides. This unique achondrite composition was apparently the result of material crystallizing from a magnesium-calcium-rich, iron-poor silicate magma under very low oxygen fugacity conditions, and it has accordingly been attributed (all be it cautiously) an origin on planet Mercury. If NWA 7325 is from Mercury then its collected fragments (and more pared fragments have been found since Ralew's initial marketplace purchase) qualify as being the rarest meteorites on Earth, and they potentially represent the only samples of the Hermean surface to be collected to date.

The third example in this list of controversial meteorites is not one specific meteorite — indeed, it is many-a-meteorite over the ages. The controversy here is not about the properties of a particular meteorite, but about who actually owns a meteorite once it has been found, and the means by which meteorites have been obtained. There are legal rules and precedents concerning meteorite finds and ownership, but as with many other aspects of the law, the rules have been wilfully bent, broken, and generally ignored if convenient or possible. Not only have many legal battles been fought over ownership, but so too have legitimate finders been defrauded and/or bullied into selling their celestial gifts. In other cases, finders have deliberately misrepresented their efforts — which is another way of saying willfully lied about where they found a meteorite.

The fall of the Crumlin meteorite in County Antrim, Ireland, on September 13th, 1902, affords one relatively harmless example of the situation (far from unique) in which a zealous museum curator found a way of coercing a purchase. The 4.2-kg Crumlin meteorite is not especially noteworthy (it is an L5 chondrite), but it did chance to fall at the same time as the 72nd meeting of the British Association for the Advancement of Science being held in Belfast, located just a few tens of kilometers away from Crumlin — an auspicious, well-timed fall, perhaps. At this meeting was Sir Lazarus Fletcher, curator of meteorites at the British Museum in London (we meet Fletcher earlier, recall, in the discussion of the Launton meteorite). Fletcher quickly made his way to Crosshill Farm and tried to close a deal for the meteorite's purchase. The landowner,

Andrew Walker, it transpired, was loathed to part with his newly fallen celestial treasure, and it seemed that a purchase deal was not going to be made. Fletcher, undaunted, however, discovered (through unidentified sources) that Walker's niece not only wielded considerable influence over her uncle, but was desirous of obtaining a musical organ. Fletcher, sensing insider influence, sent a cheque for ten pounds, from his own bank account, to Walker's niece, and sure enough, shortly thereafter the meteorite arrived at the museum in London (purchased for £105, equivalent to some US$15,000 in present day funds). In contrast, sensing a chance to promote their political agenda, the *Irish Weekly Independence and Nation* newspaper published a satirical cartoon of the British Museum, under the guise of John Bull, steeling away with the Crumlin meteorite. Bribery and politics, all rolled up in the story of one harmless meteorite. It is interesting to note that the caption to the cartoon (published on November 16th, 1902) reads in one part, "Lave down that big stone; 'tis a meteor [sic] huried from Vaynus, or Mars, or from Jupiter's moon, sure it's mine, for it fell on my land, ye bosthoon." While Martian meteorites were eventually recognized, the other source locations suggested in the cartoon caption are highly improbably — the association of meteorites with the asteroid belt is, to the modern eye, conspicuously absent.

The fall and find of the Dresden, Ontario, Canada, H6 chondrite on July 11th, 1939, affords one example of a hapless (and poor) farmer being persuaded by a knowing local to sell a meteorite at a vastly under-representative price (in the Dresden case $4 for a 40-kg meteorite). The meteorite, thereafter, was sold on for profit.

Even in the modern era, deciding who owns a meteorite can be a highly divisive and litigious issue, and the Wuxilike meteorite in Xinjiang Uyghur Autonomous Region, China (Figure 7.13) is not an atypical example of such a case. The history of the Wuxilike meteorite is convoluted: although clearly an old fall, the meteorite was discovered on pastureland by a Kazak herdsman in the 1980's, and being recognized as something special was known locally as *The Tear of Allah*. For at least 25 years the finder, Zhuman Reamazhaen, was the proud caretaker and protector of the meteorite, but in 2011

Figure 7.13. The elder son of Zhuman Ramazan poses with the Wuxilike meteorite — a possible fragment in the Aletai meteorite strewn field.

the Altay Municipal Government seized the meteorite. Zhuman and family sued for the return of the meteorite arguing that they had discovered the divine stone on the pastureland that they leased. Under Chinese law, however, agriculture land is allocated on a tenure basis and is not privately owned. Additionally, under the Chinese constitution all mineralogical resources belong to the state. The Altay government argued that the meteorite was a natural resource and won their case.

Not surprisingly, Zhuman's lawyers contested the court's decision, but then a further problem arose. In 2013 two Kazak guides took legal action against the Altay government, arguing that they were the first finders of the meteorite, and that they had reported the find to the Beijing Planetarium, receiving a reward of 5000 yuan and a certificate of honor for their efforts. By accepting the reward, however, the court ruled that the two guides had surrendered any rights to the meteorite's ownership. Three more court appeals were launched by Zhuman over the next 5 years, but it was finally ruled in 2018 that the meteorite belongs to the state and that it will not be returned to the find location.

In the mean time, a research paper published by Wang Kechao and Xu Weibiao (both of the Purple Mountain Observatory) in the *Chinese Science Bulletin* for January 2016 added to the intrigue associated with the Wuxilike meteorite, arguing that it could be chemically paired to two other meteorites: the 28-tonne Aletai meteorite found in 1898 and the 430-kg Ulasitai meteorite found in 2004. This pairing (they are all IIIE iron meteorites), the researchers suggest, indicates that the three meteorites make up the world's longest strewn field, with material being spread over a range of at least 425 km, beating out the next record holder, the Gibeon strewn field, by some 150 km. A strewn field with a major-axis many hundreds of kilometers in length requires a remarkable, low angle of entry for the parent meteoroid — remarkable, but not impossible. The researchers describe the strewn field as being a "mega-meteorite rain fall event," but a plausible atmospheric trajectory and the strewn field characteristics have yet to be fully investigated.

The final entry in this section on controversial meteorites concerns those meteorites that never actually existed. In this sense, we are not dealing with identification mistakes, but with down-right deception (which is basically a polite way for saying "lies"). The Port Orford meteorite heads our list of non-existent meteorites. In this particular case the story of a large meteorite find in the state of Oregon was invented to essentially embezzle money from the United States government. At the center of the story is John Evans, a contract explorer hired by the Department of the Interior in 1856. Evans, a medical doctor by training, set out from Port Orford and over the ensuing expedition gathered numerous geological samples. These specimens were ultimately donated to the Boston Society of Natural History in 1859, and it was then that several of the samples were recognized as being from a rare pallasite meteorite. At this time

Evans revealed that he had hammered the samples from a massive parent iron embedded in the grassy slopes below what he called Bald Mountain (now generally thought to be Johnson Mountain). Talk of recovering the main mass soon followed and Evans persuaded a number of academies to petition Congress for funds to mount a new expedition. Unfortunately, while plans were being made to recover the meteorite, the events leading to the onset of the American Civil War played out and, indeed, Evans himself died the day after hostilities began (April 13th, 1861).

Over the years many expeditions were sent to the location outlined by Evans in the hope of finding the meteorite, but nothing has ever been found. What are we to make, therefore, of the pallasite samples claimed by Evans to be from the Port Orford meteorite? Howard Plotkin (University of Western Ontario, Canada), Roy Clarke (Smithsonian Institution), and Vagn Buchwald (University of Denmark) picked up the story in the early 1990s and through a close examination of Evans's journals and travel letters discovered that a hoax had been in play. Indeed, Evans was deep in debt and brought the meteorite recovery idea into play in an attempt to garner more expedition money from Congress. The clincher in this hoax was provided by a detailed chemical analysis of the supposed Port Orford finds by Buchwald and Clarke — it turns out that the specimens are an exact match to the Imilac pallasite meteorite. It is additionally known that Evans was in the Panamanian Isthmus in 1858 and at that time Imilac meteorite fragments were being sold as natural curiosities. The rest of the Port Orford meteorite story, as the saying goes, is rumor predicated upon duplicity perpetuated down the ages.

The Norwood meteorite supposedly fell on October 7th, 1909. Preceded by a brilliant, slowly moving fireball, observed by several eyewitnesses, that lit up the sky, the meteorite apparently smashed through a set of fencing bars and then penetrated deeply into the ground. The meteorite was found the next morning by a farmhand, who claimed that the stone, when dug out, was still hot. Several local observers reported seeing the fireball and the farm owner soon deposited the meteorite in a dime museum in Boston. There was just one problem, the meteorite was clearly not a meteorite of any known class, and indeed, when investigated by astronomer Frank Very (then director of the Westwood Astrophysical Observatory) the meteorite was identified as an ordinary water-warn stone (actually an ophitic andesite porphyry).

Sensing a scam in play, Very pushed his investigations further and eventually presented his damming conclusions in the March 18th, 1910, issue of the journal *Science*. The fireball, meteorite, and broken fence posts were indeed a hoax. Very writes that the "proprietor of a cheap vaudeville show in Boston purchased the *meteorite* from a Vermont man." Thinking that the *meteorite* needed some "local color" to be of interest, however, the fireball and fall were fabricated (with the help of the farmer on whose land it was said to fall). The rock was apparently taken to the farm under the cover of darkness, the fence bars deliberately smashed, and the unwitting farmhand sent out the following morning to make the discovery. That the *meteorite* was a fake is hardly newsworthy, indeed, the "Vermont man" was able to pass it off to a profiteering conman, and such frauds are still attempted (and occasionally successful) to this very day. What is particularly intriguing with respect to this story, however, is the observations of the fireball on October 7th — it too was a hoax. Indeed, Very explains that a large skyrocket, with a delay fuse, had been attached to and then launched upward upon a balloon to a height of about one mile. The rocket, when it eventually ignited, produced the effect of a slowly falling fireball.

As a final hoax story, one of the more sophisticated scams (although propagated by an unknown antagonist) is that related to a small sample of the Orgueil meteorite. This particular meteorite fell in the south of France on May 14th, 1864, and it is a rare carbonaceous chondrite. Since the time of its fall, Orgueil has been extensively studied, but in the early to mid-1960s a fragment of this famous meteorite became embroiled in a (sometimes vitriolic) debate between research groups — one in Chicago and the other in New York. The story begins in the opening years of the 1960s, when chemists Bartholomew Nagy (Fordham University), George Claus, and Douglas Hennessy analyzed small samples of the Orgueil and the

Ivuna[2] meteorites. Reporting their findings in the March 24th issue of the journal *Nature*, Nagy and coauthors argued that they had found clear evidence of biotic material, and more specifically they argued that the material was definitely indigenous to the meteorites and not the result of terrestrial contamination. The findings were incredible, both micro-fossils and "organized elements" had been found inside of extraterrestrial matter. As with all such incredible finds, however, scepticism ran rampant — enter now the researchers from the University of Chicago. This latter group, headed up by Edward Anders, claimed that they found no similar such structures in their samples of Orgueil, and suggested that Nagy and coworkers had perhaps mishandled and unknowingly contaminated their sample. Discussing the issue of these lifelike "organized" forms at a meeting held on May 1st, 1962 at the New York Academy of Sciences, John Bernal (Birkbeck College) nicely outlined the problem at hand, arguing that "they [the touted microfossils] might be contaminations, artifacts — *jokes of nature* — or remnants of indigenous organism."[3] The problem, of course, is that the two sides could not agree on what the explanation was.

Moving into the mid-1960s, however, a new twist weaves into the story. All of the experiments up to that time had been made on samples of Orgueil that could have, at least in principle, become contaminated by terrestrial microbes, so what was required was a virgin sample, a sample of Orgueil that had never been experimented upon. Remarkably, such a sample actually existed, and it was contained within a sealed glass jar at the *Musée d'Histoire Naturelle* in Montaubin, France. It is through a 25.5-gram sample of this sealed specimen that, upon examination in Chicago, the hoax was revealed. Writing a summary of their analysis in the November 27th, 1964

---

[2]Ivuna is another well-studied carbonaceous chondrite, which fell in Tanzania on December 16th, 1938.

[3]This quotation is from the meeting summary by Harold Urey, published in the August 24th, 1962 issue of the journal *Science*.

issue of *Science*, Edward Anders and coworkers noted, "we found it contained several light brown particles, up to several millimeters in size." These turned out to be plant seeds. In addition to seeds, Anders *et al.* also discovered coal fragments and "a water-soluble protein material resembling collagen-derived glue" within their meteorite sample. The seeds turned out to be from a reed (*Juncus*) that commonly grows in the south of France, and the coal and glue are the smoking gun of deception at play. The detailed analysis presented by Anders and coworkers in 1962 revealed that what had once been thought of as a pristine and preserved Orgueil material, clearly was not. The sample studied in Chicago had clearly been tampered with, some 100 years earlier, when the specimen was placed within the sealed jar. Someone, and the perpetrator is unknown, had deliberately placed several seeds within the interior of the meteorite and then covered up their fraud by resealing the fusion crust with a glue-charcoal dust mixture.

As hoaxes go, the Orgueil one was a failure since it was not revealed until a century after the act. What is the most likely explanation for the fraudster's motivation was the debate concerning the spontaneous generation of life, which was in full rage in 1864. This idea held that inorganic matter could transform into organic matter — for example, fleas could arise from dust grains, maggots from dead flesh, and mice from corn seeds. This idea, although long held, was roundly debunked in a series of famous experiments and lectures given by Louis Pasteur during the early 1860s. The French Academy of Science even established a prize in 1862 to focus attention on the topic. It seems probable, although we will never know for certain, that the hoax was instigated in an attempt to show the spontaneous generation of plants from meteoritic material. While the original hoax failed, the seeds did not germinate (presumably from a lack of watering), the unraveling of the fraud in 1962 brought sharp attention to the problems (especially terrestrial contamination) associated with the identification of potential bio-markers within meteoritic material (as described earlier with respect to the Martian meteorite ALH84001).

7.12   The strangest meteorite stories

Just because something falls out of the sky does not mean it is a meteorite, and surprisingly many things do fall out of the sky. Indeed, as ichthyologist Eugene Gudger pointed out in his wonderful 1921 summary of *Rains of Fishes* (published in the November-December issue of *Natural History* magazine), accounts of fishes, frogs, stones, and oddities falling from the sky can be found in the literature as far back as ancient times. Aeschylus, the father of Greek tragedy, is even accounted an appropriately ironic death, in 455 B.C., by being struck on the head by a tortoise falling out of the sky.[4]

Perhaps the most well-known example of non-meteoritic material falling from the heavens is that of blue ice. Such chunks of ice, which can be many kilograms in mass, have been known to puncture holes through roofs — such as the 1971 fall through the roof of a chapel in Kensington, London (an event that resulted in the ultimate demolition of the building); or the ice that fell on multiple cars in Kelowna, Canada in 2018. These and numerous other such reports which pepper the newswires have been identified with frozen bio-waste leaking from aircraft sewage systems. The distinctive blue color that such ice fragments often show is due to the disinfectant used in aircraft septic tanks. It should be noted here that the blue ice falls are thought to be different to the megacryometeors that are discussed in Chapter 23.

Other examples of objects falling from the sky, the accounts of which are readily found within internet newsfeeds, relate to the modern use of woodchippers. The rapidly rotating blades found within these rapacious systems can throw any lose stony or metal material at very high speeds a very great distance. Indeed, numerous accounts of metal fragments being found embedded within the sides of houses and cars can be directly linked to ejection from a woodchipper, or from being flung out by a lawnmower. Other

---

[4]The story indicates the tortoise was dropped by an eagle who mistook the bald head of Aeschylus for a stone upon which to shatter the hapless Testudinidae. The well-documented falls of fishes and frogs can be readily attributed to adverse meteorological effects.

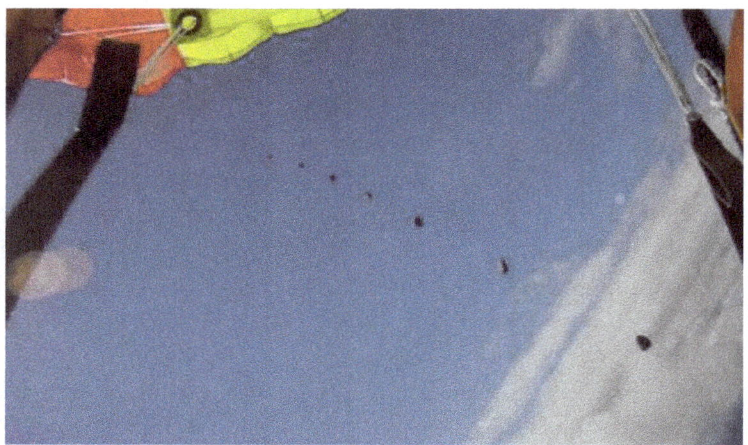

Figure 7.14. A montage of images from the camera of skydiver Anders Helstrup. The rock was most likely trapped within the parachute folds prior to opening, and it is highly unlikely that the object is a meteorite.

examples of terrestrial rocks falling from the sky and striking cars and/or buildings are more mysterious, but the mystery invariably resides not in the rocks but in how they were actually lofted into the sky.

One particularly strange story, involving an apparent near miss with a falling object, is that described and filmed by Norwegian skydiver Anders Helstrup in 2012 (Figure 7.14). It is unlikely, although not impossible, that the rock was a meteorite, since the odds of such a close call are incredibly small. Meteorite impact probabilities will be discussed in more detail in Chapter 18, but a rough estimate of the (un)likelihood of a near miss with a meteorite can be determined according to a ratio of areas. If a meteoroid is to become a meteorite then it must encounter the Earth's atmosphere, and from space this gives a target area of $A_E = \pi(R_\oplus + 100)^2$, that is the Earth's cross-section area at a height of 100 km above the Earth's surface, and where $R_\oplus$ is the Earth's radius of 6371 km. The area of a skydiver A_{SD} is about one square meter, and accordingly, the probability of being hit is of order $P_{hit} = A_{SD}/A_E \sim 8 \times 10^{-15}$ — which is an incredibly small probability. In order to determine the likelihood of a hit, however, we would need to multiple P_{hit} by the number of meteorites, of a specified size, hitting the Earth per year. Say the meteorite weighs in at 100 grams, the total number of such objects N_{100} falling to the Earth per day is of order 250. Accordingly, if a skydiver made 10 jumps per day, then of order $1/(10 \times N_{100} \times P_{\text{hit}}) \sim 140$ million years of such continued activity would be required before a hit is likely to occur.

If these odds seem small, it is even more remarkable to read in *The Marine Observer*, published by the British Meteorological Office, the following account, "November 6th, 1958. Indian Ocean. Two meteors were seen to collide ... The brighter one was travelling in a NE'ly direction while the other one moved toward the W. When the collision occurred a brilliant white flash with a bluish tinge was observed. From first to last the time elapsed was about 4$^{1}/_{2}$ sec. Sky almost cloudless and visibility excellent." In this case, given that meteors, rather than meteorite-dropping fireball events are implicated, the size of the meteoroids would be in the centimeter range, although the associated trails could be of order several meters across. There are certainly many more smaller meteoroids than larger ones striking the Earth's atmosphere each day, but if the meteor collision account is genuine, and a collision really did take place, then the observers who witnessed it saw something truly remarkable.

Additional remarkable meteorite encounters are described by Brother Potamian, from Manhattan College, New York, in an article appearing in the December 1897 issue of the *Catholic Reading Circle Review*. In this case two near miss, nautical stories are recounted. On November 17th, 1896, the German registered ship *S.S. Wilkommen*, having left New York for Dantzig had just cleared the Newfoundland Banks, when a large and brilliant meteor was seen to "plunge into the sea about 2 miles ahead of the vessel, with a loud roar and hissing sound; a few minutes later an immense tidal wave, presumably caused by the fall, struck the ship, doing no little damage." Certainly, a large meteoroid (small asteroid) impact into the sea could cause a tidal wave to occur, but one suspects that, in this case, the two events, the meteorite impact and the large wave, were unrelated.

Another account concerns the British registered ship the *Cawdor*, which while rounding Cape Horn, bound for San Francisco, on

August 20th, 1897 encountered a heavy storm, and "all hands were on deck busy engaged in trimming the vessel when suddenly out of a blinding flash leaped a huge meteorite into the boiling sea. It plunged so close to the ship that the deck was swept by the water splashed up, so for a moment the sailors thought they were going under." This same story was picked up in the February 1899 issue of *Popular Science Monthly*, where it was additional revealed that the meteor "passed between the main and mizzen masts," and that "the air was filled with sulphurous fumes." Separating fact from fiction in such stories is always difficult, but if true, the poor sailors aboard the *Cawdor* were seemingly very lucky, not only did they survive the wrath of a Cape Horn howler, but dodged a celestial *cannonball* as well.

The reverse of these two stories has also been reported on numerous occasions throughout history, where a bright meteor has been mistaken for a ship's distress flare and emergency crews have set out to initiate a rescue. Perhaps one of the most ironic such circumstances, however, was the disastrous situation concerning the sinking of RMS Titanic on April 14th, 1912. In this case nearby ships mistook distress rockets for a firework display, and surviving passengers recall seeing many meteors flashing across the sky while they awaited rescue. These meteors, in fact, belonged to the annual April Lyrids meteor shower.

The suggestion in the *Cawdor* incident that the air was filled with "sulphurous fumes" is interesting in the sense that many accounts of odors accompanying meteorite falls have been reported. Even as recently as the 2013 Chelyabinsk fireball event, eyewitness many tens to hundreds of kilometers away from the fireball track reported sensing strange odors. At large multiple-kilometer distances it is impossible to account for odors due to meteoroid outgassing and atmospheric diffusion, so such reports indicate something more psychological than actual (but see Chapter 22). Close to a recent fall, however, it is possible that odors may well be sensed, but again the observational record is sketchy at best. Some observers report that recently fallen or newly cut meteorites do have a sulphurous smell, or an odor something like burnt gunpowder, or starter fluid. Indeed, when Michel Knapp ran outside to investigate a loud crash, she not only found the Peekskill meteorite (fall date: October 9th, 1992) that had smashed through her car, but when picked up the meteorite was apparently warm to the touch and gave off a distinct smell of sulphur.

There has long been a belief that meteors, fireballs, and meteorites are portents of good fortune (recall Chapter 2), and indeed, some cultures have venerated meteorites for their medicinal properties — there are, of course, no such properties (other than perhaps a placebo effect). This being said, sulphur, which is a known (minor) constituent of meteorites, does have antibacterial properties and sulphur is a vital component of a healthy body's physiology. Nonetheless, after the fall of the Mbale meteorite on August 14th, 1992, investigators found that a paste made from crushed meteorite fragments was being applied to the skin by local inhabitants as a cure-all against AIDs. In other cultures, at other times, it has been generally held that bodily inflammations and sores could be cured by being rubbed with meteorite, or that drinking water from a glass in which a meteorite had been placed could cure any sickness.

Can meteorite falls actually start fires? This is an age-old question, but the odds are very much against any such occurrence taking place. This being said, many people who have witnessed a fall and have been able to identify meteorites within seconds of their hitting the ground have reported them as being hot or at least warm. Theoretically, it is expected that meteorites should be cold, indeed, very cold when they hit the ground — this is simply a reflection of the fact that they have been in the cold depths of space for millions of years. True the surface layer of a meteoroid is heated and ablated as it interacts with the Earth's atmosphere (recall Chapter 6), but this is very much a surface effect — there is simply not enough time during the atmospheric flight for the heat generated at the surface to diffuse into the meteoroid's interior. Additionally, the final dark flight of a meteorite, before it hits the ground, is effectively a surface cooling-off phase. And yet, accounts of hot, or at least warm, meteorites abound. How much of such accounts is driven by expectation and an exaggerated sense of touch is unclear. Upon recovery, the Mazapil meteorite was described as "a hot stone [note, it is actually an iron meteorite], which we could barely handle." Meteorites from the L'Aigle fall (1803) were described as being "scorching hot" when first picked up (see Section 2.3). In contrast, the Dhurmsala meteorite that fell in India on July 14th, 1860 was described as being so intensely cold that it "benumbed the hands."

Unsubstantiated accounts of fires being set by meteors and meteorites abound in the historical literature, with dwellings being hit and burned to the ground. The Anglo-Saxon Chronicles indicate, for example, that in Coldingham, England, in 679, a monastery was destroyed by "fire from heaven." On November 10th, 1761, a meteorite impact supposedly caused a house in Chamblan, France, to burn down [no meteorite was found, however]. The *Times* newspaper for October 30th, 1801 indicates that a mill in Suffolk was set on fire by a meteor. It has even been suggested in many popular accounts that the Great Chicago Fire of October 1871 was started by a meteorite, although there is not one shred of evidence to support this idea. Ninety-five years later, on September 17th, 1966, another bright (possibly meteorite producing) fireball was seen over Chicago, and *The Detroit Free Press* newspaper published an article suggesting that bush fires were produced by its fiery fragments. Here, the point is not that the newspaper report is entirely wrong in attributing the appearance of the bush fires to the meteor, but that the myth of meteorites being hot when they land is still prevalent and generally believed.

It seems appropriate to finish off this section with a true shaggy-dog story. Indeed, it is a story that has become an urban myth, repeated so often that it is hard to believe that it is, in fact, not a true story. The meteorite involved in the story, Nakhla, is indeed real and fell in Egypt on June 28th, 1911 — indeed, it is the first Egyptian fall to be identified. The first detailed report on the Nakhla fall was published by John Ball of the Survey Department of the Ministry of Finance, Egypt, in 1912 (Survey Department paper 25), where it is indicated that fragments of the meteorite were found over a relatively small strewn field some 5 km in diameter, and "all together about 40 stones, of a total weight near 10 kilo-grammes [sic], were collected." The fall was well observed and eyewitnesses reported details on the fireball, the fireball trail, and the occurrence of sonic booms. One farmer who witnessed the fall, however, indicated that a dog had been struck by a meteorite and instantly vaporized. Since no meteorites, or ashes, or dead dog were found at the location indicated by the witness (the location being also some 35 km away from the main strewn field), the assumption has to be that no such event took place, and that the Nakhla dog story, albeit a good story, is not just apocryphal, but entirely fabricated upon rumor. With the above being said, there is no reason to believe that the eyewitness deliberately lied — indeed, they could not profit from any such story. Ball, in fact, comments in his Survey Department Paper that the incident concerning the dog "is doubtless the product of a lively imagination." What is entirely likely, however, is that the English interpretation of what the eyewitness said was wrong. Additionally, while the farmer no doubt heard the sonic booms associated with the meteorite's atmospheric flight, so too did the dog, and accordingly frightened it ran away and hid.

Chapter 8

1969: Year of the Carbonaceous Chondrites

In 1969 I gave up women and alcohol — it was the worst 20 minutes of my life.

George Best (footballer)

All years are significant, but perhaps, with a nod towards George Orwell's *Animal Farm*, some years are more significant than others. The year 1969 is one such year — it stands out along the ever-stretching timeline of history for its social, political, and cultural impact. Indeed, echoes and rumblings of 1969 still filter down and touch us to this very day. It was in 1969 that the first automatic teller machines (ATMs) were installed to deliver cash to high-street shoppers, the first officially death from AIDS was recorded in the United States, the first Boeing 747 jumbo jet took to the skies, and the first human eye transplant was performed. 1969 saw the release of *Abbey Road* by The Beatles, the first Led Zeppelin album was recorded, and the Woodstock festival took place in a muddied field in the Catskills. This was the year that Richard Nixon became President of the United States, the Chappaquiddick affair occurred, the PLO was founded, and the Charles Manson murders captured global headlines. In terms of the greater human good, however, 1969 was the year in which PBS television was established in the United States, the first Monty Python programs were aired by the BBC in England, and Sesame Street first appeared on TV. In terms of scientific achievement, the Venera 5 probe landed on planet Venus, Mariner 6 successfully flew past Mars, and the Allende and Murchison meteorites fell in Mexico and Australia respectively.

Oh yes, and 1969 saw Robin Knox-Johnson complete the first non-stop, solo voyage around the Earth, and, of course, the Apollo 11 Moon landing took place — seeing Neil Armstrong, on July 21st, become the first human being to place a foot upon another world. On a less than scientific note, 1969 was also the year in which the UFO-documenting *Project Blue Book* came to a close. As years go, 1969 was arguably dramatic and epoch making.

8.1   Carbonaceous chondrite meteorites

A total of 68 falls and finds are recorded in the 5th *Catalogue of Meteorites* [2] for 1969, although it is the Allende and Murchison meteorite falls that are by far best remembered, and still scientifically valuable to this day. Indeed, a massive shower of stones fell at Allende, in the state of Chihuahua, Mexico on February 8th at 5 minutes past one in the morning (Figure 8.1). It has been estimated that more than 2000 kg of material has been collected from a strewn field that stretches over some 300 square kilometers. No ordinary meteorite, however, the Allende material constitutes that of a carbonaceous chondrite (CV3.2), and this material was destined to revolutionize meteoritics and significantly refine our knowledge of

Figure 8.1.   A fragment of the Allende meteorite. Much of the outer fusion crust has been lost from the particular fragment shown.

Figure 8.2.   Two small fragments of the Murchison meteorite. Note the bright inclusions (CAIs) scattered within the dark matrix material. Image courtesy of Museums of Victoria, Australia.

the early solar system. Indeed, it has been said of Allende that it is the best studied meteorite of all time.

Furthermore, if Allende was not enough for the meteoriticists of this world, on September 28th, 1969, the Murchison fall occurred, showering meteorites over the plains of Victoria, Australia (Figure 8.2). Just like Allende, hundreds of fragments of the Murchison meteorite have been collected from a large strewn field, and just like Allende, Murchison has a rare carbonaceous chondrite (CM2) composition. While carbonaceous chondrites had been collected long before 1969, and they amount to about 4.5 percent of all meteorite falls, it is probably not unreasonable to say that, at the peak of the Apollo Moon landing fervor, the arrival of Allende and Murchison meteorites could not have been better timed, and that they effectively revolutionized our understanding of the conditions that were prevalent in the primordial solar system.

Prior to 1969 only 50 carbonaceous chondrites had been identified. The first such meteorite in the group being the Alais, France, meteorite (CI1), which fell on March 15th, 1806; the first find in the group was the Weatherford (Cba — a bencubbinite) meteorite found in Oklahoma in 1926. Allende, however, beats them all in terms of total mass collected, and a request at the Astrophysics Data System webpage, under the title word "Allende," reveals a total of 1526 research papers relating to this one meteorite. There are a further 937 papers with the Murchison meteorite in their title. This volume of papers amounts to an average of about one research paper being published every week since the meteorites fell — a staggering statistic and clearly one that indicates the scientific importance of these two meteorites. The reasons behind this dramatic scientific output has everything to do with the scale of the falls — all of a sudden there was more carbonaceous chondrite material to work with than ever before.

This dramatic increase in access to material is one point, the other is the Apollo Moon Landing program. This latter project having initiated the funding and construction of dedicated research laboratories, mostly in the US, but also elsewhere in the world, capable of employing state-of-the-art experimentation on cosmic material. Indeed, Elbert King, first curator of NASA's Lunar Receiving Laboratory in Houston, heard about the fall of the meteorite on the 9th of February, headed to the fall location on the 10th of February, and was back in Houston with meteorite samples on the 12th of February. Likewise, Roy Clarke and Brian Mason from the Smithsonian Institute arrived at the strewn field on the 12th of February. These early investigations soon determined that the associated fireball approached from the southwest and was accompanied by tremendous detonations in the atmosphere to deliver many thousands of fusion-crusted meteorites on the ground. Smaller materials, weighing in at just a few grams, were found near to Rancho Polanco, and larger materials, with masses in excess of 100 kg, were found 50 km to the northeast, close to Rancho El Cairo. Within two weeks of the fall, Allende material had been distributed to 37 laboratories in 13 countries.

8.2 Composition of carbonaceous chondrites

In the initial 1970 report by Roy Clarke and coworkers [1], the Allende samples are described as being composed of three distinct components (reference Figure 8.1). About 60 percent of the volume is in the form of a fine-grained black matrix composed largely of iron-rich olivine with minor amounts of troilite, pentlandite, and taenite. The chondrules take up about 30 percent of the volume, with most of the chondrules being magnesium-rich olivine, although some chondrules are largely composed of a calcium and aluminium-rich anorthite. The third component, accounting for the remaining 10 percent of the volume, is made up of irregular aggregates rich in calcium and aluminium. These latter irregular aggregates (now described under the name of calcium-aluminium inclusions — CAIs) were not new to meteoriticists, they had been seen before in other carbonaceous chondrites, but they were soon destined to become objects of extreme interest. Indeed, the CAIs were the first-born solid materials to form in the young solar nebula, and their formation age sets the zero-moment from which the history solar system is gauged (recall Chapter 5).

The Murchison fireball first appeared at about 10:55 local time. Accompanied by loud detonations, it split into three distinct components. A sinuous smoke (actually dust) trail was left along the fireball's path, and numerous meteorites were delivered to a strewn field covering an area of about 35 square kilometers. Individual meteorite masses ranged from a few grams up to the largest specimen of about 7 kilograms, and to the surprise of the first finders, the meteorites exhibited a distinct smell — something like the smell of methylated spirits. The petrology type of CM2 indicates that the material of the Murchison meteorite underwent extensive aqueous alteration while located on the parent body, and detailed chemical analysis has revealed the presence of 74 amino acids, of which only 19 are known on Earth, within the Murchison matrix material (see Chapter 9). In all, some 100 kg of meteoritic material was collected from the Murchison strewn field, although the majority of the material now resides at the Fields Museum in Chicago and at the Smithsonian Institute in Washington D.C.

It is now well established that meteorites are fragments of asteroids, and that they have undergone dramatic dynamical evolution in order to evolve onto orbits that carry their perihelia to a distance within one astronomical unit of the Sun. The ordinary chondrite meteorites are derived from S-type asteroids, and many iron meteorite groups are from the fragmented nickel-iron cores of differentiated asteroids — others have a collisional, melt-pool origin. But what about the carbonaceous chondrites? In this case reflectance spectroscopy (see Section 12.3) indicates a close compositional link to asteroids such as 1 Ceres (that is C-type asteroids) and possibly the D-type asteroids. This association is further strengthened by the fact that the C-and D-type asteroids appear to have dark surfaces that have a significant water ice component. The C-type asteroid 162173 Ryugu (Figure 8.3), for example, has recently been explored *in situ* by the Japanese Space Agency spacecraft, Hyabusa 2 and its associated robot landers. The asteroid has a low surface reflectance of about 4 percent, an appearance that is consistent with it being a

Figure 8.3. Asteroid 162173 Ryugu as imaged by the Hyabusa 2 spacecraft. Nearly 1 km across, Ryugu is a C-type asteroid and a potential source of carbonaceous chondrite meteorites. Image courtesy of JAXA (2018).

rubble pile of loosely compacted rock,[1] with a symmetrical shape
due to a phase of rapid rotation (presumably at the time of its
formation). The current interpretation of the Hyabusa survey data
is that Ryugu is probably about 100 million years old, and that it
formed from the re-accretion of material collisionally ejected from a
much larger parent body. This formation history (and the inclusion
of some small amount of heating) may explain why Ryugu is much
drier than expected, with its surface rocks being highly depleted in
hydrated minerals.

8.3   Cometary meteorites

While a good match between the C- and D-type asteroids and
various carbonaceous chondrite meteorite groups can be made, some
meteorites, especially the CI and CM types, present themselves
as good contenders for being cometary in origin. This being said,
no meteorite with a well-defined orbit has been clearly linked to
any specific comet. Nonetheless, the characteristics of volatile-rich,
primitive carbonaceous chondrites, such as Murchison, have many
similarities in common with cometary nuclei — they have a high
water (H_2O) content, they are dark in coloration, they are fragile,
they are highly porous, they are rich in organic matter, and they have
a basic chemical composition similar to that of the Sun. At best, the
fraction of meteorites derived from comets must be small, else their
presence amongst the collected meteorites would be obvious by now.
In this respect, the CI and the CM types, the best cometary-derived
candidates, account for about 2 percent of all falls.

The suggestion that meteorites might be derived from cometary
nuclei was first considered in detail by Estonian astronomer Ernst
Opik in the mid-1960s. Opik argued that all meteorite groups (stones,
irons, and carbonaceous chondrites) had a common origin and could
be associated with comets and Apollo asteroids (see Section 12.1).

---

[1]The porosity of Ryugu, that is the ratio of empty, pore space to the volume of
solid material is estimated to be of order 50 percent — see Section 12.2.

The parent meteoroids to the meteorites, he argued, being embedded within the icy matrix of cometary nuclei at the time when the planets were first beginning to form. Opik's arguments did not carry great favor with other planetary scientists, and indeed, his arguments have now generally (but not totally) been superseded. For all this, however, Opik's suggestion that the interiors of cometary nuclei might be peppered with large, potentially (carbonaceous chondrite) meteorite-producing boulders, still holds true. Indeed, in the modern era, *in situ* spacecraft observations of cometary nuclei have revealed that while they do, indeed, have a substantial ice mass fraction, they are also inactive over the majority of their surface, with perhaps just 5 percent or less of their surface undergoing ice-driven sublimation when close to the Sun and active. Most of the surface of a cometary nucleus is an inert, de-volatived mantle of dust, rocks, and boulders (Figure 8.4), and there is no strong reason to suppose that this mantle material cannot, on occasion, be shed from the surface of an active

Figure 8.4.   Rosetta spacecraft image of the surface of comet 67P/Churyumov-Gerasimenko. The surface is dark, boulder and debris-strewn, and mostly inactive, although this image shows an active sublimation jet of dust and water ice (just below image center). Image courtesy of ESA.

cometary nucleus and ejected into interplanetary space, thereby becoming a potential Earth-intercepting meteoroid and meteorite (the possibility of finding ice meteorites is discussed in Chapter 23).

Certainly, the Earth experiences numerous annual meteor showers, when the Earth passes through the dust/meteoroid streams associated with nucleus cometary outgassing by short-period comets. These meteoroids, however, typically have characteristic sizes no larger than a few centimeters, and because of their characteristically high encounter velocities (up to 72 km/s), they are fully destroyed long before they might reach the ground. Large meteoroids presumably do exist in most meteoroid streams, but no correlation between the fall times of meteorites and the time of occurrence of meteor showers has ever been established (see, however, Section 19.2). The typical encounter velocity for material derived from a short-period (Jupiter family) comet, however, is of order 20 km/s, which is on the high side for meteorite survival — indeed, at such speeds and higher, complete sublimation and parent-body fragmentation becomes highly likely. This being said, there is at least one annual meteor shower, the December Geminids, which is associated with an asteroid: 3200 Phaethon. While Phaethon has an Apollo asteroid designation, its orbit is more typical of that of a comet, and it approaches the Sun, at perihelion, as close as 0.14 AU. Its association with a meteoroid stream clearly indicates material has been ejected from the surface of Phaethon, and interestingly Geminid meteors are distinctive with respect to their relatively high density (compared to other cometary meteoroids) and their relatively low altitude beginning heights.

Comets and asteroids are generally distinguished according to three criteria based upon 1) observations, 2) composition, and 3) dynamics. Condition 1 accounts for the fact that comets will show a coma (and possibly a tail) due to the active sublimation of ice and the release of ice-embedded dust grains. In contrast an asteroid always looks like a star-like, point source. With respect to condition 2, comets are ice-rich objects throughout their interiors, whereas asteroids, which may have ice at their surface, are not. The dynamics condition 3 is based upon what is known as the

Tisserand parameter T_J. This term is an invariant constant in the restricted three-body problem with Jupiter. Here the dynamics and motion of an assumed small body (asteroid or comet) is determined according to its gravitational interactions with the Sun and Jupiter. The parameter is given by

$$T_J = \frac{a_J}{a} + 2\cos i \sqrt{\frac{a}{a_J}(1 - e^2)} \qquad (8.1)$$

where a and a_J correspond to the semi-major axis of the small body and Jupiter respectively, and where e and i are the eccentricity and inclination of the small body. The key mathematical characteristic of the Tisserand parameter is that while the orbital characteristics (the eccentricity, semi-major axis, and inclination) of the asteroid or comet may change dramatically over time due to gravitational perturbation with Jupiter, the Tisserand parameter T_J preserves its initial value. Observations indicate that short-period, Jupiter family comets have $2 < T_J < 3$, main-belt asteroids have $T_J > 3$, while long-period comets have $T_J < 2$.

With these distinctions in place, it is observed that there is no hard boundary between those objects that we call comets and those that we call asteroids — in effect there is a continuum of objects between the two classes. These distinctions are not only dependent on formation location, but also on age. If we could follow the appearance of a single active short-period comet for many tens of thousands of years, then it would gradually transform itself into an object that looks like an asteroid, but it would preserve its initial Tisserand value betraying its origin as a comet. This evolution is thought to be the situation with respect to 3200 Phaethon (parent of the Geminid meteor shower), which even though has a high eccentricity, comet-like orbit, has a Tisserand parameter of $T_J = 4.510$, indicating an origin in the asteroid belt.

Other objects, because of their orbits, may additionally have the appearance of an asteroid, but are, in fact, comets. Indeed, there are currently seven objects that have both an asteroid and a cometary classification — these are the main-belt comets, which as their name indicates have orbits set within the main-asteroid region, but have shown occasional comet-like activity. Classic among these objects is asteroid 7968 Elst-Pizarro, which has an orbital period of 5.62 years and $T_J = 3.185$. While this object was discovered in 1979 and given an asteroid classification, it developed a distinctive tail structure in 1996 when near perihelion (cometary activity was also seen at its next perihelion passage in 2001, but not thereafter), and accordingly this earned it a secondary designation as comet 133P/Elst-Pizarro. Another similar such object is comet 107P/Wilson-Harrington, discovered in 1949, with a deduced 4.29-year orbital period and $T_J = 3.08$. The comet, however, was never recovered. Thirty years later, however, in 1979, a new asteroid (1979 VA = 4015 Wilson-Harrington) was discovered with an orbit identical to 107P/Wilson-Harrington. Following further observations and recovery observations of the asteroid, it is now recognized that the two objects, comet and asteroid, are one and the same: 107P/Wilson-Harrington $\equiv$ 4015 Wilson-Harrington.

Moving back to the question, can comets be the parent bodies to some meteorites, the answer is assuredly yes, but to date the association still requires better (that is, directly sampled) proof. Historically, it has been suggested that the Orgueil (C1) meteorite which fell in 1864 had an orbit (based upon eyewitness accounts) similar to that of a short-period, Jupiter family comet. The unique Tagish lake (ungrouped C2) meteorite that fell in 2000 has a spectrum that closely matches that of a D-type asteroid (which in turn shares many visual similarities to cometary nuclei). Additionally, the infrared spectrum of comet 162P/Siding Spring, a long-period (Oort-cloud-derived) comet that had a close flyby encounter with Mars in 2014, provided a close match to the reflectance spectrum of the Alais (CI1) carbonaceous chondrite. With respect to the question, are some meteorites derived from comets, we are in the situation of Scottish jurisprudence, where the case is neither true nor false, but (so far) not proven.

8.4 Calcium–aluminium inclusions

The carbonaceous meteorites are principally recognized as being primitive, meaning that they have not suffered any distinct heat

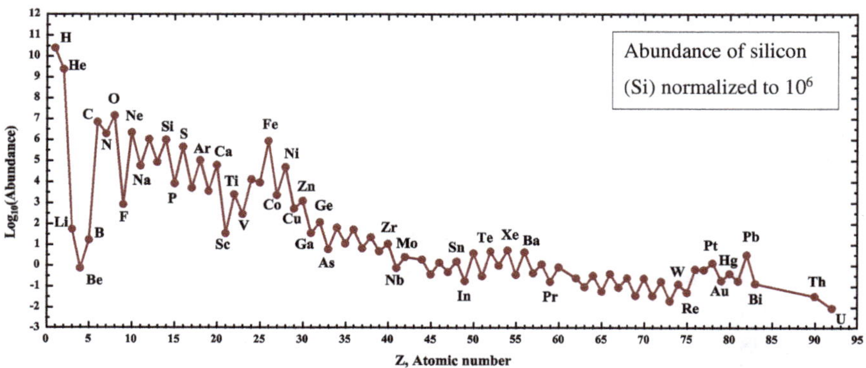

Figure 8.5.   The abundance of elements (normalized to 10^6 Si atoms) in the solar system arranged against atomic number Z. The abundance of volatile gasses, such as hydrogen and helium, that are not found within meteorites are derived from the spectroscopic analysis of the Sun's photosphere. The atomic number Z, on the x-axis, corresponds to the number of protons in a specific element's nucleus; $Z = 1$ for hydrogen, $Z = 2$ for helium, all the way up to $Z = 92$ for uranium. Image courtesy of Wikipedia Commons.

and/or pressure metamorphism, and accordingly they (along with their parent asteroids) have a composition that is representative of the very young solar nebula. Indeed, analysis of the CI meteorites in particular has enabled a detailed analysis of the initial solar system make-up. Indeed, the composition of CI chondrites is very close to that of the Sun's photosphere (as measured spectroscopically). The solar system abundance distribution (Figure 8.5) is usually normalized (by convention) to that of one million silicon (Si) atoms. Hydrogen and helium dominate with respect to their abundance (as they do in all stars and in the gas of the interstellar medium), next follows oxygen, carbon, nitrogen, and neon. Following neon is silicon, and the next most abundant elements are magnesium, iron and then, rounding off the top ten, sulphur. All other elements within the periodic table, up to uranium, can additionally be found, but in general, beyond oxygen, the abundances decrease with increasing atomic number. The zigzag, up-down appearance revealed by the abundance distribution is real, and relates to what is known as the

Oddo–Harkins rule.[2] This rule applies to atoms with an even number of protons within their nucleus and follows from the manner in which helium nucleosynthesis proceeds within stars.

The sudden availability of copious amounts of carbonaceous chondrite material in the wake of the Allende and Murchison falls, coupled with an up-surge of interest in solar system chemistry due to the Apollo Moon Landing Program, soon bore new research results. One of the first dramatic discoveries associated with Allende was published in 1971 and involved work conducted at Case Western Reserve University by Harry Green II, S. V. Radcliffe, and A. H. Heuer. Publishing their results in the May 28th, 1971, issue of the journal *Science*, Green and coworkers announced that they had found miniscule streak-like markings due to radiation damage within the chondrules found in Allende, but no similar such features in the matrix. This result implied that the irradiation exposure of the chondrules must have taken place prior to the cold accretion

---

[2]The Hodo–Harkins process, first described by Giuseppe Oddo and William Harkins in the early decades of the 20th century, the rule that even-atomic-number elements are more abundant than odd-atomic-number elements, is still not fully understood. One reason for the rule coming about builds upon the argument that since an odd-atomic-number element has one unpaired proton, it is statistically more likely to capture another proton than a nucleus in which all the protons are paired, thereby favouring the generation of even-atomic-number nuclei. This argument is encapsulated in Weizsäker's formula, which describes the nuclear binding energy in terms of the atomic number A and the proton number Z, and specifically indicates that the most strongly bound nuclei are those in which both Z and A are even, and with the equal numbers of neutrons and protons in the nucleus. Additionally, since the elements are produced via stellar nucleosynthesis within the central cores of stars, there is a tendency to favor the production of even-number nuclei. This follows the hydrogen synthesis stage, where four protons are ultimately converted into helium-4 (the nucleus containing 2 neutrons and 2 protons). Subsequent fusion reactions then tend to favor reactions built up via even-number elements — this is occasionally described as climbing the helium ladder. So, in this manner, helium fusion proceeds via the triple-alpha reaction with $3\,^4\mathrm{He} \Rightarrow\, ^{12}\mathrm{C}$, and further reactions, such as $^{12}\mathrm{C} + {}^4\mathrm{He} \Rightarrow {}^{16}\mathrm{O}$, and $^{12}\mathrm{C} + {}^{12}\mathrm{C} \Rightarrow {}^{16}\mathrm{O} + 2\,^4\mathrm{He}$, follow. Oxygen fusion then proceeds to build up silicon $^{16}\mathrm{O} + {}^{16}\mathrm{O} \Rightarrow {}^{28}\mathrm{Si} + {}^4\mathrm{He}$, which can then proceed via additional $^4\mathrm{He}$ capture events to build up those even-number elements up to iron.

and formation of the parent body. This was clear evidence for the sequential formation of structures within the solar nebula and has now become part-and-parcel of the mystery of chondrule formation (recall Section 1.2).

In many ways, the calcium–aluminium inclusions (CAIs) were the great sleeper of meteoritics. While they are present in all chondrite meteorites, having a mass fraction no larger than about 5 percent, they are especially prominent in carbonaceous chondrites. They are composed of pre-planetary grains and dust accreted and then compacted into planetesimals. They are essentially irregular in shape, can be up to several centimeters in size, and have a distinctive, whitish (sometimes pinkish) color (Figure 8.6). CAIs were first described in detail by French mineralogist Mireille Christophe Michel-Levy, in 1968, in a study of the Vigarano[3] (CV3) carbonaceous chondrite.

Figure 8.6.   A slice through a fragment of the Allende meteorite. The CAIs are easily identified in the image as white, irregular shaped blebs.

---

[3]The story of this meteorite's fall can actually begin in classic manner, "it was a dark and stormy night." Indeed, the meteorite fell near Ferrara in Italy on January 22nd, 1910, and after seeing a bright fireball through the clouds and hearing deep rumblings, the first fragments of the meteorite were collected in freshly fallen snow.

It was the fall of the Allende meteorite, however, that really brought the study and analysis of CAIs to the forefront of meteoritical research.

Mineralogical analysis indicates that CAIs are mostly composed of oxides and silicates of calcium, aluminium, magnesium, and titanium (all elements with high vaporization temperatures), but are notably deficient in all volatile elements. This composition clearly indicates that the CAIs must have formed close to the proto-Sun, where the ambient temperature was high — indeed, temperatures of order 1500 degrees centigrade are required. The fact that the CAIs must have formed in a high temperature environment, towards the inner edge of the solar nebula, but are found intermixed with the matrix material of carbonaceous chondrites, which can only form once the nebular temperature has dropped to well below 100 degrees, at first seems contradictory. This contradiction, however, is readily resolved and provides us with some fundamental information, in that there must have been radial mixing of material within the solar nebula; the problem, however, is exactly how does the radial transportation mechanism work. Current thinking[4] suggests that the mixing of material across the disk is best described by the X-wind model. Radionuclide analysis indicates that the CAIs are uniformly old, with a canonical age of 5.567 billion years, and that they all formed within a relatively short-lived window of some 100,000 years. Analysis of Allende CAIs by Christopher Gray (Caltech) and coworkers in 1973 revealed that they pose the smallest initial $^{87}Sr/^{86}Sr$ ratio of any known solar system material (recall Chapter 4). It is the combination of these last three characteristics that single the CAIs out as being the first-born solids within the solar nebula, setting

---

[4]The X-wind model was developed by Frank Shu (University of California, Berkeley) and coworkers in the mid to late 1990s. This model suggests that both chondrule and CAI precursor material follows a looping path through a protostellar disk. Material is transported radially inward until it meets the so-called X-wind region, where it is captured into a magnetocentrifugal outflow and transported outward and deposited into the outer disk (see Figure 8.7). The key idea behind the X-wind model was to explain the collimated, bipolar outflows that are associated with protoplanetary disks.

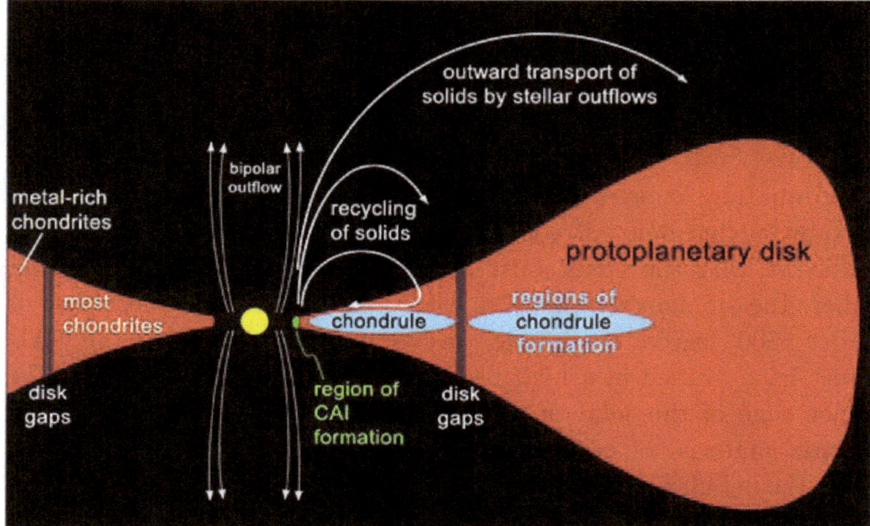

Figure 8.7. Outward model transport of solids by the X-wind stellar outflow. CAIs formed in the inner part of the disk are radially redistributed into the outer disk where they can be combined with materials that form at low temperatures. Image adapted from Van Kooten *et al.*, *PNAS.* **113**(8), 2011–2016 (2016).

the time zero for all ages to follow. The CAIs existed before the planets, and it also transpires from radionuclide analysis that they also formed several million years before chondrules. Both CAIs and chondrules were formed under high temperature conditions within the solar nebula and yet they ended up being accreted into the same parent bodies — exactly how all this worked is still being debated by researchers.

One of several important papers, based upon analysis of Allende meteorite samples, published in 1978 was that by Typhoon Lee and coworkers at the California Institute of Technology.[5] This particular research was concerned with measurements of the isotopic values of

---

[5] As a sign of their times (and a decidedly non-PC approach in ours) the authors of this paper give their address as "The Lunatic Asylum, Division of Geophysics and Planetary Sciences." The reference, of course, is intended as a humorous take on the fact that the labs at Caltech were involved in analyzing lunar material returned by the Apollo astronauts.

magnesium (Mg) in different phases of several CAIs. Importantly, what Lee and collaborators found was that the various CAIs in Allende showed large excesses of ^{26}Mg, and this they realized implicated the presence of the short-lived radionuclide ^{26}Al in the natal solar nebula. Furthermore, an earlier analysis of Allende and Murchison CAIs by Robert Clayton (University of Chicago) and coworkers in 1973 had revealed that the CAIs were notably enriched, with respect to the Earth and all other solar system materials, in their oxygen isotope abundances. This was an important discovery since those elements having the enhanced oxygen isotope ratios must have been brought to the solar nebula in the form of pre-solar grains. The fact that pre-solar grains were involved in the formation of CAIs, along with the detection of ^{26}Mg due to the decay of ^{26}Al, has resulted in the idea that the collapse of the interstellar gas cloud, that eventually evolved into our solar system (and us), was triggered by a supernova explosion. The shockwave from the supernova both triggered the collapse of the interstellar gas cloud (which contained the pre-solar grains) and implanted the active ^{26}Al into the proto-solar nebula.

That ^{26}Al is produced within supernova and thereafter injected into the interstellar medium is evidenced by observations with gamma-ray telescopes. Figure 8.8 shows an all-sky image derived from observations with the COMPTEL (the Imaging Compton Telescope aboard the Compton Gamma Ray Observatory spacecraft). The image indicates the intensity of emission at a wavelength of about 7×10^{-13} meters[6] and this emission traces out the distribution of active ^{26}Al. The emission line results, in fact, from the deexcitation of the newly formed ^{26}Mg after the decay of ^{26}Al. The image is specifically telling in that the strongest ^{26}Al emission regions are invariably associated with supernova remnants and regions in which massive stars are forming within the galaxy. At this time, it is not clear if the ^{26}Al in the proto-solar nebula was uniformly distributed or somewhat clumpy — this distribution issue has important consequences, later in

---

[6]The wavelength is derived from the energy associated with the gamma ray being 1.809 MeV.

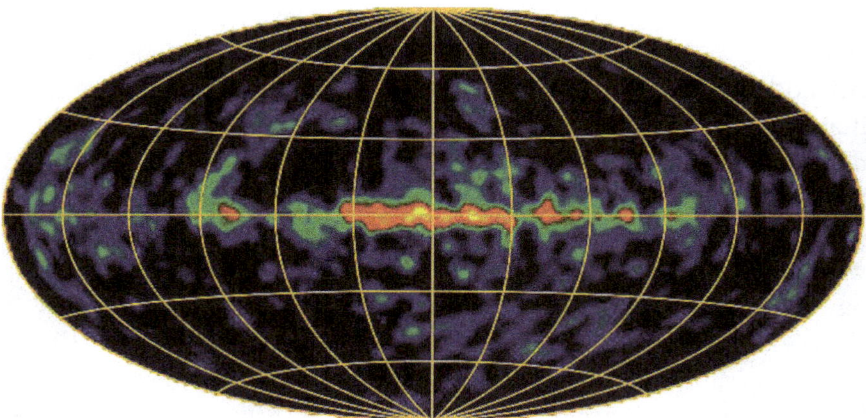

Figure 8.8. All-sky map derived with the Imaging Compton Telescope, carried aboard the Compton Gamma Ray Observatory spacecraft, showing the distribution of ^{26}Al in the Milky Way galaxy. Red indicates highest concentrations of ^{26}Al, with blue indicating the lowest concentrations. The direction to the galactic center corresponds to the image center, and the disk of the Milky Way runs horizontally along the image mid-plane. Image courtesy of the COMPTEL collaboration.

time, with respect to asteroid evolution and their ability to undergo differentiation. Indeed, Harold Urey suggested in 1955 that the decay of ^{26}Al could be more than a useful short-lived radionuclide clock, it could also be a cooker. The point here is that once bound up within the interior of a planetesimal/asteroid, the kinetic energy of the positron emitted in the decay of ^{26}Al will be transformed into thermal (heat) energy, thereby warming its surroundings (see Chapter 13 for further discussion).

Remnant magnetism in carbonaceous chondrites indicates that at the time of their formation there must have been a relatively strong magnetic field present. Indeed, a magnetic field strength of some 1 to 10 gauss within the solar nebula disk is implicated. This strong magnetic field may have been generated through a dynamo process operating within the nebula disk, or it was due to the young Sun having a much stronger magnetic field than it does in its middle-aged present. The latter scenario can, in fact, be linked to the Sun's T Tauri phase (recall Chapter 5), and the time at which the gas and dust within the solar nebula were being actively dispersed by a strong solar wind.

8.5 Water on Earth

While the carbonaceous chondrites tell us about the very first solids to form in the solar system (the CAIs), they also play an important role in determining our very existence — not just in providing important carbon compounds implicated in the origins of life (see Chapter 9). Indeed, the carbonaceous chondrites appear to have been important vehicles for delivering water to the young Earth. Oddly, we call our world Earth (intending, by inference, dryland), but in reality, we should call it Water, since water covers some 70.8 percent of its surface. The global oceans contain an estimated 1.4×10^{21} kilograms of water, and this water had to come from somewhere — the question is where? At issue is the manner and location in which the Earth formed, and specifically the way in which Earth's atmosphere has evolved. Indeed, the Earth formed in the inner part of the solar nebula where the temperature would have been far too high for water ice to exist — the Earth formed in a bone-dry environment. Not only this, the growth and formation of the Earth, via planetesimal accretion, resulted in an initial hot, molten phase which would have vaporized any initial water ice that had been incorporated into its interior.

There are a host of mechanisms by which the Earth might have derived its water supply. One, and indeed the first touted method, is that by outgassing, with water vapor being vented into the atmosphere through volcanic activity. Additionally, cometary impacts could, and no doubt did, bring water ice to the newly forged Earth, and some of the water would have been incorporated into the accreting Earth by the capture of ice-rich planetesimals. All these processes, however, face a problem when confronted with the measure of the ocean's deuterium to hydrogen ratio D/H — see Figure 8.9.

Deuterium is a heavier form of hydrogen, with a nucleus containing both a neutron and a proton, rather than just the single proton in the case of hydrogen. The D/H ratio provides a unique

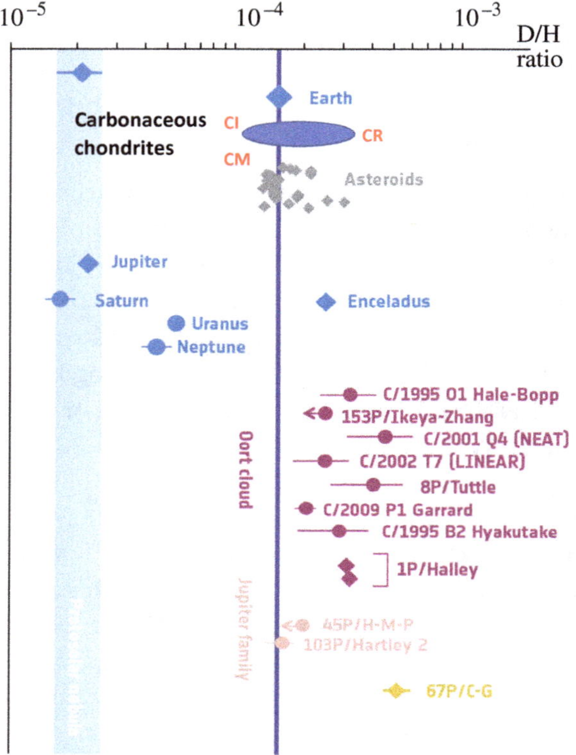

Figure 8.9.   The deuterium to hydrogen ratio for a selection of solar system objects. The wide vertical band (to the left) corresponds to the D/H ratio of the original solar nebula, while the vertical line through the Earth point corresponds to the average measure for sea water. Image adapted from and courtesy of ESA.

(but not easily read) fingerprint of the formation environment and evolution of solar system objects. For Earth, the characteristic value is $D/H = 1.5 \times 10^{-3}$, but it is an order of magnitude smaller in the case of the Jovian planets, and an order of magnitude higher in the case of Venus. The deduced D/H ratio for asteroids spans smaller and larger values to that of the Earth, while long-period comets (derived from the Oort Cloud) tend to have larger D/H ratios. The short-period, Jupiter family comets have been less well studied, but tend to have slightly higher D/H values than that derived for Earth. Indeed, of all the solar system bodies so-far investigated it is the

CI and CM carbonaceous chondrites that provide the closest match for the Earth's oceans.

Water molecules in fact account for a significant fraction of carbonaceous chondrite meteorite make-up, contributing anywhere (according to specific sample and type) from a few to near 20 percent by mass fraction. Accordingly, the late accretion of carbonaceous chondrite bodies, by the nearly fully formed and cooling-off Earth, has been proposed as a possible source for Earth's water — if this was so, then the making of the *Blue Marble* was an act precipitated by a rain of stone. The question concerning the origin of Earth's oceans, therefore, appears to relate to the late delivery of carbonaceous chondrite bodies to the Earth. There is, however, a possible explanation for the late accretion mechanism in terms of the Nice model (recall Chapter 5) and the final re-organization of planetary orbits which resulted in the late heavy bombardment[7] some 800 million years after the planets had formed.

---

[7]Evidence for the late heavy bombardment (LHB, also called the lunar cataclysm) is both complex and controversial. It is an idea that grew out of the Apollo era exploration of the Moon and the determination of Moon rock ages. Specifically, the impact-melt rocks that were returned from the various Apollo landing sites all dated to a narrow age range between about 3.8 and 4.1 billion years old. This age range sets their formation to a time some 800 million years after the planets (and the Moon) formed. Not only do the impact-melt samples returned by the Apollo missions show a specific and restricted age range, so too do the known lunar meteorites (see Section 7.9). Indeed, the feldspathic lunar meteorites, which are believed to be derived from the lunar far side, show an age clustering between 2.76 and 3.92 billion years. The data from the Apollo samples and the lunar meteorites have accordingly been interpreted as indicating a short-lived (a few hundred million years long) spike in the impact cratering rate in the inner solar system. Not all researchers are convinced that the LHB occurred, however, and argue, for example, that the age clustering is really an artifact of the material sampling process. Other researchers find evidence to suggest that there were, in fact, several very short-lived spikes in the impact cratering rate. Yet other researchers argue that impact-melt rocks older than 4.1 million years do (or did) exist, but that their radiometric clocks (recall Chapter 4) have been reset by the continuous, but steadily declining, effects of impact cratering over the age of the solar system. The debate on the LHB will no doubt continue to run for many years yet (see Section 7.9).

The debate concerning the origin of Earth's oceans is far from settled. Other theories relating to its possible emplacement are in play, and the late veneer of carbonaceous chondrite asteroids may be just one of the contributing factors behind the growth and formation of our rolling seas.

8.6   Meteorites from Antarctica

Last, and by no means least, December 1969 contained a mixed bag of good and bad news. It was the month in which the first American Vietnam war draft lottery occurred; the Rolling Stones released their *Let it Bleed* album; the first North Sea oil field was found; Shawn Carter (a.k.a. Jay-Z) was born, and Princess Alice of Battenberg died. Lost amongst the events of global news, however, was the chance find on December 21st of nine meteorites at the far end of the world, on the frozen ice fields of Antarctica. In the world of meteoritics, this number of distinct finds, in just one day, was unprecedented and seemingly miraculous. The glaciologists of the Japanese Antarctic Research Expedition, who made the finds, were not looking for meteorites, they just found them as they went about their work in the Yamato Mountain region of Queen Maud Land, but their discovery was a revolution in the making. The first find, now designated Yamato 691, was a 715-gram, 10-cm-sized enstatite meteorite. The second find, designated Yamato 692, was a 138-gram diogenite; and the third find, Yamato 693, was a 150-gram carbonaceous chondrite. The other six finds were H chondrites. The news of this remarkable discovery soon began to filter through the world of meteoritics, and Antarctica, all of a sudden and out of the blue, became a hotbed for meteorite recovery expeditions (see Chapter 14).

References

[1]   R. Clarke *et al.* Smithsonian Contributions to the Earth Sciences, No. 5 (1970).

[2]   M. Grady. *Catalogue of Meteorites (5*[th] *edition).* CUP, Cambridge (2000).

Chapter 9

Meteorites and Life

Almost in the beginning was curiosity. Curiosity, the overwhelming
desire to know, is not a characteristic of dead matter.

Isaac Asimov

Life itself is a great cabinet of curiosities. Most profound of all, however, is the mystery that surrounds the origin of life — what strange spark was it that took inanimate molecules and jiggling atoms, and transformed them into something that could be called animate and alive? We do not know the answer to this riddle yet, but in the expansive cabinet of life, there are some items that can be recognized, and which are even understood. The actual relationships and interactions that existed between the known and the unknown processes, however, remain profoundly problematic. Nonetheless, what seems to be clear is that somewhere in the mystery of it all, meteorites are important. Indeed, meteorites appear to be the very vectors by which the crucial molecular components of life were brought to the young and distinctly lifeless Earth.

9.1 Life on Earth

The final stages in the formation of the Earth were violent, hot, and energetic — no place, indeed, for life to either begin or survive. The newly formed Earth was both sterile and hostile, and it was only ripe for nurturing life after perhaps the first few hundred million years had passed, at which stage the dramatic impact resulting in the

225

formation of the Moon had occurred,[1] and oceans and land were at least available for the colonization by microbes. This colonization began fairly rapidly, as indicated, for example, by the analysis of the carbon inclusions in the 4.1-billion-year-old Jack Hills, Australia zircons [1]. Certainly, even microbial life seems to have been well established by 3.5 billion years ago, but the questions remain — how, where and when did it first appear? Unfortunately, at present there is a bewildering number of answers to these questions.

We presently have too many options to choose from to say what may or may not be true. Perhaps none of the current theories are correct; perhaps several are on the right track — we simply don't know for certain. There are many good scenarios but the data, conformational or negative, are scarce. Abiogenesis, the actual formative event resulting in life, may have occurred naturally, in house (as it were), by exploiting the chemicals and processes already present on a young Earth — this process being driven by a Miller–Urey mechanism.[2] The *essence* of life may have evolved in deep space

---

[1]The impact responsible for the Moon's formation is thought to have occurred about 50 to 80 million years after the formation of the proto-Earth.

[2]The Miller–Urey experiment is gloriously simple, and simply glorious in its design. It is predicated on the principle that if you are given lemons, then turn them into lemon juice. Designed by Stanley Miller and Harold Urey in the early 1950s, the apparatus was designed to test for the chemical origin of life. Accordingly, the experiment took the dominant atmospheric gases then expected of the young Earth — these being water vapour (H_2O), methane (CH_4), and ammonia (NH_3) — and then added energy to the mix. Figure 9.1 shows the basic experimental design. A water reservoir is heated to produce water vapor, which then rises and combines with the other gases in a collecting (atmosphere-simulating) flask. Energy is then added to the gas mixture by an electric spark — this spark mimics the natural process of lightning — and then the lower part of the atmospheric chamber is cooled allowing the water vapor to condense, and thereafter trickle into a lower U-bend trap. Given this configuration, the system is allowed to run for a day or so, and after this time the water that has collected in the U-trap is sampled and analyzed. The results indicated that many organic compounds had formed including amino acids. The original gas composition used by Miller and Urey is now known to be wrong and it does not actually represent Earth's early atmosphere, but revised, carbon-dioxide-rich gas mixtures have been found to yield similar results to the original experiment.

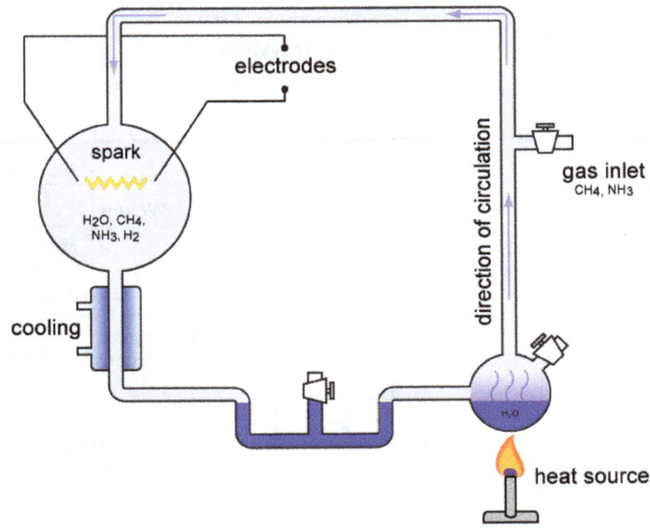

Figure 9.1.   Outline of the basic Miller–Urey experiment. Image courtesy of Wikipedia.

and have been built into the collapsing solar nebula from the very start, seeding life everywhere in the solar system from the beginning.[3]

---

[3]If the life evolved in space and was brought to Earth via some panspermia mechanism (see Section 15.2), or was simply built into the solar nebula from the very first collapse phase, the question arises as to when, rather than where, did it first appear. One way of estimating this galactic origin moment can be by looking at the evolution of biological complexity on Earth. This approach has been studied by Alexei Sharov (Laboratory of Genetics, The National Institute of Ageing, Baltimore), and as a proxy for complexity he and coworkers have analyzed the non-redundant functional genome size of various species. The idea is that the greater the complexity of a species so the larger the genome size. Figure 9.2 is a diagram based upon the numbers given by Sharov and Richard Gordon (an itinerant theoretical biologist working from Winnipeg, Canada) in a 2013 research paper: *Life before Earth* (see: https://arxiv.org/abs/1304.3381). The unexpected and intriguing result from the diagram is that while, over the age of the Earth, the complexity of life (as measured by the logarithm of the functional genome size) has increased linearly over time, it was not zero at the time that life first appeared in the terrestrial record (i.e., as inferred from the Jack Hills zircons). That complexity should increase over time is not unexpected given the action of geological transformation of the Earth's surface and Darwinian evolution, but that the complexity was so high, so early on, when life first appears is puzzling. This result Sharov and coworkers suggest is an indication that the Earth was seeded

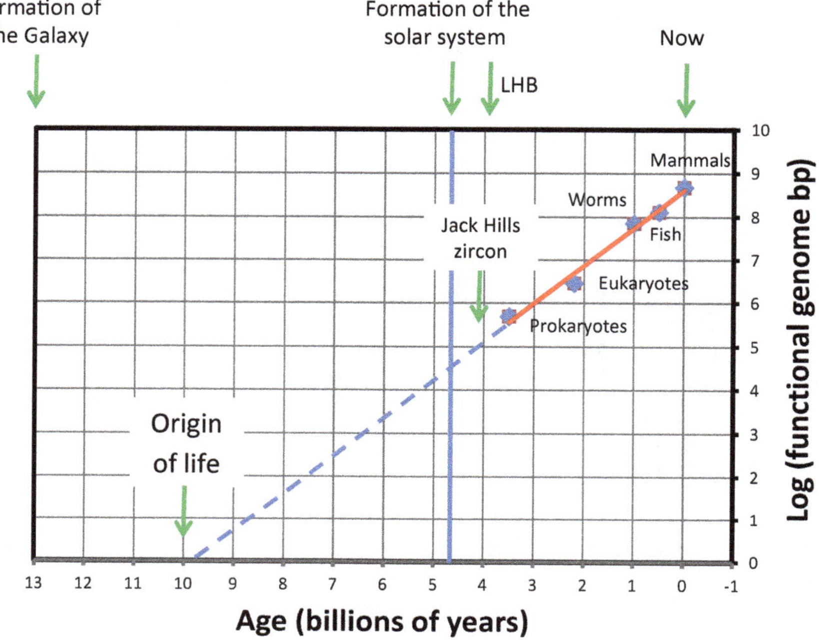

Figure 9.2.   Time before the present versus biological complexity (as measured by the logarithm of the non-redundant functional genome size). The galaxy first formed about 13 billion years before present (BP), and the origin of the Earth and solar system is set at 4.56 billion years BP. The Jack Hills zircons, containing evidence of the first biotic life on Earth, are dated to 4.1 billion years BP. The Late Heavy Bombardment started about 3.9 billion years BP. Extrapolating the linear growth line for complexity (dashed blue line) suggests life first appeared in the galaxy some 10 billion years BP.

by a low-complexity form of elementary life. Accordingly, upon finding nurturing environments on and in the newly formed Earth, this elementary life thrived — and has continued to do so ever since the proverbial day 1. Extrapolating the linear growth line backwards in the complexity diagram, however, leads to the moment at which life first evolved in the galaxy. This moment of creation, at which the fecund spark of life emerged (that is when the first set of replicating base pairs evolved), is set some 10 billion years before the present. This is at a time about 3 billion years after the galaxy first formed and about 4 billion years after the Big Bang that saw the universe itself came into existence. According to this picture the seeds of life were first sown in the galaxy 5 billion years before the Earth actually formed. Additionally, this tells us that it took *life* 5 billion years

Life may have first thrived in warm, nutrient-rich rock pools on the edges of the first continents. Life may have first thrived in the deepest oceans at the location of black smokers. Life may have first thrived in mud-rich pools located in any region where rain falls [2]. Life may have first thrived close to the hydrothermal vents associated with large impact craters. Each of these scenarios is *scientifically* possible, but none of them need be right, and none of them are inevitable. This latter point is somehow disturbing, but finds resonance with the modern-day search for biosignatures from other planets orbiting other stars than the Sun. It is entirely possible that there are planets in the galaxy that are capable of supporting life, but upon which life will never evolve.

While abiogenesis remains a topic for continued research, one of the key components in all cellular life is some form of data storage about the self, and more specifically data on how to make another self. Reproduction is key to life, and the message of successful life is passed on from one generation to the next by deoxyribonucleic acid (DNA). This highly complex molecule is composed of two chains that coil into a double helix. The molecular components of DNA carry the detailed genetic instructions for building new life — literally, new life from the old. But, for all this, it is highly unlikely that DNA was there, as the molecular director, from the very outset, rather, it is thought that ribonucleic acid (RNA) was the first code carrier. RNA is a long, single strand molecule, but it can still store and allow replication of genetic information (Figure 9.3). American biophysicist Alexander Rich (MIT and Harvard Medical School) suggested in the early 1960s that RNA probably controlled the origin and development of the first life on Earth, with DNA taking over the coding at a later stage. So, the search now for the origins of life

---

to evolve to the level of a microbe — these being the first life forms appearing and being recognized in the geological record. One additional implication of this scenario is that all life, wherever it might appear in the galaxy, has a common origin, and this is in principle a testable result — life on Mars, Europa, and Enceladus, if and when found, should have biological similarities to life on Earth. The basic chemistry of life, as such, does not evolve independently and/or in different ways.

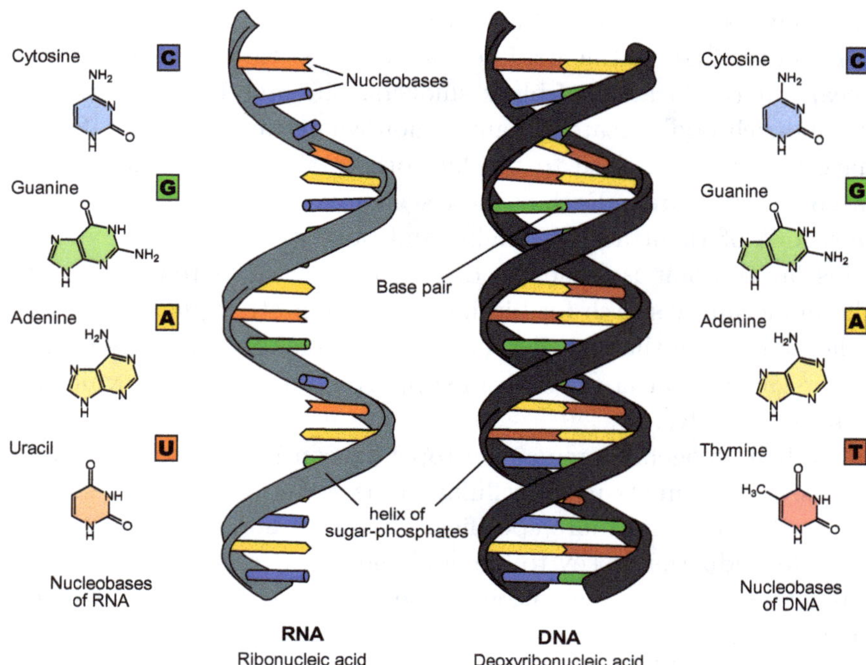

Figure 9.3.   A comparison of RNA (left) and DNA (right), showing the helices and nucleobases each employs. Image courtesy of Wikipedia.

transforms to a search for those places where the building blocks of RNA are able to form and polymerize. The key to the process is to form nucleobases, which are nitrogen-containing compounds, and there are five that are particularly important for life — adenine (A), cytosine (C), guanine (G), thymine (T), and uracil (U) — and they function as the basic units for the genetic code.[4] It is at this point that meteorites also enter into the discussion since nucleobases have been found in carbonaceous chondrite meteorites, and it is likely that they reside on small interplanetary dust particles (IDPs).

There are many possible ways in which the seed nucleobases contained within meteorites might have enabled abiogenesis on

---

[4]Technically, the bases A, G, C, and T are found in DNA, while bases A, G, C, and U are found in RNA.

Earth, but one particularly elegant way has been described by Ben Pearce (McMaster University, Canada) and coworkers [2]. The scenario described by Pearce *et al.* is based upon the seasonal, wet–dry cycle of warm little ponds (WLPs). In this scenario the alchemy (or more to the point, key chemistry steps) takes place in relatively small environments that have been enriched, or seeded, by nucleobases derived from meteorite impacts and a background of IDPs settling out of the atmosphere (see Section 7.4). Combing aspects of meteoritics, geology, meteorology, chemistry, and biology, making for a truly interdisciplinary study, Pearce and coworkers find that the wet and dry cycling that small ponds must inevitably go through acts to concentrate nucleobases, that is their abundance increases in spite of various loss mechanisms (hydrolysis, water seepage, and dissociation by UV radiation) that must be in play (see Figure 9.4). Furthermore, their model indicates that the wet and dry cycle promotes the polymerization of nucleotides, and from these nucleotides, long chains of RNA polymers are posited to eventually evolve.

Is this how life really started on Earth? Well, we shall never know for sure, and alternative models are available, but it offers a mechanism for solving the replication part of the abiogenesis problem. RNA may or may not be the way life started on Earth, the jury is still out. There is still much work to be done, not least the eventual code-carrying takeover by DNA, and the evolution of an additional metabolic cycle to go with that of reproduction.[5] In short, life is complicated.

One of the problems associated with the emergence of life is that the geological record, based upon the Jack Hills zircon ages, indicates that the key steps must have occurred at a violent, inhospitable

---

[5]It is usually assumed that metabolism (the ability to generate energy) and reproduction evolved simultaneously, but this need not be the case. The metabolism-first argument is nicely described by Freeman Dyson in his book, *Origins of Life* (Cambridge University Press, 2nd Ed., 1999). Dyson argues that life effectively began twice: once in creatures capable of metabolism, and once in parasitic creatures capable of reproduction. Eventually, the two lifeforms evolved to coexist in a single symbiotic entity.

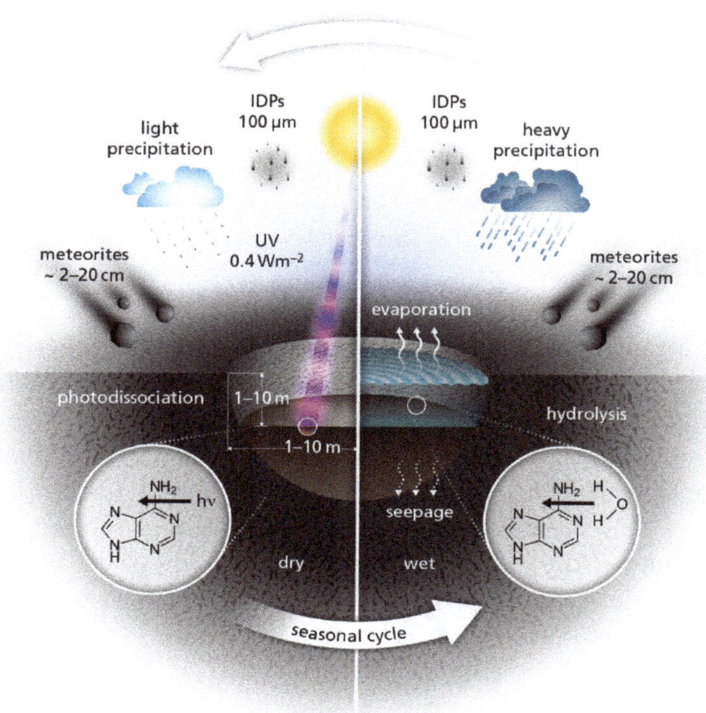

Figure 9.4.   The seasonal cycle and evolution of a warm little pond (WLP). In this depiction, the only water source is taken to be precipitation, while the water *sinks* include evaporation and seepage. The sources of nucleobases (adenine is illustrated) include meteorites and interplanetary dust particles (IDPs). Nucleobase *sinks* include seepage, photodissociation by UV radiation, and hydrolysis. Furthermore, the hydrolysis *sink* only works when the pond is wet, and the UV dissociation *sink* only operates when the pond is dry. Image courtesy of [2].

time in Earth's history. Indeed, fundamental to the delivery of the initial seed nucleotides (by meteorites and IDPs) to WLPs is the increased flux of such material associated with the Late Heavy Bombardment — a period set some 3.9 billion years ago when the meteoroid influx was estimated to be some 10 thousand times greater than it is today (see Figure 14.7). Incredibly, life found a way to survive the LHB, as well as all the other natural processes that would have readily snuffed it out.

9.2 Life elsewhere

The highly improbable emergence of life (anywhere in the universe) has caused much philosophical debate, and this debate has implications for the existence of life, intelligent or otherwise, elsewhere in the galaxy and the greater universe. Is life inevitable in our galaxy, given the composition of the interstellar medium, which shows that the raw chemical elements for life are readily available? Or is it extremely rare and hardly ever successfully realized? Some attempt at assessing the probability of life appearing has been made through the so-called Anthropic Principle. This principle is predicated on the fact that we, as humans and observers, exist, and therefore the processes that lead to the emergence of *us* cannot in their sum be so highly unlikely that we should not expect to be here. Australian astrophysicist Brandon Carter has developed the idea of the Anthropic Principle to argue that while there are highly improbable, indeed, hard but critical steps that must be navigated before life can emerge (such as the appearance of RNA and DNA), the number of such steps cannot be very large. This has led some researchers to suggest that life, for all of its improbability of emerging, is, in fact, inevitable and common within the universe [see, however, Reference 3]. Once the first few hard steps are cleared away, then the rest can be accommodated for via Darwinian evolution. In spite of much calculation, application of philosophical logic, and complex rhetoric, there is only one place in the entire universe that we know life exists — and that, of course, is Earth.[6]

The National Aeronautics and Space Administration (NASA) has, almost since its very inception in 1958, given high priority to those missions that might shed light of extraterrestrial life, both within the solar system and the greater galaxy beyond. The basic mantra that NASA mission planners have followed in such investigations is "follow the water." For indeed, wherever on Earth the thermodynamic conditions allow for water to exist, then life is

---

[6]If ever there was a rallying call to humanity to take better care of the environment, then this one statement must surely be it.

found. This mantra, taken off Earth, also allow for the identification of various other worlds where life might have evolved. These include the young Venus and Mars, as well as the moons Europa and Enceladus.

There are likely other bodies within the solar system that may have, or even do, support the existence of microbial life. One such moon being Titan, which has an atmospheric composition similar to that of the primordial Earth.[7] Given the apparent importance of meteorites in delivering the basic complex molecules that kick-started and enabled life on Earth, then there is every reason to suppose that life could have independently arisen on other planets and moons as well. Not only this, however, even if life did not evolve on other planets and moons in the solar system, it might have been seeded there via a meteorite conveyor belt. Terrestrial-born meteorites, for example, will be packed with microbes, and these same rock-bourn microbes could potentially seed life in any safe, nurturing water haven anywhere else in the solar system (see Section 7.10). The search for life (past and present) elsewhere in the solar system (and galaxy) is still ongoing, and we may hope to hear news of it soon.

References

[1] E. Bell *et al.* "Potentially bioenergetic carbon preserved in a 4.1-billion-year-old zircon." *Proceedings of the National Academy of Sciences*, **112**, 14518–2 (2015).

[2] B. Pearce, R. Pudritz, D. Semenov and T. Henning. "Origin of the RNA world: The fate of nucleobases in warm little ponds." *Proceedings of the National Academy of Sciences*, **114**, 11327–11332 (2017).

[3] National Research Council. *The Limits of Organic Life in Planetary Systems*, The National Academic Press, Washington (2007).

---

[7]Titan is a special case since its surface temperature is far too cold for water to exist in its liquid form. However, it does have liquid methane and ethane lakes (and liquid water may exist there in a eutectic form). The chemistry for life on Titan may not be the same as that on Earth (indeed, it can't be), but there is no specific reason why specialized life cannot have evolved there — if this is so, then it would represent a second abiogenesis event [3].

Chapter 10

Gathering Dust

The purpose of art is to wash the dust of daily life off our souls.
Pablo Picasso

We live under a literal rain of meteorites. Every day we actually breath them in, and we sweep them up from the floor. These meteorites are, of course, very small, and virtually invisible without the aid of a microscope — but they are there, and they are all around us. Variously called micrometeorites and more descriptively dust grains, these meteorites fall gently from the sky, being blown on the wind, and carried within raindrops. Outside your front door, right now, there could well be micrometeorites ready for collecting.

10.1 Micrometeorites

Figure 10.1 shows the flux of extraterrestrial material at Earth's surface in the mass range from 10^{-22} to 10^{22} kg. About 40,000 tonnes of material per year is accumulated by the Earth, and the greater amount of this material is in the mass range from 10^{-12} to 10^{-10} kg — that is, dust particles with sizes from just a few micrometers to a few millimeters. This latter material is derived from both asteroid and cometary sources. Although this material encounters the Earth with cosmic speeds, it is rapidly decelerated without significant ablation. Rather the material is heated to its melting point and forms a spherical bleb (Figure 10.2) that will gently settle to the ground — taking days to weeks to complete its actual descent to the ground. Occasionally, some of the extraterrestrial dust grains can reach the ground in a largely un-melted form (Figure 10.3), and

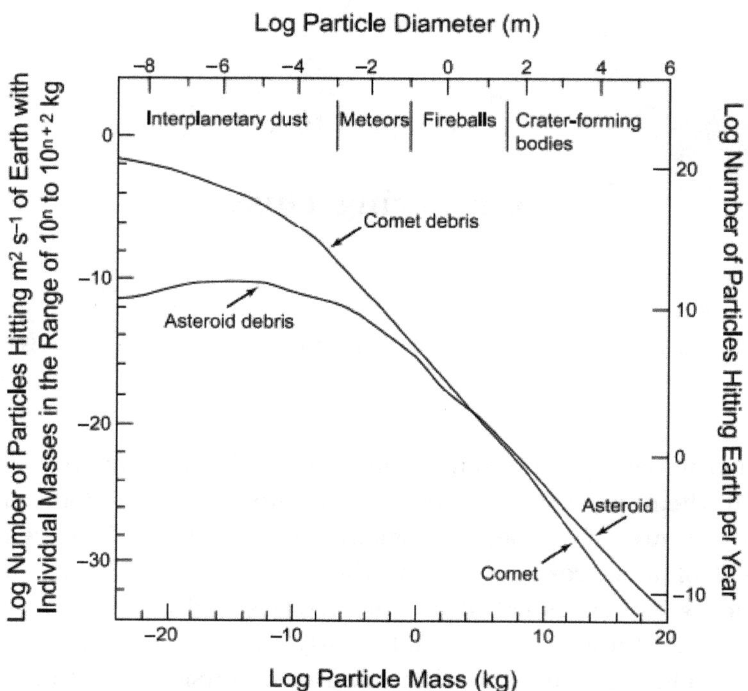

Figure 10.1.   The flux of extraterrestrial material from the smallest dust grains to large, crater-forming asteroids. The scale to the left gives the logarithm of the number of particles hitting each square meter per second of Earth's surface in the mass range from 10^n to 10^{n+2}. Image from Zolensky *et al.* [1].

these characteristically display open (that is highly porous) clusters of randomly positioned sub-micrometer particles.

Interest in this peak flux population of micrometeorites is warranted since such material may have been in part responsible for the origin of life on Earth. Indeed, micrometeorites are rich in carbon (C) and nitrogen (N), and they may have been the dominant source of organic carbon to the primitive Earth. Additionally, evaluation of the amount of H_2O and CO_2 released by the heating of micrometeorites suggests that they may have played an important and, indeed, substantial role in the formation of the Earth's atmosphere and oceans after the Late Heavy Bombardment starting some 4 billion years ago.

Figure 10.2. A typical, spherical micrometeorite. Image by Matej Pašák.

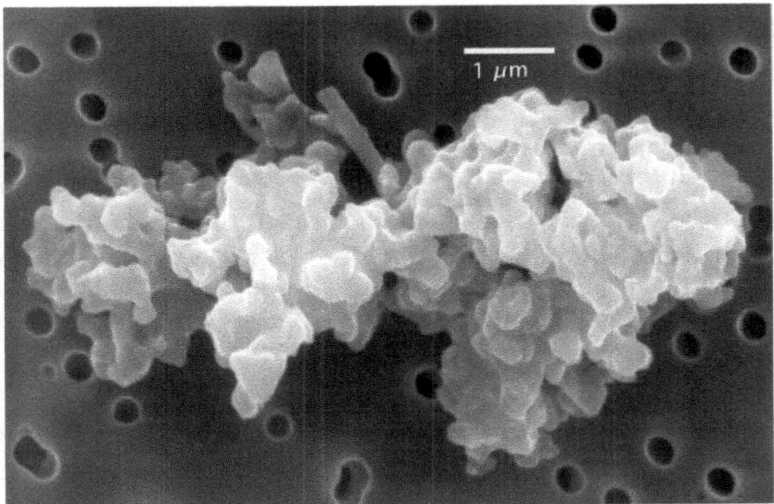

Figure 10.3. Scanning-electron-microscope image of an interplanetary dust particle. The bar to the upper right indicates a scale of one millionth of a meter. Image courtesy Donald Brownlee, University of Washington.

10.2   Atmospheric interactions

In order for a micrometeorite to survive atmospheric passage without significant ablative mass loss and/or alteration its acquired temperature must remain below the boiling point of its constituent material [2]. The temperature of a small meteoroid is determined by the transfer of kinetic energy from the direct impact of atmospheric molecules, and the energy available per unit time for meteoroid heating will be:

$$\frac{dE}{dt} = \frac{\Lambda}{2}\, S\rho V^3 \tag{10.1}$$

where Λ is the heat transfer coefficient (recall Chapter 6), ρ is the atmospheric density, V is the meteoroid velocity, and $S = 4\pi R^2$ is the surface area of the meteoroid (assumed here as being a sphere of radius R). This energy will go into the heating of the meteoroid, and its temperature T will increase per unit time as

$$\frac{4}{3} R^3 \delta C \frac{dT}{dt} = \frac{dE}{dt} - S\varepsilon\sigma\left(T^4 - T_0^4\right) \tag{10.2}$$

where δ is the meteoroid density, R is the meteoroid radius, C is the specific heat capacity of the meteoroid material, T_0 is the initial meteoroid temperature, ε is the emissivity, and σ is the Stefan–Boltzmann constant. The last bracketed term on the right in Equation (10.2) relates to the energy being radiated back into space at the surface of the meteoroid through the Stefan–Boltzmann law. Substitution of Equation (10.1) into (10.2), dividing through by the surface area S, and assuming that any terms dependent upon the micrometeorite radius R are very small (and hence negligible), the micrometeorite temperature can be expressed as

$$T^4 = T_0^4 + \frac{\Lambda}{8\,\epsilon\,\sigma}\,\varrho V^3 \tag{10.3}$$

Under the approximation of an exponential atmospheric density variation [recall Equation (6.11)], the micrometeorite velocity will decrease as per Equation (10.2), and substituting for this in Equation (10.3), the temperature as a function of atmospheric height h can be deduced, with

$$T^4 = T_0^4 + \frac{\Lambda}{8\varepsilon\sigma} V_\infty^3 \, \rho(h) \, \exp\left[-\frac{9\,H\,\Gamma}{4\,R\,\delta} \, \rho(h)\right] \qquad (10.4)$$

where V_∞ is the initial velocity, H is the atmospheric scale height, and Γ is the drag coefficient. In Equation (10.4) the radius R has been reintroduced, but must still be taken as being very small so that the approximation leading to Equation (10.3) remains true. By differentiating Equation (10.4) with respect to the height h, the density (and hence height) at which a maximum temperature is achieved ρ_{max} can be found — accordingly,

$$\rho_{\text{max}} = \frac{4\,R\,\delta}{9\,H\,\Gamma} \qquad (10.5)$$

Substituting ρ_{max} back into Equation (10.4) now enables the maximum micrometeorite temperature to be found, with

$$T_{\text{max}}^4 = T_0^4 + \frac{\Lambda\,R\,\delta}{18\,\varepsilon\sigma\,H\,\Gamma\,e} \, V_\infty^3 \qquad (10.6)$$

where $e = 2.71828\ldots$ is the base of the natural logarithms. Characteristically, for micrometeorites, $\Lambda = \Gamma = \varepsilon = 1$, and taking the density to be $\delta = 3000\,\text{kg/m}^3$, so the critical radius for the onset of vigorous ablation can be determined. Setting the boiling point for vigorous ablation to be $1500\,\text{K}$, and taking an initial velocity of $11.2\,\text{km/s}$ (Earth's escape velocity), micrometeorites (dust grains) smaller than about 40 microns across will undergo little heating alteration (recall the image and scale bar in Figure 10.3). Sizes a little larger than this limiting size will become molten (and adopt a spherical *bleb* shape), but they need not necessarily be fully ablated. However, once a meteoroid is larger than about a millimeter across then it is doomed to begin vigorous ablation at about 100-km altitude, and after briefly become a meteor, it will be fully vaporized.

It has been suggested, but never clearly proven, that there is always going to be some small residue left after ablation has ceased, and that this residue can act to seed lower-atmosphere clouds to produce precipitation. Accordingly, for example, the suggestion is that those days following the peak activity of a major meteor shower are more likely to be rainy. While the idea is compelling it is not entirely clear how the hypothesis can be unambiguously tested, and the idea remains somewhat mysterious and unproven.

10.3   Gathering meteorites

While very few people have the good fortune to find a large meteorite, or even see one fall, everyone, from the light-challenged city dweller to the remotest homesteader, is in a position to find and collect cosmic dust [3, 4]. The challenge is only in the collecting and the searching, but the latter part can be done in the comfort of one's own home. The equipment required for collecting the dust is also minimal and not expensive. What is required is a collecting system, which is basically a strong magnet, and a little patience. Indeed, key to finding micrometeorites is the fact that they will be attracted to a magnet. After this, the slightly harder part is the sampling of enough material to increase the chances of a micrometeorite being present. The material, in this case, is almost any detritus that one can gather. Some sources of detritus, however, are probably better than others.

What needs to be optimized is not so much the amount of material sampled, but the area and time over which the material being sampled was accumulated. The longer the exposure time, and the greater the area over which the material has settled, so the more micrometeorites will be found. The time component simply reflects the fact that there is a steady rain of micrometeorites to the ground, so the greater the exposure time, so the greater the number of micrometeorites. The sampled area is important, since the greater the collecting area so the greater number of micrometeorites that will be found. Figure 10.1 indicates that the flux of micrometeorites in the size range from 10^{-5} to 10^{-3} meters is of order 10^{-6} per square meter per second. Over the course of a single year, therefore, something like 10 to 30 micrometeorites will accumulate on every square meter area exposed to the sky. The area of a typical house roof is of order $100\,\mathrm{m}^2$, and accordingly a typical home will potentially accumulate over a thousand micrometeorites per year. This number, of course, is rather optimistic since many of the accumulated micrometeorites will be blown and/or washed away by wind and rain. But not all is lost — one of the best places to search for micrometeorites is within the detritus found in the eve troughs that collect and channel the rain run-off from a roof. The first task then in finding micrometeorites is to collect the material from the eve troughs (another good source of raw materials is the detritus that collects at the bottom of rain barrels).

Once this material has been collected and dried, place a strong (neodymium) magnet in a plastic bag, and then carefully scan over the collected material to extract all the magnetic material. Do this slowly and several times. Eventually, you will find material, attracted to the magnet, sticking to the outside of the plastic bag. These are your potential micrometeorites. Once you have carefully sifted through your collected material, carefully turn the bag containing the magnet inside out to safely collect the magnetic particles, which will now be on the inside of the bag. Not everything collected by the magnet will be a micrometeorite, there will be all sorts of magnetic flakes and chips in the collection — these being just inevitable industrial pollution. However, with luck, and with the use of a hand lens, or under microscope analysis, you will begin to find small, spherical blebs. These are the micrometeorites (Figure 10.4). The spherical blebs will typically have a black coloration, but white and bluish hues are also possible.

There are a whole host of good micrometeorite collecting grounds available. As already indicated, the detritus from eve troughs is a good place to look, but one could also position a series of magnets (each wrapped in a plastic bag) within the eve troughs or down spouts of your house. One can also just scan a plastic-wrapped magnet over a driveway, or a car park, or any ground that is easily traversed. If you live in region where snow regularly falls and accumulates, you can sieve through the snow run-off [4]. Basically, anywhere that water might accumulate and drag small particles to a common collecting point is a good place to try. Amateur enthusiast, Jon Larsen [3, 5], suggests approaching the managers of large, flat-topped, industrial buildings and, with permission, sampling the detritus that collects

Figure 10.4.   Micrometeorites collected from snowpack in Antarctica. Image courtesy of http://www.polarfoundation.org/.

around the roof run-off points — the advantage here is that such roofs present a large collecting area.

References

[1] M. Zolensky, P. Bland, P. Brown, and I. Halliday. "Flux of extraterrestrial materials." In *Meteorites in the Early Solar System II*. D. Lauretta and H. McSween (Eds.). The University of Arizona Press, Tucson (2006).

[2] M. Beech. "Finite-size corrections to the atmospheric heating of micrometeorites." *Monthly Notices of the Royal Astronomical Society*, **402**, 1208–1212 (2010).

[3] J. Larsen. *On the trail of Stardust*. Voyageur Press, Beverly MA (2019).

[4] M. Maurette. *Micrometeorites and the Mystery of Our Origins*. Springer, Berlin (2006).

[5] M. Genge, J. Larsen, M. Van Ginneken, and M. Suttle. "An urban collection of modern-day large micrometeorites: Evidence for variations in the extraterrestrial dust flux through the Quaternary." *Geology*, February issue (2017).

Chapter 11

Impact Cratering

Writing 5 centuries after the event, we learn from Pliny the Elder, in his *Naturalis Historia*, that Anaxagoras of Clazomene predicted the fall of a meteorite at Aegospotami in 467 B.C. This is a remarkable account not least for the fact that Pliny had no access to the writings of Anaxagoras, and that his comments are effectively based upon 500 years of rumor and handed-me-down commentaries. From all accounts, however, Anaxagoras was a remarkable man and philosopher, but it is very much the case that, in spite of his prodigious talents, he would not have been capable of predicting the fall of a meteorite — indeed, no one is. The only meteorite for which some prior warning as to its Earthly arrival is that of Almahata Sitta, which fell in the Nubian Dessert, Sudan on October 7th, 2008. The warning time before impact was less than 24 hours. The pre-impact body, asteroid 2008 TC3, was detected at 06:39 UTC on October 6th, 2008 by the Catalina Sky Survey telescope and detector system located in Arizona, and it was quickly realized that something special had been found. Indeed, the object was heading for Earth, and ground zero was predicted to be somewhere over the Middle East. Sure enough, at 02:45 on October 7th, the 4-m dimeter, 80-tonne asteroid entered Earth's atmosphere and exploded at an altitude of some 37 km with an estimated 0.9 to 2.1 kTNT equivalent energy.

The entry and explosion were seen by crew aboard distant aircraft and from Earth-orbiting satellites (Figure 11.1) — an associated infrasound signal (see Chapter 21) was detected as far away as Kenya. It took two months to organize a search for meteorites on the ground, but on December 6th a team of researchers led by Peter Jenniskens

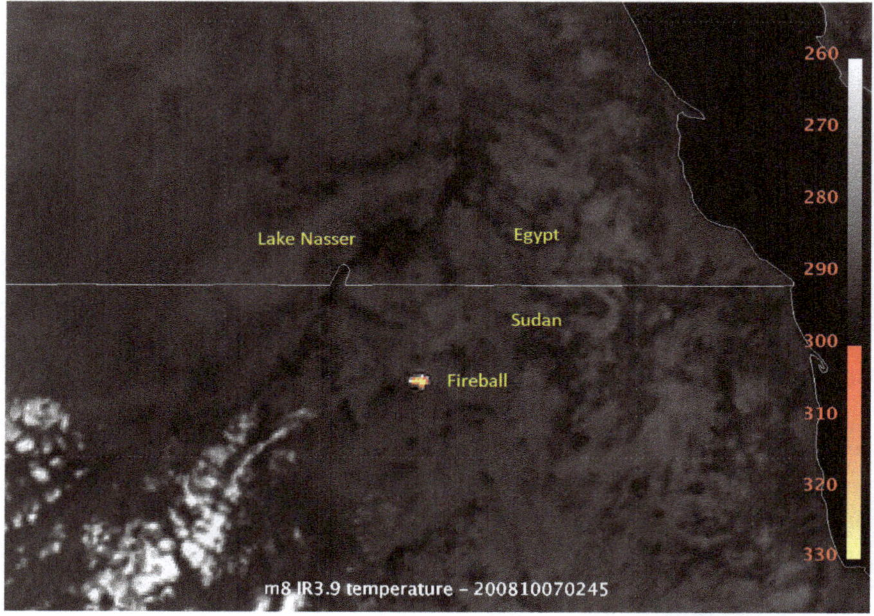

Figure 11.1.   Infrared image of the entry and detonation of asteroid 2008 TC3 over northern Sudan as recorded by the Meteosat-8 satellite. Image courtesy of EUMETSAT.

(SETI Institute, California) and Muawia Shaddad (University of Khartoum) set out to hunt for remnants of asteroid 2008 TC3. Within days of starting, freshly fallen meteorite fragments were found close to Almahata Sitta (Station 6) on the rail line between Wadi Helfa and Khartoum. In time some 600 fragments, weighing in with a total mass of 10.5 kg, were collected. Remarkably, the meteorites belong to a rare class of brecciated ureilite achondrite.[1] The meteorites are particularly rich in carbonaceous grains and were found to contain numerous nano-diamonds. These diamonds are especially interesting since laboratory analysis revealed that they must have formed slowly and at pressures greater than some 20 GPa. The pressure condition is

---

[1]The term remarkable is used since the most common meteorites to fall correspond to the H and L ordinary chondrites (recall Section 1.2).

revealing since such values are only going to be realized in planetary-sized objects (something the size of Mars or Mercury) — and this parent planet is, of course, now destroyed and no longer a member of the solar system.

Asteroid 2008 TC3 did not survive atmospheric passage to produce an impact crater — the surviving meteorites typically having a mass of just a few grams. Other small asteroids have been detected prior to encountering the Earth since 2008 TC3, but no meteorites have been recovered from these other events. The most recent crater-producing event, for which meteorites were additionally collected, is that of September 15th, 2007. At this time, without any prior warning, a meter-sized stony asteroid hit the ground at Carancas (recall Figure 6.2) in southeastern Peru near Lake Titicaca, producing a single impact crater some 13.5 meters across and 4.5 meters deep. This particular event caught the attention of the world's media since, as a result of the impact, many people fell ill after inhaling arsenic vapor — the vapor was released through the impact heating of local groundwater, which is known to contain arsenic compounds. The Carancas crater has a classical profile, being essentially circular from above and bowl (or saucer) shaped in cross-section. There were no human injuries reported when the Carancas crater formed, but in the grand scheme of things, this particular crater is just a little-un.

A more extensive fall of material than that at the Carancas crater was experienced in southeastern Russia, in the Sikhote-Alin mountains, on February 12th, 1947. In this case a complex strewn field of both iron meteorites and their associated craters were produced, with the largest crater being some 26 meters across and 6 meters deep. The bolide associated with the fall was witnessed by many observers and distinct detonations were produced as the meteoroid fragmented. A distinct smoke trail, lasting for several hours, was additionally produced (Figure 11.2). The pre-atmospheric mass of the iron meteorite is estimated to have been of order 90,000 kg, and a relatively small strewn field some 1.3 square kilometers in size was produced, from which over 23 tonnes of material has been collected. Cosmic-ray-exposure ages derived from various fragments indicate

Figure 11.2. (Left) Soviet Union stamp issued on November 20th, 1957 to commemorate the 10th anniversary of the fall of the Sikhote-Alin meteorite. The image is based upon a painting made by P. Medvedev who witnessed the fall and reveals the distinctive smoke trail. (Right) One of the many fragments from the Sikhote-Alin fall showing surface regmaglypts as well as a fragmented (shrapnel-like) appearance. Images courtesy of Wikipedia Commons.

two distinct ages of 450 million and 145 million years — the younger ages indicate the time of a significant break-up event while the parent meteoroid was still in space.

The cratering process is all about energy and ultimately the kinetic energy (KE) of the impactor. Accordingly, the initial energy available for punching through Earth's atmosphere and thereafter producing a crater in the ground is:

$$E = \frac{1}{2}M_i V_i^2 = \frac{2\pi}{3}\rho_i L_i^3 V_i^2 \qquad (11.1)$$

where M is the meteoroid mass, V is the velocity, L is the diameter, and ρ is the meteoroid density, and the i subscript indicates an initial value. The amount of energy E increases, as expected, with the mass (or size) of the impactor and its velocity. After atmospheric passage, the meteoroid mass (size) and velocity will have changed somewhat (recall Chapter 6) with the impact size and velocity being $L_{imp} < L_i$ and $V_{imp} < V_i$ respectively. Although, as shown in Chapter 6, once the size of a meteoroid exceeds more than several meters across then it is likely to retain most of its cosmic speed upon impact. The equations that follow, however, being based upon laboratory experiments, field measurements, and theoretical scaling rules, use the actual impactor size and impactor velocity as their basic parameters. A full investigation of a crater forming event would need to take into account the atmospheric deceleration, ablation, and fragmentation of the original meteoroid.

Describing the physics behind the cratering process is far from straightforward, but the basic idea is to generate a formula that accounts for the transfer of the meteoroid energy into the ground upon impact, along with the excavation of material and the physical alteration of the surrounding target matter. Clearly, there is going to be a dramatic deceleration of the meteoroid once it contacts the ground and this results in the rapid compression of material at very high pressures and temperatures. Furthermore, a shock wave develops between the compressed and uncompressed material and this propagates away from the point of initial contact into the surrounding target medium. The first-formed, so-called transient crater subsequently collapses under gravity to produce a slightly wider but less deep final crater. Close to the crater rim an ejecta deposit forms and this contains material ejected from the crater and potentially fragments (meteorites) of the impactor as well. The base of the crater further develops into what is called a breccia lens — this being composed of fragmented target material as well as infill. Figure 11.3 shows the cross-section of a so-called simple crater.

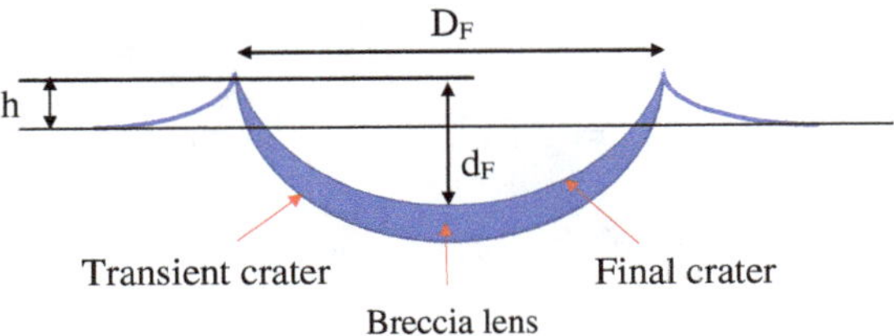

Figure 11.3.   Characteristics of a simple crater. D_F is the final crater diameter, d_F is the final crater depth, and h is the rim height above ground level.

A commonly used (although many are available) formulation for the transient crater diameter D_T is given by the parameterization [1]

$$D_T = 1.161 \left(\frac{\rho_{imp}}{\rho_t}\right)^{1/3} L_{imp}^{0.78} V_{imp}^{0.44} g^{-0.22} \sin^{1/3}\theta \qquad (11.2)$$

where the t subscript indicates a property of the target material, the *imp* subscript indicates properties of the meteoroid at the time of impact (the size is given in meters and the velocity in meters per second), g is the acceleration due to gravity ($g = 9.81\,\mathrm{m/s^2}$ for Earth), and θ is the angle of the meteoroid path to the horizon. As described in Section 20.2 the most-likely value for the entry angle is $\theta = 45°$. With an expression for the transient crater diameter in hand, the final crater characteristics are described according to a set of 6 more equations. The final crater diameter D_F (Figure 11.3) is

$$D_F = 1.25 D_T \qquad (11.3)$$

and the final crater depth d_F is given by

$$d_F = d_T + h - \Delta_{br} \qquad (11.4)$$

where d_T is the depth of the transient crater

$$d_T = \frac{D_T}{2\sqrt{2}} \qquad (11.5)$$

and h is the crater rim height

$$h = 0.07 \left(\frac{D_T^4}{D_F^3}\right) \qquad (11.6)$$

The final term in Equation (11.4) is the maximum thickness of the breccia lens, and this is given by the formula

$$\Delta_{br} = 2.8 V_{br} \left(\frac{d_T + h}{d_T D_F^2}\right) \qquad (11.7)$$

where V_{br} is the volume of the breccia lens, with

$$V_{br} = 0.032 D_F^3 \qquad (11.8)$$

The formula shown above provide for a good description of a simple crater — that is a circular crater with a raised rim showing smoothly sloping (parabolic profile) walls rising up from the crater center. Many of the craters on the Moon are of this simple kind, but on Earth two good examples (discussed further below) are the Whitecourt crater in Alberta, Canada and the Barringer (also Meteor) crater in Arizona. As the impactor energy increases, so crater morphology becomes more complex, ultimately resulting in the formation of a complex crater with a central uplift. The change over from the production of a simple crater to a complex crater (on the Moon) occurs at a diameter of about 15 km. The largest craters also tend to be accompanied by multiple ring structures about the crater center; on Earth, the Vredefort crater in South Africa is a multi-ring structure (as well as being the second oldest, at 2 billion years old, and largest, at 300 km across, impact structure on Earth). Occasionally elongated craters are observed, being more elliptical than circular, and these are formed by very shallow angle impacts. Additional equations have been developed to describe the volume of material ejected in the formation of a crater, but these will be considered in Chapter 15.

Figure 11.4 shows the variation of D_F, d_F, and h for an iron impactor when the impact velocity is taken to be 3 km/s — a characteristic impact velocity for small meteoroids. In this example plot, the crater diameter, depth, and rim height increase, as expected, with initial size. A crater that can be described under the assumptions used to produce Figure 8.4 is that of the recently described Whitecourt crater in Alberta, Canada [2]. This particular crater formed about 1100 years ago and has a diameter of 36 meters and a depth of 6 meters (Figure 11.5). The crater was formed through the impact of an iron meteoroid, as is witnessed by the abundant IIIAB iron medium octahedrite meteorites that have been discovered in the region immediately adjacent to the crater rim. Indeed, over 1200 fragments have been collected from the crater location with a total mass in excess of 50 kg of material. With an assumed impact speed of 3 km/s and taking the density of the impactor to be that of nickel-iron, and the target material to be that of consolidated soil, a final crater, with characteristics similar to that at Whitecourt, would be produced by an iron meteoroid having a diameter of 1 m. This

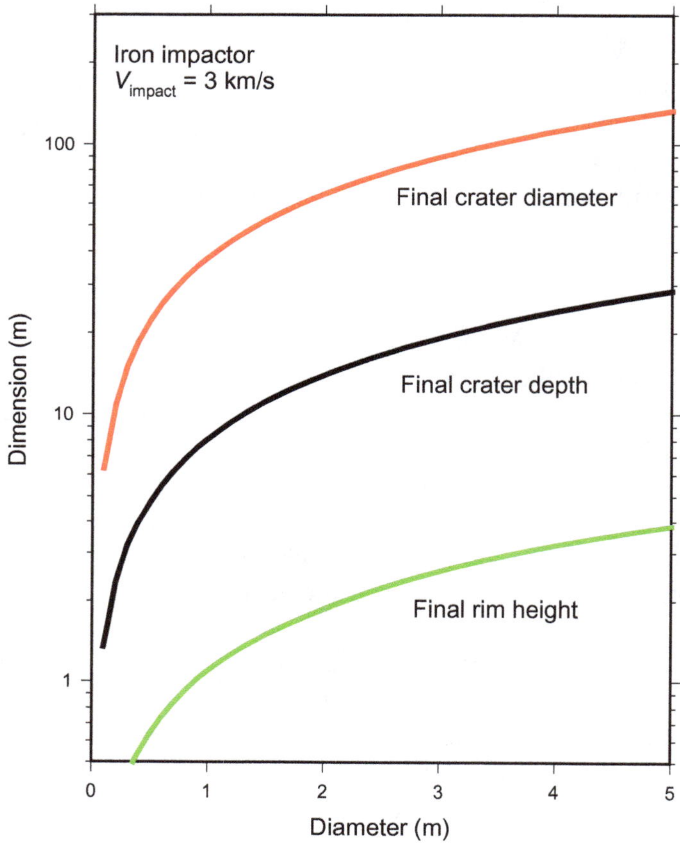

Figure 11.4.   Crater diameter, depth, and rim height variation against impactor size (a velocity of 3 km/s is adopted, and the density of the impactor and target material are taken as 9000 and 3000 kg/m^3).

indicates an impactor mass of some 5000 kg and an impact energy of order 21 billion joules (about 5 tonnes of TNT equivalent energy). The transient crater for the Whitecourt structure would have been about 28 meters across and some 10 meters deep, giving a breccia lens of a maximum thickness of 3.5 meters — and these values are in good agreement with the actual crater measurements (Figure 11.5).

A larger, older crater than that found at Whitecourt is the Barringer crater in Arizona (Figure 11.6). Named after Daniel Barringer, who acquired the land and mineral rights surrounding the

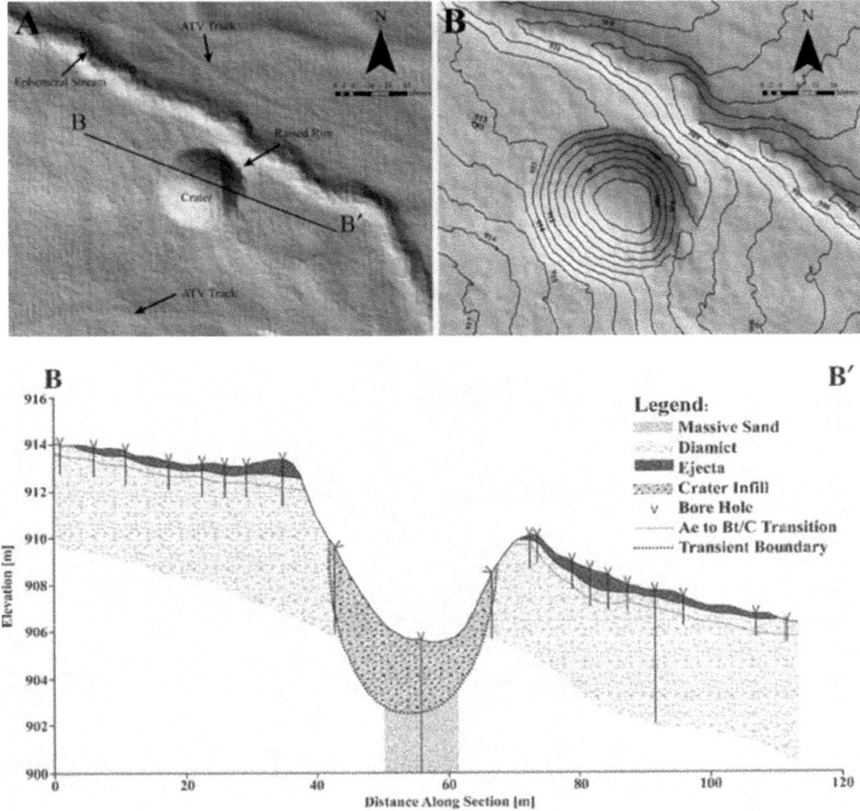

Figure 11.5. (Top) Lidar depth-ranging map (left) and contour outline (right) of the Whitecourt crater in Alberta, Canada. The line BB/ on the lidar map corresponds to the crater cross-section shown in the lower part of the figure. See Reference [2] for details.

crater in 1903, this crater is 1.186 km across and 170 meters deep, with a rim height of about 45 m above the surrounding plane. It has been estimated that the crater formed about 50,000 years ago, when an iron impactor some 50 meters across (with a mass of order 56,000 tonnes), ploughed into the region known as Canyon Diablo (the Devil Canyon). The impact energy was of order 10 megatons of equivalent TNT energy, and the impactor was mostly vaporized. The area surrounding the crater has yielded many hundreds of iron meteorites,

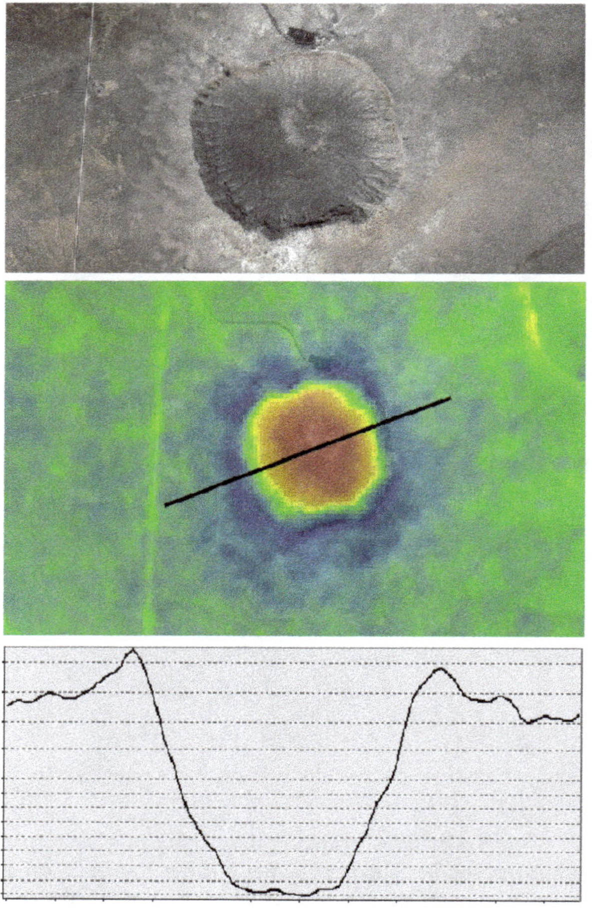

Figure 11.6.   The Barringer crater in Arizona. The top image shows the crater from above, the middle image is a color-coded elevation map, and the thick line across this image corresponds to the cross-section profile shown in the lower panel (the depth scale is exaggerated in this profile). Images courtesy of https://gisgeography.com/topographic-profiles/.

in an extensive strewn field from which some 30 tonnes of material has been extracted. The largest Canyon Diablo meteorite fragment ever to be collected, weighing in at 639 kg, was actually found just below the crater floor by Samuel Holsinger, the drilling engineer hired by Barringer in what turned out to be a fruitless attempt to mine the iron-impactor material. Although 40 times larger than the Whitecourt crater, the Barringer crater still maintains a simple-crater parabolic profile (as seen in the lower image in Figure 11.6). The Barringer crater is also the largest terrestrial crater for which a strewn field of meteorites is known to exist (the Canyon Diablo strewn field) — iron meteorites (literally, fragments of the impactor) have in fact been found at distances out to about 10 km from the crater.

Indeed, the energy involved in producing a crater bigger than about a kilometer or so across is so large that essentially none of the impactor will survive to be collected in the form of meteorites, although the composition of the impactor (stony or iron) can sometimes be deduced by the chemical analysis of shock-altered rocks in the crater region. In the production of very large craters the signature of the impact can be detected not just locally (as a crater) but globally as a distinct chemical anomaly. The classic example of such an Earth-perturbing impact is that which saw the formation of the Chicxulub crater on what is now the Yucatan Peninsula in Mexico. This particular impact occurred some 66.03 million years ago, and its formation triggered a global climate crisis.

The Chicxulub crater (Figure 11.7) is over 150 kilometers in diameter, some 20 kilometers deep, and was formed at a time corresponding to the Cretaceous–Paleogene (K–Pg) boundary. Indeed, this geologically distinct boundary change was largely caused by the production of the crater, and most infamously it precipitated the extinction of the dinosaurs (on land) and the ammonites (at sea) [6]. The crater has a complex morphology and is the only known terrestrial crater with a peak ring structure. The crater is now buried under about a kilometer of sediments and is not visible upon the surface ground — this being said, it has been extensively studied by drilling (where core samples are collected) and via gravitational anomalies.[2] Being so large, the energy required to produce the Chicxulub crater could only have been delivered through the impact

---

[2]The gravitational anomalies come about by the redistribution of material between the crust and mantle, with the different density rocks producing a small, but measurable differences in the local gravitational acceleration.

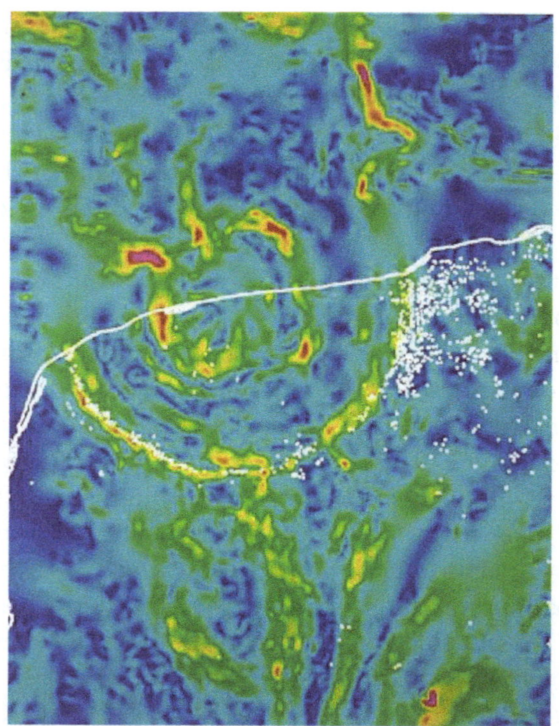

Figure 11.7.   Gravity anomaly map of the Chicxulub (from the Mayan world meaning *tail of the devil*) crater, showing it to be a complex (eroded) crater with a central peak ring (smaller circular arcs left of image center). The white dots correspond to sink holes (cenotes) associated with the location of the crater rim — the outer crater rim diameter is some 180 km. The white line across the image center indicates the present-day land–sea boundary, the crater now being buried beneath about a kilometer of sediment. Image courtesy of Wikipedia Commons.

of a multi-kilometer-sized asteroid or comet — it is still not clear which kind of impacting body was involved. The material excavated in the production of the Chicxulub crater produced a massive cloud of hot ash and steam that spread upwards and outwards to eventually enveloped the entire globe, while the accompanying blast wave flattened trees and wrought unconstrained destruction at distances many thousands of kilometers away from ground zero. Additionally, since the impact occurred in a shallow sea region, the mega-tsunami produced would, upon reaching shore, produced massive devastation well inland. The impact literally changed the Earth, resulting in a profound change in the climate and in the killing off of numerous land and marine species. There is a certain irony in this devastating action, since earlier in Earth's history it was such impacts (from asteroids and comets) that seeded life and produced the life-enabling oceans (see Chapter 9).

The dust and ash placed into Earth's atmosphere by the Chicxulub impact had a significant and detrimental effect upon the propagation of sunlight to the ground. Indeed, not only did it halt photosynthesis in plants, it caused a strong cooling effect that lasted for many decades.[3] The environmental stress imposed by the impact effectively destroyed the food chain and triggered an ecological disaster from which few, previously dominant, animal species were able to adapt and survive. The great die-off at the K–Pg boundary is distinguished in the geological record as a thin (just a few centimeters thick) clay boundary, and it is within this boundary clay layer that the chemical signature of the Chicxulub impactor resides. The signature is in the form of an anomalous increase in the abundance of the element iridium in the boundary clay. And, this iridium anomaly was noted and described in what is now a classic research paper by Luis and Walter Alvarez, with Frank Asaro and Helen Michel (A^3M) published in the June 6th, 1980 issue of *Science*. Entitled *Extraterrestrial cause for the Cretaceous-Tertiary extinction: experimental results and theoretical interpretation*, the paper presented the results of a study measuring the iridium abundance across the K–Pg boundary at outcrops in Denmark, Italy, and New Zealand. In each case the iridium abundance showed a significant increase within the clay layer. The numbers vary from one location on Earth to another, but characteristically the abundance in Earth's crustal rocks is of order 0.05 nanograms of iridium per gram of material.[4] In the boundary layer clay, the abundance is

---

[3]This cooling effect is similar to that invoked in the Nuclear Winter Scenario — see Reference [7].

[4]Iridium is a siderophilic (iron-loving) element and is accordingly rare in Earth's crust, but abundant in Earth's iron core.

characteristically 50 nanograms of iridium per gram of material. The question addressed in the *Science* paper by Alvarez *et al.* was, why is there an enhancement of a factor of 10^2 to 10^3 within the clay layer — and this is where the extraterrestrial impactor comes in.

The iridium abundance in meteorites is known to be much higher than that of Earth's crustal rocks — indeed, for chondritic material the abundance is characteristically of order 500 nanograms of iridium per gram of material. Accordingly, it was argued by A^3M that the iridium enhancement in the K–Pg boundary clay layer is the fall out of the iridium originally contained within the impactor.[5] A^3M estimated that the total amount of iridium within the global clay layer amounts to some 750,000 kg of iridium. The baseline abundance of iridium in the solar system is given by Katherina Lodders[6] as 4.69×10^{-10} kg per kg and accordingly, the mass of the Chicxulub impactor is $M_{\text{chic}} \approx 750,000/(4.69 \times 10^{-10}) = 1.6 \times 10^{15}$ kg. Assuming a rocky asteroid impactor, with a density of $\rho_{\text{stone}} = 3000 \, \text{kg/m}^3$, the implied radius of the impactor is 5 km. It is unlikely, however, that any asteroid impactor is monolithic, so the bulk density ρ_{bulk} should be reduced in accordance with a porosity measure $0 < P < 1$ (see Section 7.7), with $\rho_{\text{bulk}} = (1 - P)\rho_{\text{stone}}$. Estimates place P as high as 0.5 in some asteroids, and this implies the Chicxulub impactor could have had a radius as large as 10 km. Accordingly, the total iridium mass deposited as estimated from the K–Pg boundary layer indicates an impactor (stony asteroid) with a diameter between 10 and 20 km.

---

[5] The debate concerning the exact mechanism(s) behind the mass extinction event at the end Cretaceous continues to rage. It was probably not just one event that was responsible for the die off, but a combination of events including the extensive volcanic activity associated with the formation of the Deccan Traps in India, continental break up, and the Chicxulub impact along with associated smaller impact. Three additional craters with the same deduced age as Chicxulub are the Silverpit crater (20-km diameter), the Boltysh crater (24-km diameter) and the Shiva crater (500-km diameter). It should be noted, however, that an impact origin for the Silverpit and Shiva craters is not universally agreed upon and/or unambiguously demonstrated. Interestingly, however, if an impact origin for the Shiva crater is established, then it and not the Chicxulub impactor would have been the main instigator of the K–Pg extinction event.

[6] See: https://arxiv.org/ftp/arxiv/papers/1010/1010.2746.pdf.

If the impactor was an iron asteroid, then the diameter could have been as small as 7 km across. If the impactor was a comet then the implied diameter would be of order 20–25 km across.

The Earth Impact Effects program [3] can be used to gauge the effects of the Chicxulub impact. Here we shall adopt as parameters an impactor diameter of 15 km, an impact velocity of 25 km/s, an impact angle of 45°, and a density of 1500 kg/m^3 corresponding to porous rock, and assuming a target density of 2500 kg/m^3 (something like sedimentary rock). With these parameters, the energy involved in the impact amounts to 5.3×10^{23} joules — 127 million megatons of TNT equivalent energy. The transient crater will be about 75 km across and 26 km deep, with the resultant final crater being some 130 km across and about 1.5 kilometers deep, and of order 5210 cubic kilometers of target material will be melted and vaporized. About half of this material will be ejected into Earth's atmosphere. If such an event occurred in the modern era, then even at 1000 km away from the impact location, the fall-out effects would be lethal. Just 8 seconds after impact the thermal radiation generated would ignite trees, grasses, and clothing, and produce third-degree burns upon any hapless observers. And then, $3^1\!/_2$minutes after impact, seismic waves would produce dramatic shaking and structural damage to most standing structures — indeed, the effects would be equivalent to a magnitude 10 earthquake.[7] Some 8 minutes after impact small debris fragments would begin to rain down from the sky, and about 50 minutes after impact a tumultuous blast wave will hit, with wind speeds of order 240 m/s and an overpressure of 162 kPa — such conditions will utterly destroy any buildings that might have survived the earlier seismic shaking. At greater distances than 1000 km from the impact location the direct effects will be moderated, closer they will be worse, but no matter how far away the observer might be from ground zero, the effects of the impact will sooner or later be felt.

---

[7] An earthquake of this magnitude has never actually been recorded. The strongest known earthquake event was that in Valdivia, Peru in 1960, which measured magnitude 9.5 on the Richter scale.

Not only will the physical ground effects within the first few hours following impact be devastating, the effect of the impact upon the atmosphere will be equally catastrophic. Material ejected from the crater, with speeds less than Earth's escape velocity of 11.2 km/s, will soon begin to fall back to the surface, and upon being heated, through atmospheric friction, will become incandescent and ignite numerous and extensive firestorms. These firestorms will produce vast quantities of soot and smoke, which will mix with the smaller debris dust particles to choke off sunlight from reaching the ground. Night will literally fall. To say the least, the immediate and short-term (that is, on timescales of months to years) consequences of injecting some 2000 km^3 of target material into the atmosphere, and then adding to this mix massive amounts of smoke and soot particles, will be profound. The mass of material vaporized will be of order 5.2×10^{15} kg of material, and this will, over several days and weeks, spread throughout the entire atmosphere.

The amount of attenuation of sunlight caused by the fine-grained dust and soot lifted into the atmosphere can be estimated through the extinction formula $I = I_0 \exp(-\tau/\cos Z)$, where I is the intensity of sunlight on the ground, I_0 the normal intensity of sunlight under no extinction conditions, τ is the optical depth, and Z is the solar zenith angle [7]. The optical depth is the key quantity, and it can be expressed as $\tau = \sigma\mu L$, where σ is the specific cross-section area of particles (usually expressed in cm^2/g), μ is the mass concentration in g/cm^3, and L is the optical path length in centimeters. For fine dust particles and soot, σ is characteristically $\sim 5 \times 10^4$ cm^2/g, and the μ term will be of order the mass of material vaporized and ejected into the atmosphere divided by the volume of atmosphere to a height of about 30 km (about the middle of the stratosphere). Accordingly, $\mu = M_{\text{vap}}/(1.537 \times 10^{25})$ — units of g/cm^3 — where M_{vap} is the material density (~ 2 g/cm^3) multiplied by the amount of material vaporized in cubic centimeters. Let us assume that just 1% of the dust and ash ejected in the crater formation event ends up being rapidly dispersed around the globe, then $M_{\text{vap}} \sim 5.2 \times 10^{16}$ g of material, and $\mu \sim 3.4 \times 10^{-9}$ g/cm^3. With these characteristic numbers, the optical depth of the dust cloud will be of order 500.

Needless to say, at such an optical depth the Earth's surface will be plunged into deep darkness, with the Sun's rays being incapable of penetrating to the ground. Such conditions will not prevail for more than a few days, however, and as more material falls back to Earth, so the mass loading of the atmosphere will be reduced, and the optical depth will correspondingly decrease. Nonetheless, detailed simulations of dust-cloud effects suggest that the Earth would remain in complete darkness for at least several weeks to a month or so, and that the light levels will be below that at which photosynthesis can operate for some 6 months to one year. Indeed, for a dust-loaded atmosphere photosynthesis will cease once the opacity exceeds $\tau \sim 25$ [8], and this will be achieved if just 0.05 percent of the material vaporized in the formation of the Chicxulub crater were uniformly dispersed into the atmosphere.

This situation is made much worse, in the sense of climate change, by the fact that not only will sunlight be unable to penetrate to the ground, but more sunlight than normal will be reflected back into space (that is, it will increase the atmospheric albedo), the consequence of which is to induce significant global cooling. Evidence for the increased reflection effect was found during the compositional analysis of Chicxulub peak ring rocks which revealed a deficiency in the calcium sulfate component that is normally found in the deep crustal (target) rocks. This absence of calcium sulfate is interpreted as a vaporization effect and the formation of sulfur dioxide, the latter gas being dispersed into the atmosphere. The relevance for the climate is that sulfur dioxide will increase the atmospheric albedo, and indeed a number of geoengineering proposals (suggested in order to combat global warming now) are based upon the injection of sulphate aerosols into the stratosphere either by aircraft, balloons, or rockets [9].

That several important turning points in Earth's history have been predicated upon a giant impact event, such as Chicxulub, is now uncontroversial, although not all of the bigfive mass extinction events can be attributed to cosmic catastrophe.[8] The evidence for extinctions and significant global forcing is certainly well substantiated with respect to the Chicxulub impactor, and more recently the global cooling associated with the Younger Dryas period from 12,800 to 11,550 years ago has been linked to a large impact cratering event (the Hiawatha crater in Greenland — recall Figure 4). Also, in recent times, the highly eroded Yarrabubba crater in Australia has been dated to an age set some 2.229 billion years ago, and is associated with the end of a deep, snowball Earth glaciation epoch. In this case it is thought that a 7-km diameter asteroid impacted into a thick-ice surface layer, producing a 70-km diameter crater which released significant quantities of water vapor (a potent greenhouse gas) into the atmosphere.

A recent study by Garth Collins (Imperial College, London) and coworkers published in *Nature Communications* on May 26th, 2020 has simulated the formation of the Chicxulub crater under varying impact conditions (Figure 11.8). The impactor is taken to be a stony asteroid some 17 km across, impacting the atmosphere with a speed of 12 km/s. Various impact angles were adopted in the simulations, and a best fit to the observed crater structure was obtained when the impact angle was between 45 to 60 degrees above the horizon — this is quite a steep angle of entry, and interestingly, it is the angle range over which the greatest amount of climate change, greenhouse gases, such as carbon-dioxide and sulfur, are released. This further suggests a perfect storm condition, in which a maximum global perturbation effect was induced, resulting in enhanced nuclear-winter conditions.

---

[8]The five biggest mass extinction events are (i) end-Ordovician (439 Myr BP), (ii) Frasnian–Famennian boundary (364 Myr BP), (iii) the end-Permian (250 Myr BP), (iv) end-Triassic (200 Myr BP), and (v) end-Cretaceous (65 Myr BP). Only the end-Cretaceous mass extinction has been associated with an impact event (the Chicxulub impact), while the others are the result of climate change, sea-level variation, volcanism, and ocean anoxia. Some researchers have argued that we are presently entering into an epoch of a 6th major extinction. This extinction event, however, is being driven by human activity and the onset of human-generated global warming.

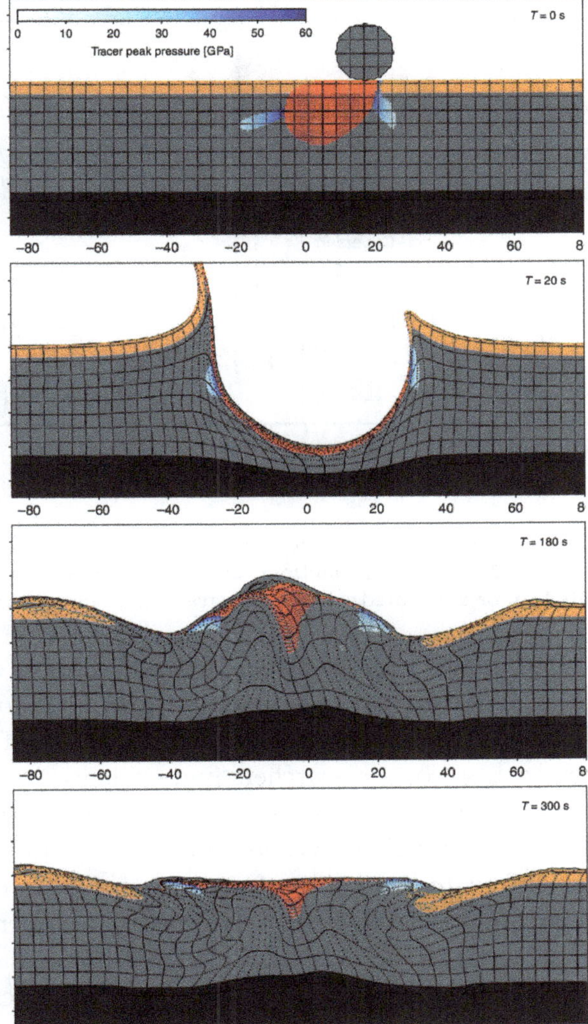

Figure 11.8. Time series development of the Chicxulub impact structure as modeled by Garth Collins and coworkers. The four panels cover the first 5 minutes of crater formation. The top panel indicates the moment of impact, when a 17-km diameter stony asteroid impacts the ground. The second panel shows the opening up of the transient crater 20 seconds after impact. The final two panels illustrate the substrate rebound phase and the formation of the peak ring structure. The top panel indicates the color scheme for pressure alteration, while the red-colored regions correspond to melted rock. The horizontal scale is measured in kilometers. Image courtesy of Garth Collins/Imperial College, London.

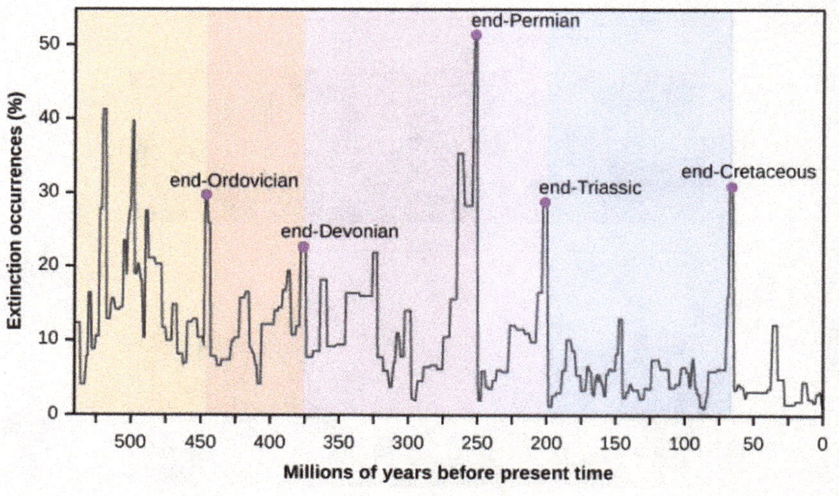

Figure 11.9.   The extinction record as a function of time over the past 550 million years. Of the *Big-5* mass extinctions, only the end-Cretaceous extinction can be attributed to, or associated with, a massive impact cratering event.

Is there a periodicity to extinction events? This is a question that has vexed biologists, astronomers, and geologists for the past many decades. According to how the statistical analysis of extinction record is organized, so some researchers find evidence for a 25 to 35-million-year periodicity, whereas other researchers find no evidence of periodicity at all (Figure 11.9). While some of the major extinction events can be linked to the times of massive cratering events, the majority cannot, and it is far from clear what physical mechanism might drive the apparent periodicity if it actually proves to be real. One suggestion for explaining a 35-million-year periodicity is that of astronomical forcing, the period being related to the oscillatory motion of the solar system above and below the galactic plain. It is also about the same time interval at which the solar system encounters a galactic spiral arm. The argument here is that at the galactic plain and within spiral arms the solar system's Oort cloud[9]

---

[9]Named after the Dutch astronomer Jan Oort, who first proposed the existence of a vast cloud of cometary nuclei surrounding the solar system, out to a distance

of cometary nuclei is more likely to be gravitationally perturbed by giant molecular clouds. These perturbation events will scatter millions of cometary nuclei, a fraction of which will move inwards, towards the Sun, thereby initiating a period of enhanced planetary impacts. Mass extinctions, under the astronomical forcing hypothesis, are accordingly comet-impact driven, rather than asteroid-impact driven.

While it remains unclear if there is a periodicity to major extinction events, it is a 100 percent certainty that the Earth will be hit by either a large cometary nucleus or asteroid at some time in the future. At present, however, we are 100 percent unsure as to when this impact will take place.

11.1   The Hoba giant: Something's missing

When there is a boundary, there is always the question of its rigidity: is it a soft or hard boundary, and can the boundary move under different physical conditions? One such boundary in meteoritics relates to the largest mass that can survive atmospheric passage to the ground. In essence this is the question concerning the largest possible meteorite that can exist on Earth. As seen in other sections (recall Chapter 6), if the impacting meteoroid is very large then it is likely to produce an impact crater and be fully destroyed in crater excavation process. If it is very small then it will be completely destroyed in the atmosphere and seen as a meteor. Somewhere between these limiting processes is an upper boundary that allows a meteoroid to survive both atmospheric passage and impact. This upper limit will be determined by the size, shape, composition, and velocity of the incoming meteoroid, and accordingly it is a moveable boundary that we are looking for. In terms of record holders (recall Section 7.1), the Hoba meteorite in Namibia is currently recognized as the largest, most massive meteorite. Indeed, at the time of writing, it has held this record for the past century. This statement, of course,

---

of order 100,000 AU, in 1950. It is estimated that the Oort cloud could contain as many as 10 trillion cometary nuclei.

and in light of the current discussion, is predicated by the rider: largest and most massive meteorite discovered *so far*.

The Hoba meteorite was discovered in 1920 although there is some considerable confusion by whom and how. Peter Spargo (University of Cape Town) indicates that it was found by Jacobus Brits who, while out hunting, "saw this strange rock [with only the upper surface visible], sat on it and examined it" [10]. At about the same time, Michael Hanssen, who also lived on the Hoba West farm, claimed the find in a letter to the Transvaal Museum, Pretoria in September 1921. It is generally given, in modern accounts, that the meteorite was discovered while breaking ground for a new field, but there seems to be no actual statement of this by the finders. However, while short accounts of the meteorite's existence, and prodigious mass, circulated for a number of years, it was an excursion organized during the International Geological Congress in 1929 that brought it to worldwide prominence (Figure 11.10) — the first detailed account of its properties being published by Leonard Spencer[10] (British Museum of Natural History) in 1932. After various attempts to buy and move the meteorite to various museums, it was eventually declared a National Monument in 1955, and is now on display at the center of a sculpted amphitheater.

Hoba is a nickel-rich ataxite (IVB, Ni $= 16.44\,\mathrm{wt\%}$) meteorite with an estimated mass of 60 tonnes. Cosmic-ray-exposure date indicates that the collision resulting in the formation of the Hoba parent meteoroid occurred some 200 million years ago, and it is estimated that the terrestrial residency time is some 80,000 years. Its approximate shape is that of a square slab with dimensions of $2.95 \times 2.84 \times 0.9\,\mathrm{meters}$. It shows no significant cracks or obvious fracture planes, and it appears to be of a highly homogeneous structure — even on scales as small as a few tenths of a millimeter (Figure 11.11).

---

[10]Spencer is seen standing on the Hoba meteorite (second from left) in Figure 11.10. In his 1932 description of the meteorite, Spencer noted that "a dozen people can walk round on the level surface."

Figure 11.10. Geologist attending the International Geology Congress, held in Pretoria, South Africa, in 1929, inspect the Hoba meteorite during a special excursion.

Figure 11.11. Etched slab-section of the Hoba meteorite showing its uniform (Widmanstätten-pattern-free) interior structure. The parallel dark bands were most probably produced through low-velocity shock and heating alteration. The lower straight edge of the slab is about 16 cm long. Image courtesy of the Smithsonian Institution.

Perhaps the most surprising feature of the Hoba meteorite is that it is not associated with any distinctive impact structure. No evidence of any crater-like features, even eroded ones, can be found in its vicinity. This characteristic, along with it being clear that Hoba is essentially a mono-ferric mass, indicates that it must have landed relatively gently into the Kalahari limestone upon which it now rests. The Hoba meteorite is a boundary object: if the parent body had been much larger then it would have produced a distinct crater, possibly being vaporized in the process; much smaller, and it would not be the record holder that it is. Fortunately, there is enough physical data available to constrain the fall circumstances of the Hoba meteorite [11], and these allow for an estimate of the properties of the parent meteoroid and how it interacted with the Earth's atmosphere.

In order to produce a non-cratering event with an object the mass of the Hoba meteorite, multiple special circumstances have to be in place at the same time. The meteoroid must have a slow entry velocity, at a very shallow angle to the horizon; it must have very few internal fractures and a high tensile strength. Preferably, as well, it should present a stable profile to the oncoming airflow that additionally generates a high drag coefficient, and maybe even provides some lift — indeed, something like a flat slab with a large square front profile is ideal. Such a shape has a drag coefficient twice that of a sphere and it will also provide a small amount of lift. A high drag coefficient is good, since it will act to decelerate the incoming meteoroid more rapidly (recall Chapter 6), thereby quickly robbing it of its cosmic velocity and setting up the conditions for a relatively gentle landing. Asteroids have a whole range of shapes and profiles, but remarkably, one asteroid, 2017 BQ6, has been discovered that is essentially the big brother of the Hoba meteorite (Figure 11.12). This particular asteroid passed close by the Earth on February 17th, 2017 and was imaged with the Goldstone radar telescope at which time its slab-like profile was revealed. Unlike Hoba, which is an iron meteorite, it appears that 2017 BQ6 is a stony object, and its size is estimated to be about 150 meters across.

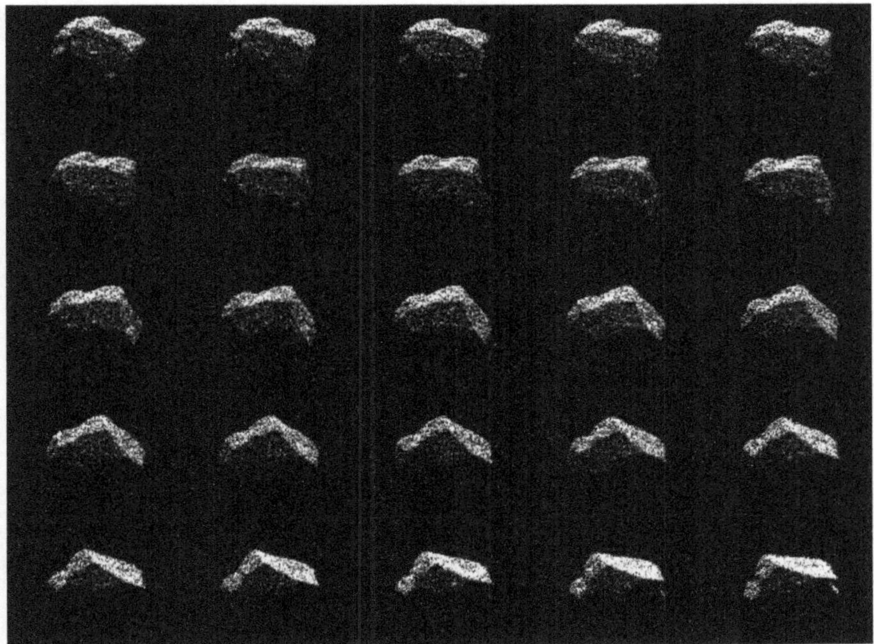

Figure 11.12. A sequence of radar images showing the rotation of the slab-like asteroid 2017 BQ6. The radar observations reveal an angular asteroid about 150 meters across, with a rotation rate of 2.15 hours. Image courtesy of NASA/Goldstone Space Communications Complex, California.

The landing conditions, such that no substantive crater is produced and in which the meteoroid is not fragmented and/or vaporized, are determined by the impact speed and the shock conditions (described by the so-called Hugoniot equations [12]) between the target material and the impactor. A detailed numerical study [11] reveals that a meteoroid with the physical characteristics of the Hoba meteorite will survive impact onto a limestone target provided the impact speed is smaller than about 300 m/s. At such an impact speed, only a small crater will be produced — a crater that would soon be eroded — and the impacting body will not undergo catastrophic disruption (it may well undergo some structural deformation, however — recall Figure 11.11). To understand the

Hoba meteorite, therefore, is to find ways in which the parent body can be decelerated from its cosmic velocity to just a few hundred meters per second upon impact.

The way to do this is effectively straightforward and requires a large surface area, a long travel path through the atmosphere and a low initial velocity. In terms of the initial velocity, the lowest encounter speed is that of the Earth's escape velocity: 11.2 km/s. Such encounter speeds do occur, but they are more commonly higher, and this reveals one reason (amongst many) for massive meteorites being relatively rare. The meteoroid encounter angle is discussed in Section 20.2, where it is revealed that the most likely angle is 45° to the horizon. Smaller encounter angles are certainly possible, but they are less likely to occur — this is a second rarity condition for producing large meteorites on Earth's surface. The results from a detailed analysis [11] indicate that for an assumed slab-like meteoroid, with physical characteristics of an ataxite, a Hoba-mass meteorite can be produced by a 115-tonne meteoroid (having a forward, stable-orientation, face with dimensions of 2.95 × 2.84 meters) if the initial velocity is taken as 11.2 km/s and the entry angle to the horizon at 500 km (starting) altitude is 16 degrees.[11] Such a meteoroid would impact the ground at a speed of order 220 m/s, after traversing a ground path of some 2900 km through Earth's atmosphere (Figure 11.13). At this impact speed a crater some 20 meters across and about 5 meters deep would be produced — a relatively small crater that, after 80,000 years (the estimated terrestrial age of Hoba), would no longer be detectable. At angles just a few degrees smaller than 16° the meteoroid will skip out of the Earth's atmosphere; at angles just a few degrees steeper than 16° the meteoroid will be destroyed upon impact. If the impact angle is taken as 45° the impact velocity is about 620 m/s.

The point of the analysis described is not to necessarily say that this is the exact scenario by which the Hoba meteorite managed to

---

[11]This particular calculation adopts an ablation coefficient of $\sigma = 10^{-8}\,\mathrm{s}^2\,\mathrm{m}^{-2}$ — see Chapter 6.

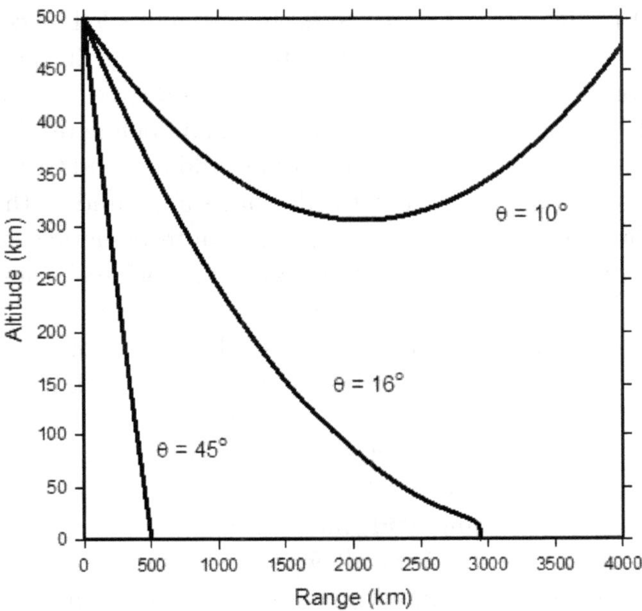

Figure 11.13.   Height versus range for a series of calculations adopting an initial mass of 115 tonnes and an entry velocity of 11.2 km/s. The curves are labeled according to the entry angle to the horizon (at 500 km altitude). The profile labeled 16° results in a Hoba-mass meteorite impacting the ground at a speed of about 220 m/s following a flight path of some 2900 km. For an entry angle of 10° the meteoroid skips out of the atmosphere.

survive atmospheric passage, but that it is one possible way that it could have done so, and that other scenarios can no doubt be made to work — the art of the *game*, however, is to fix as many of the physical parameters as possible, the ground mass, physical size and shape, the material density, the material yield strength, and so on, and then produce a range of models that include possible values for the unknown parameters: initial mass, initial velocity, entry angle, and ablation coefficient (recall Chapter 6).

A similar study to the one above, but concerning the Tunguska event (see Section 21.2), has recently been published by Daniil Khrennikov (Siberian Federal University) and coworkers in the February 4th, 2020 issue of the *Monthly Notices of the Royal Astronomical*

Society. In this case, the 1908 fireball is modeled according to the passage of a 100-m diameter asteroid, with an initial speed of 20 km/s, passing at a skimming angle of 11.5 degrees to the horizon through Earth's atmosphere. The numerical modeling by Khrennikov *et al.* suggests that such an iron asteroid would descend to a minimum height of some 10–15 km in altitude, have a ground path of some 3000 km, and eventually sail back out into interplanetary space.

While successful Hoba-meteorite-production scenarios can be developed, the modeling also allows for the study of maximum possible meteorite-delivering scenarios. In terms of ataxites like Hoba, the maximum ground mass might stretch as high 100 tonnes — a cube of this mass would have side dimensions of about 2.5 meters. Above this mass it is increasing less likely that the cosmic velocity can be slowed enough to not produce a substantial crater and a single ground mass. This being said, meteorites even more massive than 100 tonnes may yet exist if one allows for atmospheric fragmentation. Such scenarios have been investigated by Natalia Artemieva (Russian Academy of Sciences) and Phil Bland (Curtin University, Australia), who find that ground masses up to 10,000 tonnes can be realized for iron meteorites — a cube with such a mass would have side dimensions of 11 meters across. Such bodies are not going to be produced very often, perhaps at a rate of one per ten-million-year time interval, but the end result is that meteorites much more massive than Hoba are possible, and, given luck with respect to preservation and recovery, some such record breakers may eventually be found.

11.2 Stranger matters

The standard model of particle physics, developed over the past century, has proved remarkably successful in describing the structure of matter and the forces that hold that same matter together (Figure 11.14). For all this, particle physicists, ever-chasing deeper and deeper mysteries, are not convinced that it is the full picture. The universe is stranger (or so it seems) than the already strange physical theory developed to describe it. However, in decreasing size

Elementary Particles

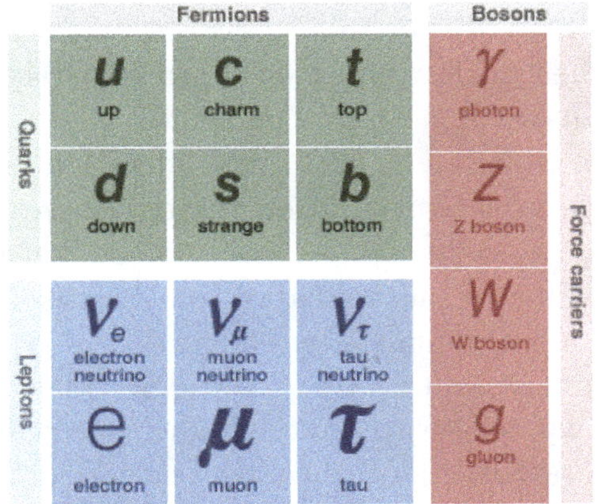

Figure 11.14. The standard model of elementary particles. Fermions are composed of quarks, of which there are six flavors. Ordinary matter is composed of fermions composed of up and down quarks. The electron in ordinary matter is one of the 6 lepton types. The forces that hold matter together are carried by bosons: the strong force between quarks in atomic nuclei is mitigated by the exchange of gluons, the photon transmits the electromagnetic force, and the W and Z bosons are involved with the interaction of the weak nuclear force. Why there are 3 generations (or families) of particles is presently unknown.

of constituents, matter is made of atoms, and atoms are composed of a nucleus, made of protons and neutrons, surrounded by an electron cloud. Electrons are elementary in that they are not thought to have any substructure. Protons and neutrons, however, are made of quarks. Ordinary matter (generation — or family — I) has its protons and neutrons made of up and down quarks. While protons and electrons carry positive and negative charge e, the quarks have a fractional charge: up quarks have a charge of $+\frac{2}{3}e$, while the down quark has a charge of $-\frac{1}{3}e$. Accordingly, a proton is composed of two up and one down quarks, and a neutron is composed of one up quark and two down quarks.

The standard model requires that the quarks within the proton and neutron must remain bound and as such no free quarks are predicted to exist. However, there is always wriggle room, and it has been suggested that under certain extreme conditions quarks may be liberated from their nuclear prisons and produce fractionally charged particles. The question now, of course, is, if such matter exists, where might it accumulate and can it be identified with present-day detectors. The answer to the latter question is yes, if fractionally changed matter exists it can be detected, and a good candidate material for the former accumulation medium is iron meteorites. Iron meteorites are especially favored since they typically have the longest cosmic-ray-exposure ages. The point being that the longer a meteoroid is exposed to space, the higher the probability that it might *pick up* a rare fractionally charged interloper.

One experiment to look for fractionally charged particles in small fragments of meteorites has been carried out by researchers at the Blackett Laboratory, Imperial College, London [13]. This group investigated the magnetic levitation properties of milligram cubes of the Hoba, Forsyth County (type IIA iron) and Murchison (carbonaceous chondrite) meteorites. The idea being that a small sample containing fractionally charged particles would have measurably different levitation properties to those samples that have none. No evidence for fractionally charged particles was found, however. A second group of researchers, headed by Peter Kim at Stanford University [14], has performed a modified Millikan oil drop experiment to look for fractionally charged particles in powdered samples of the Allende meteorite — but again no evidence for any exotic particles was found. While more refined and extensive searches for fractionally charged particles will no doubt continue into the future, the use of meteorites as particle detectors is being additionally expanded to include the search for the elusive dark matter.

Dark matter does not fit into the standard theory. Its universal existence is indicated, however, on the basis of astronomical observations, and indeed, dark matter contributes a greater proportion than ordinary matter to the universe's matter content (85% versus 15%). The problem, however, is describing exactly what dark matter is, and how might it be detected in the laboratory. One possible way of investigating dark matter is to look for evidence of nuclear recoils (essentially damage tracks) within ordinary matter caused by an interaction with a dark matter particle. This is where meteorites come into play since they have very long exposure times, thereby enhancing the possibility of having undergone some dark matter recoil event. In addition to meteorites, other so-called paleo-detectors have been proposed to search for dark matter interactions, and these include rocks extracted from several kilometers below the Earth's surface (where they are shielded from any cosmic ray interactions) or ancient marine evaporates such as halite crystals [15]. The task, however, will be a hard one, and part of the problem is to know exactly what effect one is looking for (since it is not known what dark matter is), and to then separate out any signal from other noise sources (e.g. cosmic ray). For physics, the future is definitely dark, but meteorites may yet prove to be a vital part in illuminating the new physics beyond the standard model.

One of the profound mysteries of our universe is that it is matter dominated. That is, there is not an equal (and potentially explosive) mix of matter and antimatter. The standard model allows for the existence of antimatter in the sense that any particle can have an antiparticle doppelganger — the positron, for example, is an exact match to the electron except for its positive, rather than negative, charge. Models of the early universe predict that equal quantities of matter and antimatter should be produced, but somehow (and somewhen) the universe became matter dominated. There are no antimatter stars or antimatter planets or asteroid (as far as astronomical observations can tell). But, for all this, there may yet be antimatter meteoroids. The problem with respect to antimatter meteoroids is that of survival, since matter–antimatter interactions release tremendous amounts of energy. Indeed, the interaction releases enough energy to make their destruction very obvious at gamma ray wavelengths — and no such events have been detected. If two equal mass particles of matter and antimatter interact, then the energy released will be $E = 2mc^2$, where m is the particle mass and c is the speed of light. If just one gram of matter/antimatter undergoes annihilated then the energy released will be of order 1.8×10^{14} joules, which is about twice the energy yield of the first atomic bomb, Trinity.

Turning these same ideas to a meteoroid, the rate of annihilation mass loss of a spherical antimatter meteoroid of radius r at height h in the atmosphere is given by the equation

$$\frac{dm}{dt} = -\frac{\sigma_{an}}{\sigma_{el}}\pi r^2 \rho(h) V \qquad (11.9)$$

where m is the meteoroid mass, $\rho(h)$ is the atmospheric density at height h, and V is the meteoroid velocity. The multiplier on the right-hand side of Equation (11.9) gives the ratio of the annihilation cross-section of an atom by an antiatom σ_{an} to the elastic collision cross-section σ_{el}. The latter term essentially allows for the fact that the atom and antiatom need not collide head on in order to interact. It is generally assumed that $\sigma_{an} = \sigma_{el}$, but this has yet to be demonstrated experimentally, and accordingly, Philip Papaelias and Alexandros Apostolakis (both at the National University in Athens, Greece) have suggested that if $\sigma_{an} < \sigma_{el}$ then the interaction of an antimatter meteoroid with the Earth's atmosphere could be distinguished from that of an ordinary matter meteoroid of the same initial mass [16]. The observational test, if $\sigma_{an} < \sigma_{el}$, would be to look at the mass–height variation, with the antimatter meteoroid being more massive than the same initial mass (and same initial velocity) ordinary matter meteoroid as a function of height. This, of course, would be a very difficult comparison to make, but presumably not impossible.

For all this, however, it is generally understood that there are no primordial antimatter meteoroids in the solar system (or the universe) at the present epoch. This being said, a number of non-standard particle physics models do allow for the continuous formation of antimatter and even antimatter nuggets. Needless to say, perhaps, these alternative theories are highly speculative and currently completely unproven. That there is no observation evidence to suggest that antimatter meteoroids exist, might simply mean that there are no such entities as antimatter meteoroids — full stop. Absence of evidence is not evidence of absence, however, and a role for this contraterrene matter may yet play out. Indeed, among the strangest of the many strange explanations[12] for the Tunguska event of 1908 (see Section 21.2) is that given by Philip Wyatt (Florida State University) who suggested in a short letter to the journal *Nature* on April 26th, 1958 that the great Siberian explosion might have resulted from the destruction of an antimatter meteoroid;[13] this idea was further developed in some detail by Ken Bullough (University of Sheffield) in a 1995 publication. Given the estimates of the energy involved in the Tunguska explosion (some 15 megatons of TNT equivalent energy), then just $1/3^{rd}$ of a kilogram of antimatter would need to be annihilated in the atmosphere.[14]

---

[12] Among suggested explanations for the Tunguska event is that by A. Jackson and M. Ryan (both at the University of Texas) who suggested in *Nature* **245**, 88–89 (1973) that it was caused by the passage of a small black hole through the Earth. This suggestion was made at the time when Stephen Hawking first suggested that such objects might exist. Author Bill De Smedt (2004) invokes this same idea (with a twist) in his sci-fi, mystery, thriller called *Singularity*. There is currently no observational evidence to indicate that small-mass black holes actually exist.

[13] Author Isaac Asimov adopted this idea in a 1989 short story entitled *The Mad Scientist*. In this case self-aggrandized physicist professor Martinus Augustus Dander (even his initials are MAD) explains the Tunguska explosion in terms of a Russian matter–antimatter experiment gone badly wrong.

[14] It seems worth noting that what is generally credited as being the first science-fiction story to feature antimatter is that by Jack Williamson (written under the pseudonym Will Stewart) in *Astounding Science Fiction* for July 1942. The term antimatter had not actually been coined at that stage, and the term contraterrene matter (or seetee for the verbal contraction of CT standing for contra and terrene) was used. Williamson's story, called *Collision Orbit*, is set in the main asteroid belt region of the solar system, and it is taken that some asteroids are made of antimatter, and that there are antimatter meteoroid streams. Interestingly Williamson described the use of antimatter to provide an explosive thrust reaction to change the orbit of an asteroid. Some 59 years later in life, Williamson wrote *Terraforming Earth* (published in 2001), a novel which sees life on Earth destroyed by an asteroid impact, with the only human survivors being those cloned and nurtured by robots on a lunar base in Tycho crater. These lunar survivors/revivors eventually initiate the Herculean task of terraforming the asteroid-blasted Earth, making it fit for human habitation one more time. This science-fiction novel has resonance with the present-day idea of making humans a multi-planet species (in order to guard against the extinction of humans on Earth), and with the ideas of colonizing and terraforming Mars, along with geoengineering Earth in order to combat climate change.

References

[1] P. Jenniskens, M. H. Shaddad, and S. P. Warden. "The impact and recovery of asteroid 2008 TC3." *Nature*, **458**, 485–488 (2009).

[2] J. Borovika and P. Spurny. "The Carancas meteorite impact — encounter with a monolithic meteoroid." *Astronomy and Astrophysics*, **495**, L1–L4 (2008).

[3] G. Collins, H. J. Melosh, and R. A. Marcus. "Earth Impact Effects Program: A web-based computer program for calculating the regional environmental consequences of a meteoroid impact on Earth." *Meteoritics and Space Science*, **40**, 817–849 (2005). See also the interactive webpage: https://www.purdue.edu/impactearth.

[4] R. S. Kofman, C. D. K. Herd, and D. G. Froese. "The Whitecourt meteorite impact craters, Alberta, Canada." *Meteoritics and Space Science*, **45**, 1429–1445 (2011). See also: C. D. K. Herd *et al.* "Anatomy of a young impact event in central Alberta, Canada: Prospects for the missing Holocene impact record." *Geology*, **36**, 955–958 (2008).

[5] H. J. Melosh and C. S. Collins. "Meteor crater formed by low-velocity impact." *Nature*, **434**, 157 (2005).

[6] J. Brittain. "Iridium at the K/T boundary — the impact strikes back." *Astronomy and Geophysics*, **38**(4), 19–21 (1997). See also: A. Hildebrand. "The Cretaceous/Tertiary boundary impact (or the dinosaurs didn't have a chance)." *Journal of the Royal Astronomical Society of Canada*, **87**(2), 77–118 (1993).

[7] R. P. Turco, O. B. Toon, T. P. Ackerman, J. B. Pollack, and C. Sagan. "Nuclear Winter: Global consequences of multiple nuclear explosions." *Science*, **222**, 1283–1292 (1983).

[8] R. P. Toon, R. P. Turco, and C. Covey. "Environmental perturbations caused by the impacts of asteroids and comets." *Reviews in Geophysics*, **35**, 41–78 (1997), Fig. 9.

[9] P. J. Rasch *et al.* "An overview of geoengineering of climate using stratospheric sulphate aerosols." *Philosophical Transactions of the Royal Society A*, **366**, 4007–4037 (2008).

[10] P. Spargo. "The history of the Hoba meteorite — part 1: Nature and discovery." *Monthly Notices of the Astronomical Society of Southern Africa*, **67**, 85–94 (2008).

[11] M. Beech. "Towards an understanding of the fall circumstances of the Hoba meteorite." *Earth, Moon and Planets*, **111**, 15–30 (2013).

[12] H. J. Melosh. *Planetary Surface Processes* (Cambridge University Press, Cambridge, 2011).

[13] W. G. Jones *et al.* "Searches for fractional electric charge in meteorite samples." *Zeitschrift fuer Physik C — Particles and Fields*, **43**, 349–355 (1989).

[14] P. Kim *et al.* "Search for fractional charged particles in meteorite material." *Physical Review Letters*, **99**, 161804 (2007).

[15] S. Baum *et al.* "Searching for dark matter with paleo-detectors." https://
arxiv.org/abs/1806.05991. Accessed March 2020.

[16] P. Papaelias and A. Apostolakis. "Mass–height relation for antimatter
meteors." *Earth, Moon and Planets*, **49**, 1–13 (1990). See also: K. Bullough.
"Interactions of antimatter with the atmosphere." *Journal of Atmospheric
and Terrestrial Physics*, **57**, 1533–1551 (1995).

Chapter 12

The Impactors

In order for an object to hit the Earth, be it a meteorite, asteroid, or comet, its orbit at some point must cut through the ecliptic (the plane of the Earth's orbit) at a heliocentric distance of 1 AU. With reference to Figure 4.2 and Equation (4.2) this requires that

$$r = 1\,\mathrm{AU} = \frac{a\left(1 - e^2\right)}{(1 - e\cos\omega)} \tag{12.1}$$

and this condition depends upon the values of the specific object's semi-major axis a, its orbital eccentricity e, and argument of perihelion ω. If for a specific object its orbital values of (a, e, ω) satisfy condition (12.1) then sooner or later an Earth impact will occur — presupposing, that is, that the orbital parameter set (a, e, ω) never changes. In reality, of course, this latter condition is never satisfied, since a close approach to the Earth, say, during one specific encounter will result in a gravitational perturbation of the orbital elements, and the new set of orbital elements need not satisfy condition (12.1). It is for this reason that predicting a future impact with the Earth is a far from easy task. For all this, however, that the Earth will be hit by meteorite-producing meteoroids, asteroids, and comets in the future is a certainty — unfortunately, we just don't know where and when. The main corpus of objects of interest with respect to impacts are the near-Earth objects (NEOs) — a group of objects comprised of both asteroids and cometary nuclei.

12.1 Near-Earth asteroids

Various subgroups of near-Earth asteroids (NEAs) are recognized and are arranged according to the extent of perihelion and aphelion

279

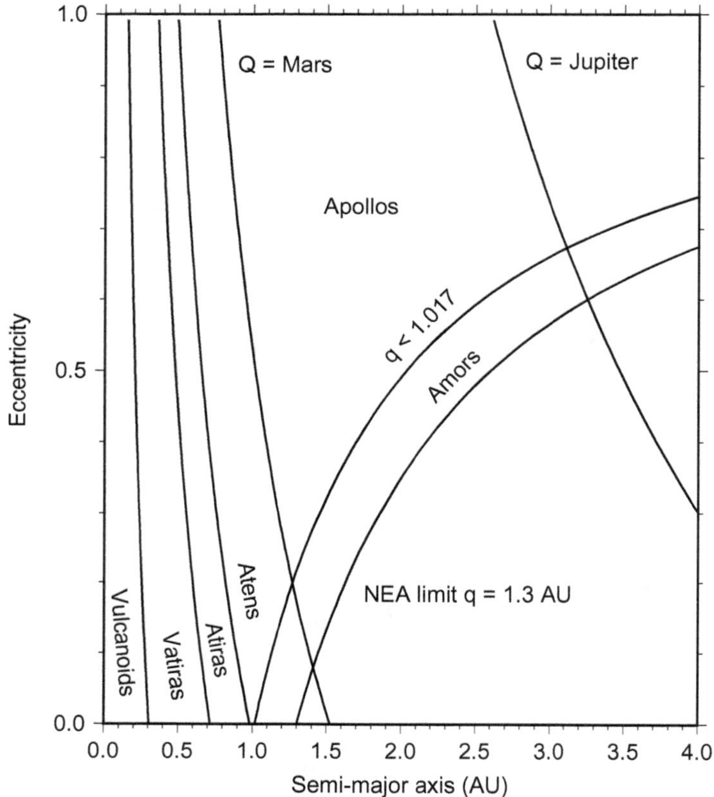

Figure 12.1. Asteroid classification domains in the eccentricity e versus semi-major axis a diagram. Near-Earth asteroids are defined as those asteroids with perihelia $q = a(1-e) < 1.3$ AU. Curves delineating the aphelia $Q = a(1+e)$ limits for Mars and Jupiter are also shown, along with those aphelion limits defining the Atens, Atiras, Vatiras, and Vulcanoid asteroid groups.

distances. In general, NEAs are those asteroids that have perihelion distances smaller than $q = 1.3$ AU — that is, at their closest point to the Sun they are interior to the orbit of Mars, but at this limit they do not necessarily cross Earth's orbit (Figure 12.1). Asteroids with $q < 1.017$ AU having semi-major axis $a > 1$ AU are called Apollo asteroids, after the classic example in the group, asteroid 1862 Apollo. Asteroids with $1.017 < q$ (AU) < 1.3 are called Amor asteroids (after the largest known example in this group, 1221 Amor).

Asteroids with a semi-major axis $a < 1\,$AU and aphelion distances $Q > 0.983\,$AU are called Atens (after asteroid 2062 Aten, the first such object found in the group). These three groups of asteroids and those comets that cross the Earth's orbit constitute the NEOs. Additional asteroid groups are recognized, however, according to their aphelion distance. Those objects that have $Q < 0.983\,$AU are called the Atiras, and these objects orbit entirely inside of the Earth's orbit (Earth's perihelion distance is 0.983 AU); those objects that orbit entirely within the orbit of Venus are called Vatiras, and those that orbit interior to the orbit of Mercury are called Vulcanoids. At present there are about 20 known Atiras, just one Vatira (2020 AV2), and no Vulcanoid asteroids.[1] Atens, Atiras, Vatiras, and Vulcanoids are the most difficult asteroids to look for, however, since, from the Earth, they will always be located close to the Sun, and this is a region generally inaccessible to survey telescopes.

The first NEA to be discovered was 433 Eros (Figure 12.2) in 1898, moving along what is now classified as an Amor-type orbit. When at its perihelion, Eros can approach within 0.13 AU of the Earth, and this is sufficiently close that its parallax can be measured directly from Earth. A worldwide program to study the parallax of Eros was launched in 1900, and this resulted in the definitive determination of the astronomical unit when measured in kilometers — indeed, this value held strong until the late 1950s, at which time radar ranging of the planets Mercury and Venus enabled the determination of a more accurate measure. Eros was the first asteroid to be studied in detail by a dedicated, orbiting spacecraft

---

[1]It is not entirely clear whether any Vulcanoid asteroids should be expected to exist at the present time — but the odds are against them. That there was a primordial population of such objects seems quite likely, since the region between 0.1 and 0.2 AU provides a highly stable orbital domain in which few gravitational perturbations and/or resonances are at play. Being so close to the Sun, however, results in rapid orbital decay, and then evaporation by the Sun, due to the Poynting–Robertson effect and Yarkovsky drift (see Section 13.3). Collisions between Vulcanoid asteroids would also have been common and because of their high orbital velocities relatively destructive. For a review of possibilities concerning the present-day existence of Vulcanoid asteroids, see Reference 1.

Figure 12.2.   First-found NEA 433 Eros as imaged at a distance of 200 km by the NEAR-Shoemaker spacecraft on February 29th, 2000. Image courtesy of NASA.

(The Near-Earth Asteroid Rendezvous — Shoemaker[2] spacecraft), and the first asteroid upon which a spacecraft successfully landed (the NEAR-Shoemaker spacecraft at the end of its mission). The second NEA to be discovered was 719 Albert (another Amor-type) in 1911, although at that time its orbit was only poorly determined and it remained lost until rediscovery in 2000. Currently (and the numbers increase almost daily) over 20,000 NEOs have now been detected, of which about 100 are comets, and nearly 2000 have been classified as potentially hazardous asteroids (PHAs). Of the asteroids

---

[2]The Shoemaker appellation was added after the launch of the spacecraft, and honors geologist Eugene Shoemaker who tragically died in a car accident while studying impact craters in Australia in 1997.

detected, some 12,400 are Apollo-type, 1700 Aten-type, and 8200 are Amor-type.[3]

The orbital dynamics exhibited by NEAs have been studied in great detail, and it is clear that they are relatively short-lived objects. With lifetimes of order a few millions of years, it is accordingly clear that there must be a reservoir of such objects that is capable of steadily replenishing those NEAs that are lost. The loss mechanisms include ejection into interstellar space through gravitational scattering during a close encounter with a planet (especially Jupiter); collision, thermal, and/or tidal disruption by the Sun; and finally by hitting a planet. The reservoir that feeds into the NEA population is the main-belt asteroid region situated between Mars and Jupiter. Specifically it is those objects in the main-belt region that chance to drift into one of the chaotic resonance zones (delineated by the Kirkwood gaps — recall Chapter 3), as a result of Yarkovsky drift (see Section 13.3) or a past, random collision, that feed into the NEA population — these objects slowly acquiring an increasingly large orbital eccentricity and a concomitant smaller (that is, closer to the Sun) perihelion distance. As the perihelion distance decreases so terrestrial-planet-crossing orbits begin to come about. Once the perihelion distance is smaller than 1 AU, of course, then an Earth impact is possible.

12.2 Numbers, structure, and composition

The first of what have become known as the asteroids (technically now, small solar system bodies), 1 Ceres, was discovered on January 1st, 1801 by Giuseppe Piazzi working from the Palermo Astronomical Observatory in Sicily. It was an inevitable discovery and many astronomers, from across Europe, were actively involved in the search for what was then an expected planet situated between Mars and

---

[3]This data is from the Center for Near Earth Object Studies (CNEOS): https://cneos.jpl.nasa.gov/stats/.

Jupiter.[4] In a letter written some three weeks after his discovery Piazzi commented that, "I have announced this star as a comet, but since it is not accompanied by any nebulosity and, further, since its movement is so slow and rather uniform, it has occurred to me several times that it might be something better than a comet." Piazzi no doubt hoped that he had discovered a planet, but within relatively short order three more such objects, orbiting between Mars and Jupiter, were discovered. William Herschel suggested that the name asteroid be applied to this new group of objects, since they are star-like in appearance. The observational distinction between asteroids and comets is still, to this day, based upon whether a tail or coma is seen. A comet will support these latter features through the sublimation of its surface ices.

In situ, spacecraft observations of asteroids reveal a whole range of profiles, from near spherical to highly elongated (Figure 12.2), and highly fragmented. These observations further betray the fact that

---

[4]The Celestial Police were formed by Baron Franz Xavier von Zach with the specific aim of finding the long-sought-after planet between Mars and Jupiter. That a planet *should* exist in this region was conjectured by Johannes Kepler in the early 17th century, but the real stimulus to up the search was provided by William Herschel's discovery of planet Uranus in 1781. The key aspect of this planetary discovery was that its deduced orbital radius was very close to that indicated by the Titius–Bode rule. This mathematical rule has a long and often tortuous history all of its own, but it builds upon a simple numerical formula that gives the sequential orbital radii of the planets (in astronomical units) as: $a(n) = 0.4 + 0.3 \times 2^n$, where the power-law index varies as $n = -\infty, 0, 1, 2, 3, 4$, and so on. The start of the sequence with $n = -\infty$ looks odd, but it gives the orbital radius of mercury as 0.4 AU; the orbit of Venus corresponds to $n = 0$, Earth to $n = 1$, Mars to $n = 2$, and Jupiter $n = 4$. The discovery of Uranus was consistent with the prediction for $n = 6$, and accordingly the apparent gap at $n = 3$, giving $a = 2.8\,\text{AU}$, began to look increasingly suspicious. Suspicious in the sense that a planet had been missed. When Ceres was found by Piazzi, its orbital radius was found to be 2.6 AU, and accordingly it was initially thought to be the fabled $n = 3$ *gap* planet. The rapid discovery of three more asteroids, however, soon indicated that the situation was much more complex than the Titius–Bode rule allowed for. It is still not universally agreed upon that the Titius–Bode rule is some truly remarkable and fundamental law, or just a formula that provides a lucky sequence of numbers that chances to match that of the orbital radii of the planets.

NEAs (and asteroids in general) have a complex, ongoing collision-dominated history. Indeed, the asteroids have been interacting with each other, colliding and chipping each other away since the solar system formed 4.5 billion years ago. Within one century of Piazzi's discovery of 1 Ceres, that is, by January 1901, some 302 additional asteroids had been discovered; by January 2001 this number had risen to 26,000, and by January 2020 the total number of asteroids with known and cataloged orbits has climbed to over 54,500 objects. Ceres, the first-found, remains the largest asteroid with a diameter of some 934 km and is large enough to have remained spherical since its initial formation — this technically placing it within the dwarf planet category of objects. With a deduced mass of 9.4×10^{20} kg, Ceres constitutes about 30% of the entire mass of the main-belt asteroid region, which contains an estimated 3×10^{21} kg of material.

Observations clearly show that there are many more smaller asteroids than larger ones in the main-belt asteroid region, and that the variation can be reasonably well approximated by a power law in the radius R, with the number of asteroids $N(R)\, dR$ in the size range R to $R + dR$ being described by

$$N(R) = N_0 \left(\frac{R}{R_0}\right)^{-q} \tag{12.2}$$

where N_0 and R_0 are constants and where q is the size distribution index. The observations indicate that $q \approx 3.5$, and this is a telling result, since this value corresponds to that expected of a collisionally interacting population of objects. This derived value of q also indicates that most of the mass of the asteroid belt must reside in just a few of its largest bodies, while most of the surface area resides in the smaller objects.[5]

---

[5] Given a size distribution function $N(R) = C_1 R^{-q}$, where C_1 is a constant, the mass distribution function will vary as $N(M) = C_2 M^{-p}$, where C_2 is a constant and $q = 3p - 2$. This follows on the basis that the mass varies as the cube of the radius and that the density is constant: accordingly, $M = kR^3$, where $k = 4\pi\rho/3$ is a constant, and $dM = 3kR^2 dR$. Accordingly, given that $q = 3.5$, so $p = 1.8$. The distribution function for the surface area will vary as $N(\Sigma) = C_3 R^{1-q}$, where C_3 is a constant.

The irregular, shard-like shapes of asteroids and their often highly cratered surfaces are clearly indicative of their being a collisionally evolved population, and indeed, it is the population of the smaller fragments, produced during collisions, that feeds into the flux of meteoroids capable of producing meteorites on Earth. The surface of asteroid 101955 Bennu is literally littered with shattered boulders and fragments that have been no doubt produced during past collisions (Figure 12.3). Future collisions between Bennu will result in numerous surface fragments being ejected into interplanetary space, along with sub-surface material — and these will feed into the smaller population of objects and potential meteorite-producing meteoroids. About 15% of NEAs are observed to have accompanying moons, and these bodies are presumably either spun-off material that has coalesced into a larger object, or the collisional ejection of material that has remained in orbit about the impacted asteroid — the moon literally being a chip off the old block.

Figure 12.3.   Surface detail of asteroid 101955 Bennu as imaged by the OSIRIS-ReX spacecraft from 3.6 km above its surface. The large rock in the upper left-hand corner is about 15 meters across — it was a similar sized meteoroid that produced the Chelyabinsk fireball and meteorites. Image courtesy of NASA.

In addition to sporting rubble pile surfaces, it is also thought that asteroids, in general, have a rubble pile interior. That is, asteroids (above a certain size) are not solid monolithic blocks: rather they are composed of numerous building blocks, with large amounts of smaller material infill, all held together by their mutual gravitational attraction and electrostatic forces. The smallest possible rotating rubble pile system is that composed of just two components. If these two components are assumed to be of equal size, with radius R, then the closest that they can orbit about each other, their surfaces just touching, is at a separation of $2R$. Given such a configuration, it is a straightforward matter to determine the minimum orbital rotation period, and this will correspond to the maximum spin rate — if the system spins any faster then it will fly apart.

To determine the spin limit we need to make use of Kepler's 3rd law (recall Chapter 3). The mass of each lobe in our system, assuming that they have the same density, will be $M = (4\pi\rho/3)R^3$, and accordingly, Kepler's 3rd law gives the orbital period P as:

$$P^2 = \frac{4\pi^2}{G}\left(\frac{a^3}{M_1 + M_2}\right) \tag{12.3}$$

where the constant k has been substituted for in terms of the constant $4\pi^2/G$ in Equation (12.3), where G is the gravitational constant, and the sum of the masses involved in the system M_1 and M_2. Given equal size, equal mass components, just touching spheres $a = 2R$, Equation (12.3) reduces to:

$$P^2 = \frac{4\pi^2}{G}\left(\frac{8R^3}{2 \times 4\pi R^3 \div 3}\right) = \frac{12\pi}{G\rho} \tag{12.4}$$

Under the assumptions made, the actual size of the components cancels out and the minimal orbital period reduces to an expression related to the density only, with P (hr) $= 208.83\rho^{-1/2}$. Taking a characteristic density for a stony material of $4000\,\text{kg/m}^3$, so the shortest orbital period of a two-component spinning rubble pile is $P = 3.3$ hours.

The observed values of asteroid spin rates is shown in Figure 12.4 and the data indicates that for objects larger than about 500 meters

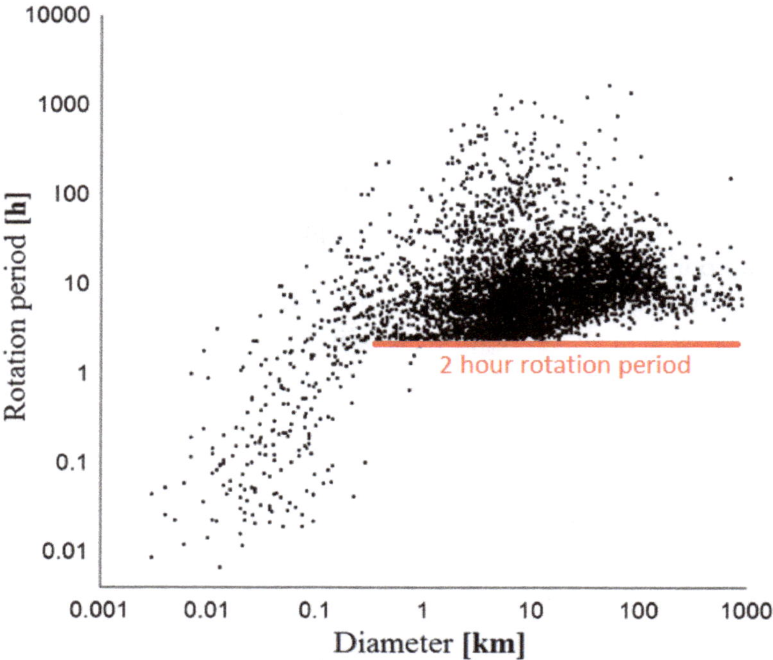

Figure 12.4.   Measured rotation rates (in hours) for some 6000 asteroids with sizes ranging from a few meters to that of Ceres (the largest asteroid at 934 km across). Note the distinct cut-off limit at about 2 hours to the rotation periods of large asteroids. Only objects smaller than abut 500 m across rotate more rapidly than this cut-off rate. Image adapted from Isabella E. Johansson *et al.*, *Proc. SPIE*, **9226**, 922607 (2014).

across there is a spin-rate limit at about 2.2 hours (just a little smaller than the 3.5-hour idealized limit just determined). This observation of a spin limit is taken to indicate that all large asteroids (larger than 500 m or so) are rubble piles, and only asteroids smaller than about 500 m across are monolithic, which allows them to spin more rapidly, indeed, up to the limit of its material strength.

That asteroids are rubble piles helps explain the oddly low values deduced for their bulk density. The bulk density is derived from the ratio of the asteroid's mass and volume. The former of these quantities can be determined by measuring the deflection of

Figure 12.5. Asteroid 253 Mathilde as imaged by the NEAR-Shoemaker spacecraft in June 1997. The 20-km diameter crater Damodar is seen here in partial shodow. Image courtesy of NASA.

a spacecraft's path during a fly by encounter,[6] and the latter can be estimated from the shape model derived from the observed light curve (see Chapter 17). Inevitably, when mass and volume data are available, the deduced densities are lower than would be expected if asteroids were made of solid stone and/or iron components. On its way to 433 Eros, for example, the NEAR-Shoemaker spacecraft encountered asteroid 253 Mathilde (Figure 12.5) — flight-path deviations revealed a mass of 1.033×10^{17} kg while images of the asteroid were used to deduce a volume of 7.8×10^{13} m^3. These numbers imply a bulk density of 1324 kg/m^3, a value just a little above that of water ice and entirely inconsistent with the fact that Mathilde is a C-type asteroid (see next section). Indeed, given the

---

[6]The mass can be determined via Kepler's 3$^{\rm rd}$ law if the spacecraft is placed into orbit about the asteroid.

C-type asteroids are thought to be the parent bodies of carbonaceous chondrite meteorites, which have a laboratory measured density of order $3300\,\text{kg/m}^3$, the apparent density is about half that which might be expected. This result is explained, however, if the interior of Mathilde, and asteroids in general, contain many void (low density) spaces. The void volume fraction can be expressed in terms of the porosity P, where P = volume of void space/total volume = $1 - \rho_{\text{bulk}}/\rho_{\text{material}}$. For Mathilde, taking the spacecraft-observation-derived bulk density and comparing this to the laboratory material density of carbonaceous chondrite meteorites, it appears that $P(\text{Mathilde}) = 0.6$. This indicates that some 60 percent of the interior of Mathilde is essentially void space — making it a loosely assembled rubble pile.

This low porosity helps explain one of the mysteries of Mathilde — that mystery being that it actually still exists. Indeed, given the presence of a massive 20-km wide impact crater (officially named Damodar) it would be expected, if Mathilde was a solid body, that such an impact would have resulted in its complete disruption. The fact that the impact did not destroy Mathilde is explained by the high porosity, with the void spaces between its interior rubble-pile components allowing for the energy of the impact to be cushioned and safely absorbed without being transmitted throughout the entire body. Fittingly, perhaps, it was Eugene Shoemaker (just three weeks before his tragic death) who first suggested that Mathilde had survived the massive impact because it was composed of a loose aggregation of rocky material. Similarly, high porosity values have been deduced for other asteroids beside Mathilde; for example, asteroid 433 Eros has a porosity of about 0.3, asteroid 101955 Bennu has a porosity of about 0.4, while asteroid 162173 Ryugu has a porosity of about 0.5. Large asteroids, such as 1 Ceres and 4 Vesta, have relatively low porosities, the deduced values both coming in with $P < 0.1$, indicative of the fact that they have been able to withstand the numerous impacts that they must have suffered since they first formed.

12.3 Asteroid taxonomy

The asteroids display a diverse range of surface composition —
some asteroids have surfaces rich in carbon compounds and hydrated
minerals, other surfaces are entirely composed of refractory silicates
and/or metals. This diversity of surface composition is classified
according to the observed reflectance characteristics in the optical
and infrared parts of the electromagnetic spectrum (see Table 12.1).
The largest class of asteroids (accounting for some 40% of the
total number) are the dark, low-albedo, C-type. The spectra of
such asteroids (see column 2 in Table 12.1) is flat for wavelengths
longward of 0.4 μm and is similar to that expected for carbon-rich
material. There are absorption bands in the ultraviolet part of the
spectrum shown by C-type asteroids, which indicates the presence
of water in the form of hydrated silicates. The next major group
of objects are the S-type or stony asteroids. These asteroids tend
to be fairly bright, with albedos between 0.1 and 0.22 (see column
4 of Table 12.1). The spectrum of S-type asteroids shows a strong
absorption feature in those wavelengths shorter than 0.7 μm, and this
is thought to indicate the presence of iron oxides. Absorption features
near 1 and 2 μm are suggestive of assemblages of iron and magnesium
bearing silicates mixed with metallic nickel-iron. The D and P-type
asteroids account for about 5 to 10% of the total number of asteroids,
and these show a distinct linear infrared spectrum increasing, being
brightest at longer (red) wavelengths. This spectrum is explained by
the presence of organic compounds. It is generally thought that the D
and P-type asteroids are the most primitive, that is least chemically
altered since formation, of all the asteroids — even the C-types.
The M-type asteroids are generally bright, with albedos between
0.1 and 0.18, with relatively featureless spectra and are thought
to be predominantly composed of nickel-iron. The V-type asteroids
have basaltic surfaces and spectra very similar to that of the HED
meteorites (see Chapter 13).

It is a curious fact that no large asteroid group shows spectral
characteristics that are a perfect match to the ordinary chondrite

Table 12.1.   The asteroid taxonomic system. The taxonomic types (column 1) are based upon the characteristics of the reflected sunlight in the infrared part of the spectrum (column 2). The main spectral features are listed in column 3, and the range of reflectance characteristics (the albedo) is given in column 4. Column 5 indicates the compositional properties. Table from J. Bodnarik *et al.*, IEEE Nuclear Science Symposium Conference Record (2011). DOI:10.1109/NSSMIC.2011.6154374.

Major Taxonomic Types	Reflectance Spectrum (0.4–0.9 um)	Spectral Features	Visible Albedo	Suspected Composition
D (D, T)		Relatively featureless spectrum Steep red slope	0.02–0.06	Primitive carbonaceous Organic-rich compounds Hydrated minerals

C (C, B, F, G)		Slight bluish to slight reddish slope
		Shallow to deep absorption
		   blueward of 0.5 μm
		Hydrated asteroids with absorption
		   at 0.7 and 3.0μm
			0.03–0.10	Hydrated minerals
				Silicates
				Organics

X (E, M, P)		Slightly reddish spectrum E: absorption features at 0.5 and 0.6 μm	E: 0.18–0.40 M: 0.10–0.18 P: 0.03–0.10	E: Enstatite-rich M: metallic, Nickel-Iron P: Carbonaceous, Organics

S (S, Q, A, K, L)		Moderately steep red slope
		   ($\lambda < 0.7\mu$m)
		Shallow to deep absorption
		   at 1.0 and 2.0 μm
			0.10–0.22	Stony composition
				Magnesium Iron silicates

V		Moderate to steep red slope ($\lambda < 0.7\mu$m) Very deep absorption at 1.0μm	0.20–0.60	Volcanic basalts Plutonic rocks

meteorites — these meteorites being the most commonly found on Earth. The small group of Q-type asteroids affords the best spectrum match to ordinary chondrites, but it is generally agreed that the abundant S-type asteroids are the real parent bodies. The spectral differences between these asteroids and ordinary chondrites is a consequence of space weathering, which is driven by cosmic ray impacts, collisions with other asteroids, solar-wind sputtering and implantation, and irradiation.

Plotting the relative number of asteroid types against heliocentric distance reveals the existence of distinct zones of dominance (Figure 12.6). The high albedo E-type asteroids are exclusively found at the inner edge of the main-belt asteroid region, and while the S-type asteroids tend to be more numerous in the inner regions of the main belt, the C-type dominate in the outer regions. The D and P-type asteroids are exclusively found in the extreme outer part of the main belt and among the Trojan asteroids that are co-orbital with Jupiter (recall Chapter 3). This distribution of asteroid types is essentially primordial, the distribution being qualitatively consistent with the temperature structure expected in the solar nebula (recall Chapter 5), with the more primitive asteroid groups

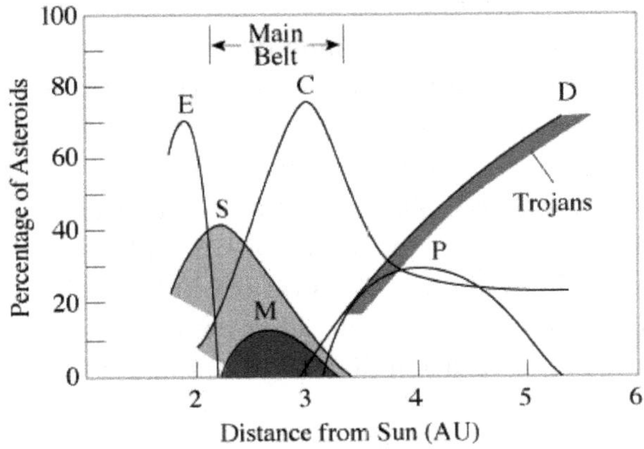

Figure 12.6. The relative abundance of asteroid types across the main-belt asteroid region to the orbit of Jupiter. Image courtesy of *NASA's Cosmos*: https://ase.tufts.edu/cosmos/.

being located further from the Sun. For all this, however, it is also clear that some considerable dynamical processing must have additionally taken place to explain the full distribution — especially the Trojan asteroid group. This latter group is potentially explained by the Nice model (recall Chapter 5), but there are still many issues that have as yet to be solved.

12.4   The comets

A small fraction, perhaps of order a few percent, of the NEOs are cometary nuclei. These bodies are chemically distinct from asteroids, and they are evolved from a different set of solar system reservoirs. Structurally, cometary nuclei are described under the dirty-snowball model introduced by American astronomer Fred Whipple in 1950. This model invokes water ice as the main binding constituent of cometary nuclei, and embedded within the ice are refractory grains and minerals. In this model, cometary activity is controlled by sublimation outgassing of the ice as a consequence of solar heating. This outgassing drags the previously embedded dust grains away from the nucleus, feeding them into an associated meteoroid stream.

Appropriately, the first cometary nucleus to be directly imaged was that of 1P/Halley (Figure 12.7). Such images, along now with images of various other nuclei (recall Figure 2.5), confirm Whipple's basic model, but there are still many details concerning the internal structure that need to be resolved. One of the startling results from the *Giotto* mission to Halley's comet was the realization that only a small fraction of its surface was undergoing active outgassing — no more than about 10%. This observation has been confirmed with respect to other active comets, and indicates that repeated perihelion passages and sublimation mass loss results in the build up of a dark, insulating, inactive rubble mantle. That cometary nuclei are able to develop a dark, outer crust indicates that some NEAs may well be old, inactive cometary nuclei.[7]

---

[7]One such example is 3200 Phaethon, which is sometimes described as an active Apollo asteroid. This object was the first asteroid to be discovered through spacecraft observations (the IRAS spacecraft in 1983), and is the parent body to the annual Geminid meteor shower.

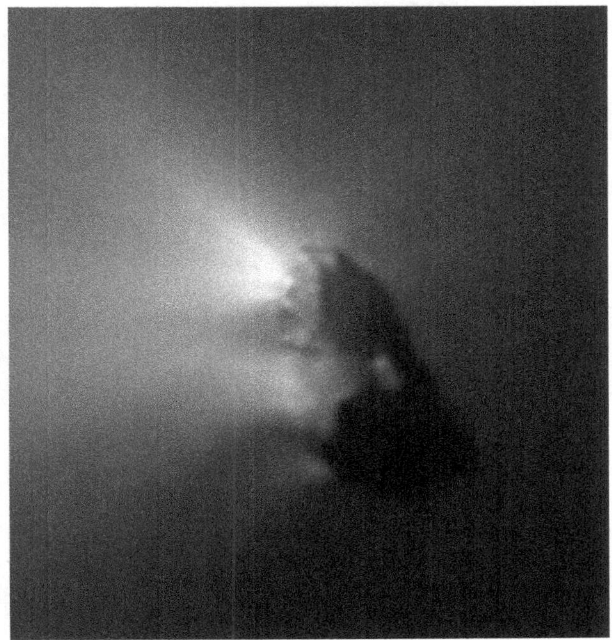

Figure 12.7. The nucleus of comet 1P/Halley as revealed by the *Giotto* spacecraft in 1986. The nucleus is about 15 km long and 8 km wide. Active sublimation jets are visible on the Sun-facing side of the nucleus. Image courtesy of ESA.

As with asteroids, the deduced density of cometary nuclei is invariably lower than expected. For Halley's comet, the derived mass and volume were 2.2×10^{14} kg and 3.7×10^{11} m^3, giving a bulk density of some 600 kg/m^3. Given water ice has a density of 917 kg/m^3, the deduced bulk density suggests a porosity of $P \approx 0.35$ for Halley's comet. This again suggests that cometary nuclei are essentially rubble-pile-like structures assembled of numerous subnuclei.

Active comets have a relatively short lifetime of just a few millions of years, with loss mechanisms including planetary impacts, thermal and tidal disruption by the Sun, as well as ejection from the solar system via gravitational scattering. Given the short lifetimes of active comets, there must exist reservoirs of cometary nuclei within the solar system, with nuclei leaking from these regions to replace the nuclei being lost (Figure 12.8). For comets, two reservoirs are

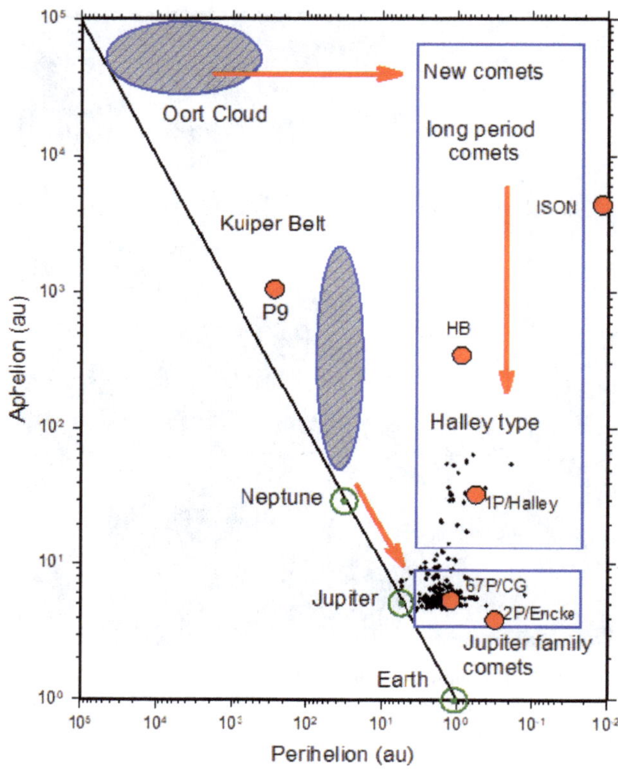

Figure 12.8. Scale of the solar system and cometary reservoirs. The small black dots correspond the perihelion/aphelion points of well-studied comets. Cometary nuclei perturbed from the Oort cloud enter the inner solar system on highly elliptical orbits as long-period comets. If such a long-period comet is gravitationally perturbed by a close interaction with one of the Jovian planets, then it will possibly become a Halley-type comet. In contrast, the Jupiter-family comets are derived from the Kuiper belt. In this case, fragments produced between collisions of Kuiper-belt objects are gravitationally perturbed and steered inwards towards the Sun by interactions with the outer planets to an eventual capture by Jupiter. **Labeled objects key**: short-period comets 1P/Halley, 2P/Encke, and 67P/Churyumov-Gerasimenko; long-period comets HB = C/1995 O1 (Hale–Bopp) and C/2012 S1 (ISON); P9 indicates the estimated perihelion and aphelion distances of hypothetical planet nine.

known, the Oort cloud, as described by Dutch astronomer Jan Oort in 1951, and the Kuiper belt, as described by Gerald Kuiper[8] again in 1951. The outer boundary of the Oort cloud, at about 100,000 AU from the Sun, essentially defines the limit of the solar system — that is the assembly of those objects influenced by the Sun's gravitational influence. While often depicted as a sphere, its boundary is dynamic and controlled by the presence of other stars in the solar neighbourhood. Indeed, it is the passage of near-by stars that gravitationally perturb cometary nuclei from the Oort cloud reservoir into the inner solar system — these will constitute the long-period comets.[9] Since the Oort cloud is a sphere-like halo about the solar system, one of the defining characteristics of long-period comets is that they can approach the Sun at any angle to the ecliptic, along highly elliptical orbits. Short-period, or Jupiter-family comets, in contrast, move along orbits confined towards the ecliptic plane. The source region for these comets is the Kuiper belt, with the icy fragments produced during the collisions of Kuiper-belt objects, evolving via gravitational interactions with the outer planets to orbits that are dynamically controlled by Jupiter — these then constitute the Jupiter family of comets, which typically have orbital periods less than 20 years.

12.5 Impact risk assessment

Our everyday lives and activities are all predicated upon risk assessment. When we cross the road, for example, we intuitively know

---

[8]Kuiper did not actually predict the existence of the outer solar system disk of icy bodies that is named after him. The disk is also known as the Edgeworth–Kuiper disk (or belt) in recognition of the ideas published by Irish astronomer Kenneth Edgeworth in 1943. The first Kuiper-belt object (or trans-Neptunian object, TNO) was discovered by Jane Luu and David Jewett in 1992.

[9]Long-period comets have orbital periods greater than 200 years, but there is technically no physical meaning behind this number. The main distinction between long and short-period comets is their source region. A third cometary group — the Halley family — is also recognized, and this group is composed of those long-period (Oort cloud derived) objects that have been captured into shorter period orbits by gravitational interactions with the outer (massive) solar system planets. These cometary families are, in fact, best distinguished according to their Tisserand parameter — recall Equation (8.1).

it is a good idea to look to see what traffic might be coming. We could just walk across the road, and we might not be hit by a car, but it is far better (that is, safer and potentially far less painful) to make some assessment of the traffic situation. The impact risk from NEOs is based upon similar factors to those that we might apply when crossing a road — that is, we want to know what the probability of a specific object hitting us is. This probability will be based upon the number of cars on the road and the distance across the road. Likewise, we would also be interested in knowing something about the traffic and how big the various cars are — is it a road being crossed by small Smart cars or is it a highway for massive Ford trucks. Additionally, the speed of the cars is of interest — we might survive an impact with a small car traveling at a low speed, but certainly not an impact from a large car travelling at great speed. In terms of NEO impacts, rather than the risk to an individual person being of interest, it is the regional and/or global effects that are of interest, and this is generally taken to mean those objects hitting the Earth with energies in excess of 1 megaton of TNT equivalent energy.

The primary properties of interest are the probability of impact $0 \leq P_{\mathrm{imp}} \leq 1$, and the size of the impactor D. Clearly an object with a very low probability of impact is not an object to worry about, and it accordingly carries a low risk. In contrast an object with a high or certain probability of impact is an object of interest. Indeed, once $P_{\mathrm{imp}} \sim 1$ the question becomes how big is the object and how fast is it moving at the time of Earth encounter. As described in Section 7.5, the larger the impactor so the bigger the resultant crater on Earth, and likewise, the faster an object is moving, so the larger the impact crater for a given size — these two quantities, of course, determine the kinetic energy of the impact $E = \frac{1}{2}MV^2$, and the mass M is determined by the size of the impactor and its composition.

An impact probability is calculated according to how well the orbit of a specific Earth-approaching object is known, and this will vary according to how long the object has been monitored. An object that has only been observed over just a short arc of its orbit, will have less certain orbital parameters (that is there will be observational uncertainty errors) than an object that has been observed over many orbits. Likewise, an object that passes close to the Earth (or another planet) on one specific orbit will undergo significant gravitational perturbations that will change its future impact probability. Accordingly, impact probabilities are typically applied over some specified time interval.

The background probability that the Earth might be hit by, say, a long-period comet can be readily determined since these objects have orbit inclinations that are randomly distributed around the sky. This means that, in general, long-period comets pass through a sphere of radius 1 AU about the Sun at any angle. The Earth has a cross-section area $\pi R_\oplus^2$, where $R_\oplus$ is the Earth's radius, on the surface of the sphere of radius 1 AU, and the impact probability for a long-period comet P_{LP}, will accordingly be of order the ratio of the Earth's cross-section area divided by the surface area of a sphere of radius 1 AU, with

$$P_{\mathrm{LP}} = 2\,\frac{\pi\,R_{\mathrm{Earth}}^2}{4\,\pi\,(1\,AU)^2} = 9.07 \times 10^{-10} \qquad (12.5)$$

The factor of 2 in the calculation for P_{LP} is due to the fact that the comet will pass through the sphere of radius 1 AU twice: once before perihelion and once after perihelion. Observations of long-period comets indicate that there is approximately one new such object passing perihelion per year (giving a flux $f_{\mathrm{LP}} = 1\,\mathrm{yr}^{-1}$), so the characteristic time interval between the Earth being hit by a long-period comet will be $T_{\mathrm{LP}} \sim 1/f_{\mathrm{LP}}P_{\mathrm{LP}} \sim 1$ billion years.

In 2002 Steven Cheseley and coworkers at the Jet Propulsion Laboratory [2] introduced what is called the Palermo Scale P to assess impact probabilities.[10] The scale is defined according to the calculated impact probability P_{imp}, the time over which the impact

---

[10]The Palermo attribute was given in recognition of the fundamental research on asteroids carried out at the Palermo Observatory in Italy. It was also the location of a conference in 2001 at which Chesley and coworkers presented their ideas. Giuseppe Piazzi was additionally working from the Palermo Observatory when he discovered the first asteroid 1 Ceres.

is possible T, and the background impact frequency f, with

$$P = \log_{10} \left[\frac{P_{\text{imp}}}{f\,T} \right] \qquad (12.6)$$

where the background frequency of impacts is determined by the approximation formula

$$f\,(\text{yr}^{-1}) = \frac{3}{10}\,E^{-4.5} \qquad (12.7)$$

where the energy E is expressed in megatons of TNT. The manner in which the scale is calculated means that a value of $P = 0$ indicates that the hazard is equivalent to that of the natural background. A value of $P = 2$, in contrast, would indicate a 100 times greater risk than a random background impact. Negative values for P are possible, and objects with $P < -2$ are considered to pose little impact risk. Objects with $-2 < P < 0$ are deemed to be of interest and requiring closer study.

Only a very few asteroids have received an initial Palermo Scale value of $P > 0$, and none of these have retained that value after further analysis — subsequent positional analysis and orbital determination, in each case, leading to an improved, but reduced, value for the impact probability. As of writing, the object with the highest Palermo Scale value is asteroid (29075) 1950 DA, with $P = -1.42$. This 1.1-km diameter object has, as far as current observations indicate, a 1-in-8300 chance of hitting the Earth in 2880, with an estimated impact energy $E = 75{,}000$ megatons.[11] The important point at this stage is that the impact probability may well change (possible to a higher value, but much more likely to a lower one) as a better understanding of close-encounter conditions and future planetary perturbations becomes available. The orbital period of 1950 DA is 2.21 years, and accordingly it will round the Sun a further 389 times before any potential impact in 2880. Observations indicate that 1950 DA has a relatively high density, and this suggests

---

[11]The predicted impact date is (March 17th) St. Patrick's Day, 2880, which would seem like a good excuse, in that year at least, to partake of a few extra draughts of Guinness — just in case.

that it is probably an iron-rich asteroid. An Earth impact from an object like 1950 DA would certainly result in a high death count and global climate change effects.

While the Palermo Scale is based upon the outcome of an object's impact probability relative to the background frequency of terrestrial impacts, the Torino Scale[12] is more often used as a guide to impact risk. Introduced by Richard Binzel (MIT) at a conference in Turin in 1999, this scale divides the 2-D diagram of impact energy versus impact probability into a discrete set of risk domains (Figure 12.9). Here the idea is that objects with a low probability of impact and a small relative size (impact energy) pose a much lower risk than those objects with a high impact probability and a high associated energy of impact. Any object that plots within the white region of the scale (Figure 12.9) is considered to have no associated risk — essentially the probability of an impact is very small and below that of a random chance hit. Objects plotting in the green and yellow zones (numbers 1 through 4) are deemed to be objects of interest, requiring careful monitoring — essentially these objects might pass very close to the Earth at a future date, and a collision might produce regional,

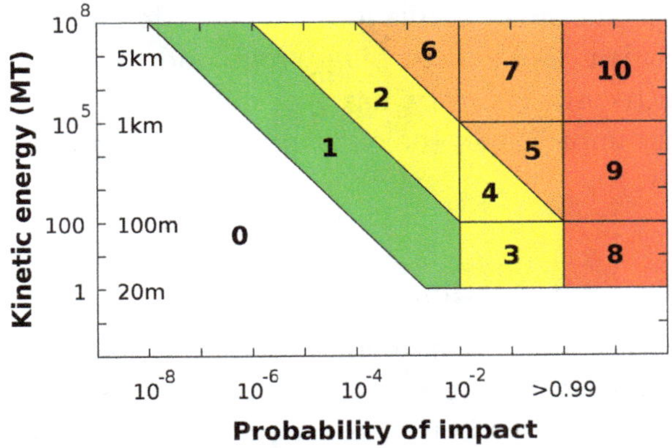

Figure 12.9. The Torino Scale. The severity of the risk (white through to red, zero to ten) is determined according to the probability of impact and the size of the impactor.

---

[12]See: https://cneos.jpl.nasa.gov/sentry/torino_scale.html (accessed June 2020).

but not global, damage (zones 3 to 4). The orange zone (numbers 5 to 7) indicates those objects for which Earth impact is highly likely, and capable of causing large-scale to global devastation. The red zone (numbers 8 to10) is reserved for those objects that will definitely hit the Earth, with effects that will vary from regional (red 8 and red 9) to global, causing climate catastrophe (red 10). If the Chicxulub impactor (recall Chapter 11), for example, had been detected prior to its impact with Earth, it would have been a red 10 object. The Chelyabinsk impactor (if its orbit had been known of prior to Earth encounter) would have rated as a white 0 object. The impactor that produced the Barringer crater (Figure 11.6) would have rated as a red 8 object — this same rating, red 8, would additionally apply to the Tunguska impactor. Writing as of 2020, there are no objects with a Torino Scale rating higher than white 0.

Various estimates have been made concerning the odds of dying as a result of an asteroid impact, and these range from of order 1 in 5000, to 1 in 100,000. Depending upon which of these odds one wishes to apply it does transpire that your chances of dying as a result of an impact are higher than dying in an earthquake, or from being struck by lightning, or a shark attack. This being said, you are more likely to die in an airplane accident (odds of 1 in 30,000) or a vehicle accident (with disturbingly high odds of 1 in 90). No one, of course, in recorded history, has ever died as a direct consequence of an asteroid impact, and the odds are calculated according to best guess models.

12.6   Impactor deflection

Is an impact the inevitable outcome of the detection of a Torino Scale red 9 or red 10 object? The answer here is, not necessarily [3]. But to overturn and reduce the impact probability away from certainty, the key element will be lead time. If, for example, a long-period comet is detected that is heading straight for an Earth impact, given what would be a typical lead time of months, then there is essentially nothing that could be done. If, on the other hand, a near-Earth asteroid is discovered with a predicted impact time set, say, at least many tens of years into the future, then there is a chance that a collision could be avoided. There are a whole number of procedures that, in principle, can be used to change the path of a potentially Earth-impacting asteroid. The options essentially range from the Hollywood-style impulsive explosion approach (see Section 25.2), in which the asteroid is broken apart into many smaller components, to the more subtle, slow-and-steady acceleration approach in which the radiation pressure of sunlight is utilized (effectively using the Yarkovsky effect — see Section 13.3). The point, however, is that all of the potentially available approaches will take years to set in place and implement — hence the importance of lead time. While the ability to change the orbit of an asteroid has not, as yet, been demonstrated, the technology required to perform the deflection is well within the realm of human engineering skill. Likewise, while the full population and orbit of ever Earth-approaching asteroid is not currently known, the technology required to complete the inventory is in place. For the first time in human, even Earth, history, the outcome of an impending collision is no longer an inevitable impact.

References

[1] M. Beech and L. Peltier. "The Vulcanoid asteroids: Past, present and future." *American Journal of Astronomy and Astrophysics*, **5**, 28–41 (2017).

[2] S. Chesley *et al.* "Quantifying the risk posed by potential Earth impacts." *Icarus*, **159**, 423–432 (2002).

[3] S. Pack *et al.* "Optimization and decision-making framework for multi-staged asteroid deflection campaigns under epistemic uncertainties." *Acta Astronautica*, **167**, 23–41 (2020).

Chapter 13

Exploring Vesta

Main-belt asteroid 4 Vesta was discovered by Wilhelm Olbers on March 29th, 1807. It was the last of the first four — a time span of 38 years would need to pass before the next asteroid, 5 Astraea, was to be discovered by Ludwig Hencke (in 1845); indeed, Hencke was looking for Vesta when he made his discovery. Vesta is the second largest and second most massive of all the asteroids, only bested by 1 Ceres. Additionally, it is also the only main-belt asteroid that intermittently achieves naked-eye brightness (if one knows where to look and when). Although the name asteroid had been introduced by William Herschel in 1802 as a descriptive name for those objects located between Mars and Jupiter, Vesta was initially thought to be a planet or perhaps a large fragment from a disrupted planet. This initial planetary status betrays a tangled history, starting with the discovery of the first asteroid 1 Ceres, by Giuseppe Piazzi on January 1st, 1801. Indeed, Ceres was initially heralded as the long-sought-after planet suspected to reside in the large orbital gap between Mars and Jupiter. The existence of this planet being predicated on the apparently "empty spot" betrayed in the Titius–Bode rule at $n = 3$. Ceres, since 2006, has been reclassified as a dwarf-planet, although the definition of what makes an object a planet, dwarf or otherwise, is far from being agreed upon. Indeed, from a geological perspective, 1 Ceres, as a differentiated body (see below) could quite reasonably be called a planet. Likewise, 4 Vesta, since it too is a differentiated body, can also quite reasonably be called a planet. Technically, however, since the new naming policy introduced by the International Astronomical Union in 2006, 4 Vesta is a small solar system body, losing out on a dwarf planet classification due to the

Figure 13.1.   Image of 4 Vesta taken by the DAWN spacecraft in 2011. The surface betrays a complex cratering, erosion, and resurfacing history. The up-lift peak located at the southern pole, making the whole asteroid look decidedly non-spherical, rises some 23 km above the floor of the Rheasilvia impact crater (see below). Image courtesy of NASA.

fact that it is not spherical. Indeed, the south polar regions of 4 Vesta are dominated by two massive impact basins and a mountain-like up-lift feature (Figure 13.1).

Asteroid 4 Vesta orbits the Sun along a low eccentric ($e = 0.08874$) orbit with a semi-major axis of 2.362 AU, making it orbit closer to the Sun than 1 Ceres. It takes 3.6 years to complete one circuit about the Sun, spins once on its axis ever 5.342 hours, and has an average diameter of 525 km. The essential profile and shape exhibited by 4 Vesta is that of a gravitationally relaxed oblate spheroid. That is, it has the sphere-like shape expected of a rotating object held together by its own self-gravity. The exception to this basic shape rule, however, being the large concavity in the southern polar region (more on this later). It is this southern polar concavity that denies a dwarf planet status for 4 Vesta — this and the fact that it has not cleared the region of its orbit of smaller objects. Indeed,

Vesta orbits the Sun with a whole cohort of Vestoid asteroids — these followers literally being chips of the old block (see below).

Detailed physical models of 4 Vesta based upon its size, mass, rotation rate, and composition indicate that it is a differentiated body. That is, its interior has been rearranged from its primordial largely uniform structure to one divided according to density. The densest material has settled out and migrated towards the center, forming a metallic, nickel-iron core about 200 km across (Figure 13.2). Above the core sits a rocky mantle region, about 80 km thick. The mantle is predominantly composed of intermediate-density olivine. Finally, on top of the mantle is the outer, crustal layer, again about 80 km deep. The outmost, regolith layer of 4 Vesta shows the scars leftover from numerous impacts (Figure 13.1)

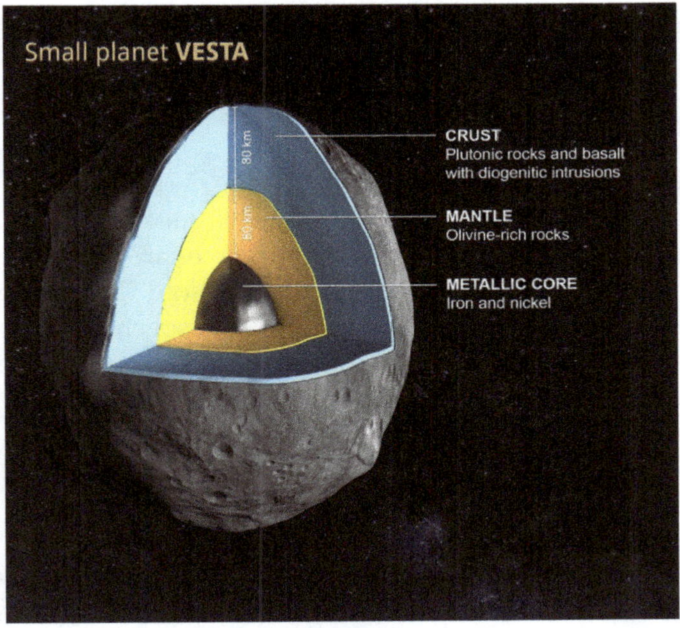

Figure 13.2. The internal structure of Vesta. Internal heating by the decay of [26]Al has resulted in a differentiated, core-mantle-crust configuration in which the densest material is situated in the core, and the lowest density material is in the upper mantle and crust. Image courtesy of NASA.

and a composition consistent with that due to substantial basaltic resurfacing.

If internal differentiation and basaltic resurfacing is going to occur, then an object requires an internal heat source, and accordingly, the question arises, where did 4 Vesta derive its internal heat from? This heat source arises as a direct consequence of the formation process, one of accretion and then radioactive decay. In the first case it is the conversion of the kinetic energy of motion of the accreting particles, acting to build up the mass of 4 Vesta, that is partially converted into internal heat, and in the second case it is the decay of the relatively short-lived radionuclide ^{26}Al.

The internal heating of 4 Vesta due to accretion can be estimated according to the conservation of energy. This conservation law dictates that the total amount of energy (in a closed system) remains constant, and when an object is growing by accretion the energy to be considered is spread between the kinetic K and potential energy U. Conservation of energy requires that

$$E_{\text{total}} = (U + K)_{\text{initial}} = (U + K)_{\text{final}} \tag{13.1}$$

The minimum accretion model imagines 4 Vesta to form by the coming together of two hemispheres each of mass $M/2$, where M is the total mass of 4 Vesta. The two hemispheres are taken as initially being a distance R_{initial} apart and when coalesced their centers are a distance R apart. To begin with, the two hemispheres that are going to coalesce to make our 4 Vesta model are taken to be at rest, and accordingly the initial kinetic energy is zero. The total initial energy is therefore

$$E_{\text{initial}} = U_{\text{initial}} = -G\frac{M}{2}\frac{M}{2}\frac{1}{R_{\text{initial}}} \tag{13.2}$$

At the accretion stage, when the two hemispheres are imagined as being coalesced to form 4 Vesta, the energy will be spread between the potential energy and the kinetic energy of impact. Accordingly,

$$E_{\text{final}} = U_{\text{final}} + K_{\text{final}} = -G\frac{1}{4}\frac{M^2}{R} + K_{\text{final}} \tag{13.3}$$

Invoking conservation of energy and substituting for Equations (13.2) and (13.3) now give the kinetic energy of impact as

$$K_{\text{final}} = G\,\frac{M^2}{4}\left(\frac{1}{R} - \frac{1}{R_{\text{initial}}}\right) \approx G\,\frac{M^2}{4R} \qquad (13.4)$$

where in the last step on the right-hand side we adopt the approximation that $R_{\text{initial}} \gg R$. The final step in this approximation is to argue that a fraction f of the accretion energy K_{final} goes into the internal heating of 4 Vesta. The temperature rise ΔT within the interior of 4 Vesta, due to the deposition of energy E_{heat} within its interior, can be estimated through the specific heat capacity C, which gives

$$C\,M\,\Delta T = E_{\text{heat}} = f\,E_{\text{final}} \qquad (13.5)$$

which upon rearrangement and substitution for the mass M in terms of the density ρ and radius R, gives

$$\Delta T = G\,\frac{\pi}{3}\,f\,\frac{\rho\,R^2}{C} \qquad (13.6)$$

In Equation (13.6) all the terms on the right-hand side are known. Taking the density of 4 Vesta to be $3456\,\text{kg/m}^3$, the radius of 4 Vesta to be 2.625×10^5 m, and the characteristic specific heat of the material to be $900\,\text{J/kg/K}$, so the accretion formation of 4 Vesta results in an internal heating of about $\Delta T = 20\,\text{K}$ (if we take the unlikely — best — situation that $f = 1$). This is clearly not enough energy to heat the interior in any substantial way, and certainly not enough to drive the differentiation process. So, if the energy of accretion is not going to enable the density sorting and metamorphism of the interior of 4 Vesta, another heat source must be found, and this is where the decay of ^{26}Al appears to be crucial.

It was seen in Chapter 4 that various radionuclides can be studied in order to determine various timescales in the life of a meteorite — its formation age from the decay of rubidium-87 into strontium-87, for example, and the meteoroid ejection timescale from the build up of argon-40 from the decay of potassium-40. Radionuclides, however, are not only clocks, they are also internal cookers. Indeed, the alpha (helium-4 nuclei) and beta (electron) particles emitted in the radioactive decay process carry kinetic energy with them, and this energy is eventually deposited in the surrounding rock, thereby raising the internal temperature. The question of importance, of course, is how much energy can a particular radionuclide supply per kilogram of material and for how long. Fortunately, such quantities can be evaluated in the laboratory, and if radioactive decay generates heat at a rate of Q joules per kilogram per second, then the energy generated per second in a freshly accreted body will be $E_{\mathrm{RD}} = QM$, where M is the total mass, and the characteristic time interval over which the energy source is active corresponds to about 2 half-life times $(T_{1/2})$ for the particular radionuclide being considered.

For uranium-235 $Q(^{235}\mathrm{U}) = 4.6 \times 10^{-11}$ J/kg/s, and $T_{1/2} = 704$ million years; for potassium-40, $Q(^{40}\mathrm{K}) = 3.7 \times 10^{-11}$ J/kg/s, and $T_{1/2} = 1.25$ billion years. So, while it is seen that radionuclides such as $^{235}\mathrm{U}$ and $^{40}\mathrm{K}$ do not produce much energy per kilogram of material, it is the case that they can produce this energy for a very long time. The core of the Earth is still being kept warm by the decay of these radionuclides — even after 4.5 billion years. In contrast to these long-lived radionuclides, however, aluminium-26 $(^{26}\mathrm{Al})$ follows the dictum of live fast and die young. For this radionuclide we have, $Q(^{26}\mathrm{Al}) = 2.5 \times 10^{-8}$ J/kg/s, and $T_{1/2} = 730,000$ years. Incredibly, active $^{26}\mathrm{Al}$ produces over 500 times more energy per second per kilogram of material than $^{235}\mathrm{U}$, but the catch is that it only does this for perhaps a few million years. The effective heating timescale of $^{26}\mathrm{Al}$ is just $1/1000^{\mathrm{th}}$ that of $^{235}\mathrm{U}$. But, in terms of 4 Vesta, a rapid internal heating phase, due to the energy released in the decay of $^{26}\mathrm{Al}$, following its relatively cold assembly is exactly what is needed to drive the differentiation process.

The decay of aluminium-26 is a beta-plus decay, and the end stable (daughter) product is magnesium-26. That is, one of the protons in the nucleus of the parent $^{26}\mathrm{Al}$ decays into a neutron and emits a positron e^+ and an electron neutrino ν, giving: $^{26}\mathrm{Al} \Rightarrow {}^{26}\mathrm{Mg} + \mathrm{e}^+ + \nu$. Since, it is argued, that $^{26}\mathrm{Al}$ was relatively well mixed into the young, just beginning to collapse, solar nebula, the first-formed planetesimals would have contained a ready and built-in internal heating source. The energy available for heating via ^{26}Al decay per second will be $E_{\mathrm{RD}} = QM$, where M is the mass of the planetesimal. For 4 Vesta, $M = 2.59 \times 10^{20}$ kg, and accordingly E_{RD} (^{26}Al) $\approx 6.5 \times 10^{12}$ joules per second $= 6.5$ billion kilowatts. Running this heat source for one half-life time of 730,000 years provides a total energy supply of $E_{\mathrm{supplied}} = E_{\mathrm{RD}}T_{1/2} = 1.5 \times 10^{26}$ joules. As with our earlier calculation for accretion heating, the heating due to ^{26}Al supplied over, say, two half-life times will result in a temperature change ΔT of

$$E_{\mathrm{supplied}} = QM(2T_{1/2}) = MC\Delta T \qquad (13.7)$$

which evaluates to $\Delta T \approx 1300\,\mathrm{K}$. Now, this latter increase in the internal temperature is more than enough to drive the metamorphism of the interior of 4 Vesta and allow the differentiation of materials to come about.

At this stage the internal heat source that enabled 4 Vesta to undergo differentiation has been identified as ^{26}Al, and that this heat source is essentially available for perhaps 2 million years. The initial picture for the early history of 4 Vesta is that it will have formed via accretion in some 10^4 to 10^5 years, and then undergone a stage of internal heating for some 2 million years (a few half-life times of ^{26}Al), during which the differentiation and material metamorphism phase will have been set in process. As time moved on, however, the internal heat was radiated into space at Vesta's surface. The cooling timescale t_{cond} for heat loss can be estimated by dividing the surface area (which varies as the radius squared) by the thermal diffusivity K(the units for K being m^2/s). This latter term is the ratio of the thermal conductivity k divided by the thermal heat capacity per unit volume of material: $K = k/C\rho \approx 10^{-6}$ for characteristic values of k, C, and ρ. Accordingly, the cooling timescale is of order

$$t_{\mathrm{cond}} \sim \frac{R^2}{K} \qquad (13.8)$$

Inserting characteristic values into Equation (13.8) indicates a cooling-off timescale for 4 Vesta of some 2 billion years. After this time 4 Vesta is effectively a geologically dead world and all of its initial internal heat will have been radiated into space. Given this cooling-off timescale, 4 Vesta has, for the past 2.5 billion years, accumulated surface modifications only as a result of impacts from other asteroids. It is these latter impacts that have resulted in the HED meteorites and the formation of Vestoid family of asteroids.

The HED meteorites are achondrite meteorites. They are not uncommon meteorites, as such, and while they account for about 5% of all meteorite falls, they account for about 60% of all achondrite falls. The HED label combines three distinct meteorite types, with the H representing howardites, the E corresponding to eucrites and the D to diogenites. The howardites, named in honor of British chemist Edward Howard, are a regolith breccia, formed from eucrite and diogenite clast material. Furthermore, the brecciated structure indicates that they must have been formed via impact gardening of the parent asteroid's outer layers. The eucrites are basaltic in structure and usually contain calcium-poor pyroxene and calcium-rich plagioclase (anorthite). The name eucrite is derived from the Greek *eukritos*, meaning 'easily distinguished,' and indeed, the eucrites are the most commonly found of all the HED meteorites. The diogenites are also igneous in origin, but unlike the eucrites, they have plutonic characteristics, meaning that they formed slowly within magma chambers located deep below the surface crust. The crystals contained within diogenite meteorites are largely magnesium-rich orthopyroxenes, with small amounts of plagioclase and olivine crystals also being found. The name diogenites is given in honor of the Greek philosopher Diogenes of Apollonia, who first suggested in the 5th century B.C. that stones (that is meteorites) might fall from the sky (recall Section 2.1). It is the basaltic/plutonic nature of the HED meteorites that accords them a link with asteroid 4 Vesta. Indeed, 4 Vesta, along with its associated family of (Vestoid) fragments, are unique amongst the main-belt asteroids in that they show the same reflectance characteristics, in the infrared part of the electromagnetic spectrum, as the HED meteorites.

13.1   Reflectance spectroscopy

The infrared spectrum of 4 Vesta in the wavelength range from 0.5 to 2.5×10^{-6} meters is shown in Figure 13.3. This spectrum is composed of reflected sunlight and a series of superimposed absorption features. Importantly, the absorption features can be used to identify the compositional make up of the asteroid's surface material. In particular there is a deep, and relatively narrow absorption line centered at about 0.9 microns, and another, much broader absorption line feature centered at a wavelength of about 2 microns. Independent measurements of the infrared reflectance characteristics of various rocks and minerals within the laboratory have been used to identify these two absorption features with pyroxene — and this, as seen earlier (Section 1.2), is a major component of eucrite meteorites. Figure 13.4 shows a comparison between the infrared reflectance spectrum of 4 Vesta and the laboratory derived reflectance spectra of sample HED meteorites. The key, double-dip characteristic of the pyroxene bands centered at wavelengths of 0.9 and 2 microns are evident in all 4 spectra, and it is this commonality that identifies 4 Vesta as the parent body to the HED meteorites.

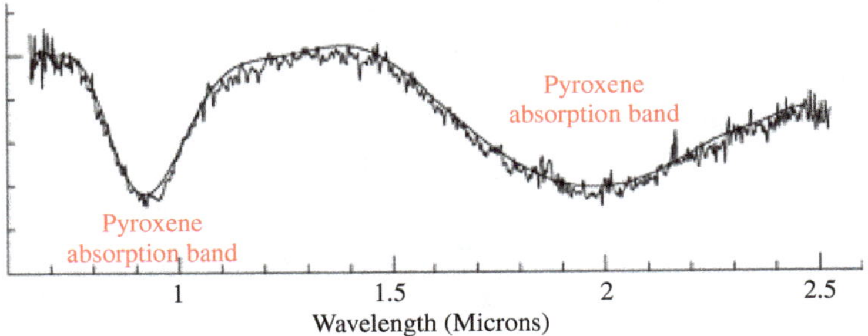

Figure 13.3.   The infrared spectrum of 4 Vesta. A strong, relatively narrow, absorption feature centered at a wavelength of 0.9 microns is indicated, along with a second, broader absorption feature centered at about 2 microns. These spectral characteristics are highly distinctive and limited to 4 Vesta, its associated Vestoid family members, and few additional objects within the main-belt asteroid region.

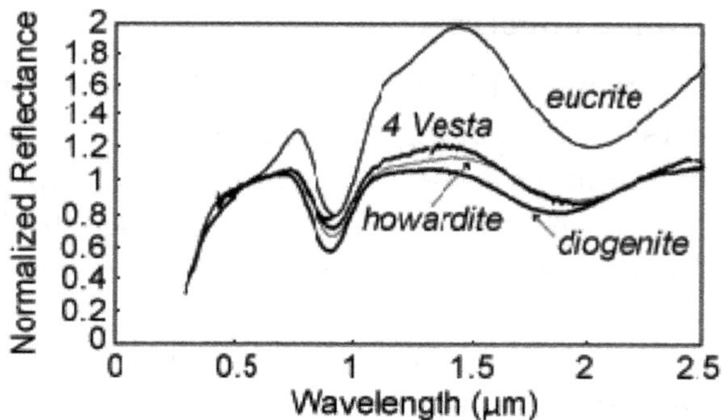

Figure 13.4. Comparison of the infrared reflectance spectra of 4 Vesta and representative samples of HED meteorites. Matching the shape and reflectance characteristics across the full spectrum provides for a broad compositional analysis.

That the reflectance spectrum of 4 Vesta is distinctly different from those produced by the other (predominantly stony, carbonaceous, and iron) asteroids is illustrated in Figure 13.5. The stony (S-type) asteroids, along with the ordinary chondrite meteorites, show a distinct absorption feature at about 1 micron, but unlike the HED meteorites no second absorption feature is seen at 2 microns. Iron (M-type) asteroids and iron/stony-iron meteorites show a relatively featureless spectrum that increases smoothly across the wavelength region from 0.5 to 2.5 microns. The spectrum of carbonaceous (C-type) asteroids and carbonaceous meteorites show flat, relatively featureless spectra. The enstatite (E-type) asteroids and the aubrite meteorites are again relatively featureless and increase in near linear manner across the wavelength region from 0.5 to 2.5 microns. While the enstatite and aubrite spectra are similar in appearance to the spectra of M-type asteroids and iron-rich meteorites, the former objects are much more reflective than the latter ones. The E-type asteroids reflecting some 50 to 60 percent of the incident sunlight, compared to about 10 percent or less being reflected by the M- and C-type asteroids — with similar sunlight reflectance percentages being exhibited by the aubrite and carbonaceous chondrite meteorites. The basic spectral forms exhibited by the various asteroid types and meteorite groups allows

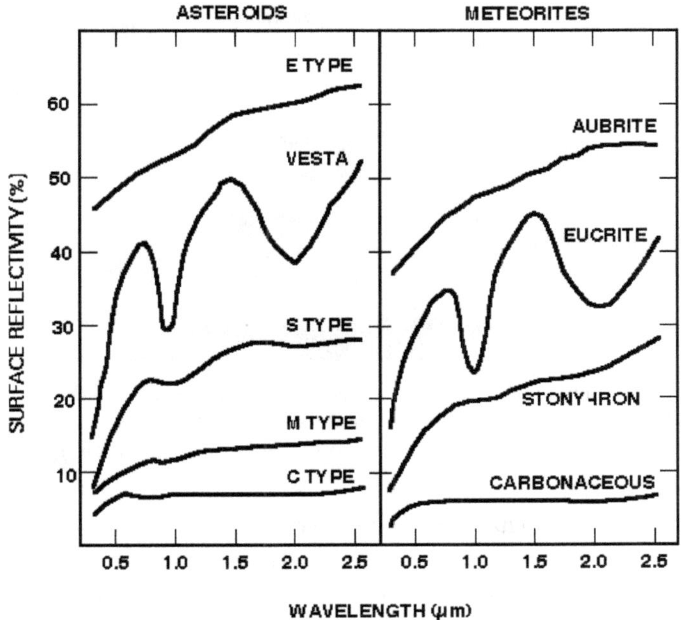

Figure 13.5. Characteristic reflectance spectra of several asteroid types and meteorite groups. Image courtesy of C. Chapman.

for potential source asteroids to be identified, but it is only the HED meteorites (with their highly distinctive reflectance spectra) that can be traced to a unique parent asteroid (that is, 4 Vesta).[1]

Given the distinct compositional characteristics of the HED meteorites, the reflectance spectra match with 4 Vesta reveals that the latter's surface must be largely different from that of the more common stony (S-type) asteroids, and that its surface must be deeply cratered (a result technically available before the 'ground-truth' images from the DAWN spacecraft — Figure 13.1 — became available). This is so, since the eucrite meteorites are representative

---

[1]This being said, there are eucrite meteorites that do not appear to be from 4 Vesta. The Bunburra Rockhole meteorite, which fell on July 21st, 2007 in Australia, for example, has a eucrite designation, but its oxygen isotope signature is significantly different to that indicated by eucrites believed to be derived from 4 Vesta.

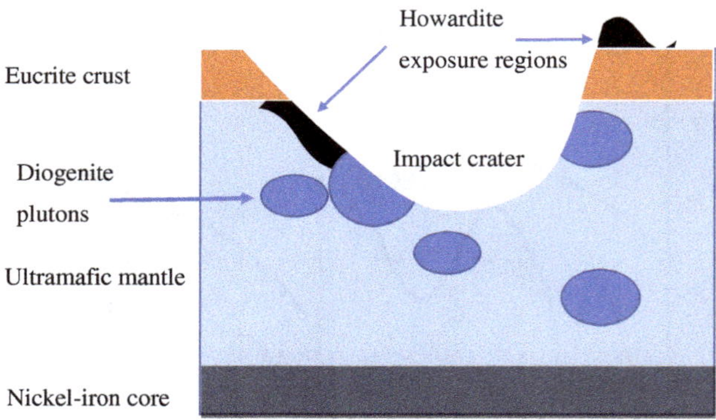

Figure 13.6.   Schematic surface stratigraphy of 4 Vesta and the HED source regions. The formation of deep impact craters is required to reveal the sub-crustal, diogenite plutons and also to produce the howardite breccia regions.

of basaltic material that has been extruded at the surface of 4 Vesta, while the diogenite meteorites indicate, in contrast, that deeper, upper-mantle, plutonic-dome regions must have become exposed. The howardite meteorites, being a brecciated mixture of eucrite and diogenite material, further indicate that prolonged and vigorous impact excavation and surface modification must have occurred on 4 Vesta. These properties of the HED meteorites allow for a basic stratigraphy of 4 Vesta to be developed, and such a stratigraphy is illustrated in Figure 13.6.

Not only can 4 Vesta be identified, through reflectance spectroscopy, as the most probable parent body to the HED meteorites, the DAWN spacecraft imaging camera has identified the most likely source location from which the meteorites are derived. This region is the south polar concavity (Figure 13.7) of Vesta. Surface morphology and crater count analysis indicates that the south polar concavity is the result of at least two massive impacts. It has been estimated that something like 1% of the volume of 4 Vesta was excavated and ejected into space during the formation of the 500-km diameter Rheasilvia basin. Furthermore, crater count analysis suggests that this impact basin formed in the relatively recent past, about 1 billion years ago. A second, massive impact basin, the 400-km diameter Veneneia structure, is also evident in the DAWN imaging data, and

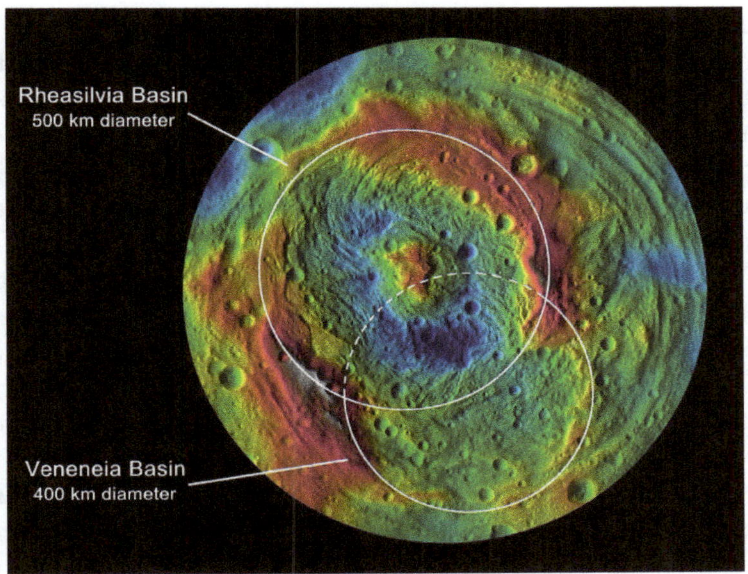

Figure 13.7. False color altitude map of the southern hemisphere of 4 Vesta. The Rheasilvia crater is outlined, along with an even older impact basin: Veneneia. Red represents the highest elevations, blue the lowest. Image courtesy of NASA.

this particular collision event is estimated to have occurred about 2 billion years ago. Prior to the occurrence of these two impact events, a time interval greater than half the age of the solar system, 4 Vesta would have had a spherical profile and accordingly would have qualified in those aeons as being a dwarf planet (along with 1 Ceres).

Combining the heating and cooling mechanisms associated with 4 Vesta, with the DAWN spacecraft observations, HED meteorite crystallization, and cosmic-ray-exposure ages, a detailed timeline for the geological evolution of 4 Vesta can outlined. Formation ages for eucrite and diogenite meteorites range from 4.56 to 4.40 billion years ago, indicating a crystallization phase about 10 million years after the formation of the CAIs at a time some $T_0 = 4.567$ billion years ago. Prior to the crystallization of eucrite material, 4 Vesta must have undergone dramatic internal evolution. Formation models for planetesimal's growth and accretion suggest that it would have taken between 1 and 2 million years (after T_0) for 4 Vesta to form.

At this time active ^{26}Al was incorporated into its interior, and over the next several half-life times (about 2 million years) the decay of this radionuclide enabled internal differentiation to take place. By about 5 million years after T_0, 4 Vesta had evolved a core-mantle structure (Figure 13.2), and over the next 5 million years the mantle would have largely crystalized to form the eucrite mantle and diogenite plutons.

From about 10 million years after T_0, 4 Vesta would have entered into a cooling-off stage. As argued for earlier, via Equation (13.8), the cooling off will not be complete until about 2 billion years after T_0. During this extended time span, 4 Vesta would have undergone continued impact alteration at its surface. Such impacts would result in thermal metamorphism and begin to expose deeper crust and upper mantle layers, allowing for the formation of the howardite breccias. Such impacts would also produce crustal melt pools that would reset the crystallization ages and allow for the formation of the younger eucrite meteorites — some eucrites, for example, have apparent formation ages set several hundred million years after T_0. About 2.5 billion years after 4 Vesta formed, it suffered a large impact that resulted in the formation of the Veneneia impact basin at its southern pole. A further billion years on, at a time some 3.5 billion years after T_0, another large impact event occurred, resulting in the formation of the Rheasilvia impact basin. It seems most likely that these massive impacts were responsible for the formation of the Vestoid family of asteroids. Measurements of cosmic-ray-exposure (CRE) ages of HED meteorites indicate distinct clustering at 22 and 39 million years ago, and these times presumably indicate two relatively recent impact and ejection epochs on 4 Vesta.

13.2 The HED conveyor belt

British writer and philosopher Bo Fowler described, in his 1998 book *Scepticism Inc.*, the journey to the top of Mt. Everest of a supermarket shopping trolley that had been fitted out with an artificial intelligence chip. When the shopping trolley was asked by a reporter how it had managed such a fantastic feat, the trolley replied "slowly." The same "slowly" can, in many ways, be applied to the HED meteorites.

At issue is what process enables them to reach Earth since 4 Vesta is not located close to any of the resonance escape hatches that have been identified in the main-belt asteroid region. Vesta essentially sits about mid-way between the 4:1 and 3:1 mean-motion resonance locations, and its orbit does not carry it close to the ν_6 secular resonance. In order to reach one or other of the meteorite escape hatches, a fragment ejected from the surface of Vesta has to undergo some considerable degree of orbital migration. How such HED-meteorite-producing meteoroids achieve this is hinted at by the existence of the Vestoids.

Indeed, the Vestoids have reflectance spectroscopy characteristics that link them to 4 Vesta, but as Figure 13.8 indicates they have an extended range in their orbital semi-major axis. Identified as ejecta from 4 Vesta (most probably associated with the formation of the Rheasilvia impact basin about 1 billion years ago), the Vestoids have evolved a range of orbital characteristics due to differential gravitational perturbations, additional collisions, and radiation pressure. Indeed, it is the latter mechanism that is thought to be particularly important, and the mechanism at play is usually described as the Yarkovsky effect — described in more detail at the end of this section. For the present it is simply stated that this effect can act to increase and/or decrease the semi-major axis of an object's orbit (the effect varies according to the orientation of the object's spin axis relative to its orbital plane — see below). The Yarkovsky effect depends upon the physical size of the object in question, with the timescale of variation in the semi-major axis being slower the larger the object is. Likewise, the slower the object spins, so the slower is the change in the orbital semi-major axis. The orbital change is also related to the solar flux $(L/4\pi r^2)$, which decreases with increasing heliocentric distance r, as well as the actual shape of the object.

While there are many parameters that go into describing the Yarkovsky effect, it has, nonetheless, been detected at play in the orbital evolution of many main-belt asteroids. For the HED meteorites, it would appear that the Yarkovsky effect is the key

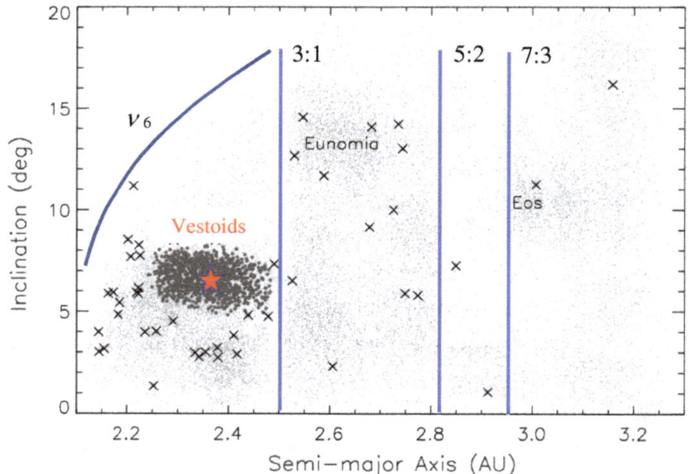

Figure 13.8.   Grey dots = Vestoids, crosses = basaltic asteroids, small dots = main-belt asteroids. Vesta is the large red star. The 3:1, 5:2, and 7:3 mean-motion resonance locations and the approximate locus of the ν_6 secular resonance are shown in the diagram, and these will act as the important escape hatches for material leaking outward and inward from the Vestoid family. The families associated with 15 Eunomia and 221 Eos are also shown in the diagram and are additional source locations for basaltic meteorites. Image adapted from Moskovitz *et al.*, "The distribution of basaltic asteroids in the main belt," *Icarus*, **198** (2008).

component in their orbital evolution, and that it directs them, ultimately, to either the 3:1 or ν_6 resonance locations, at which juncture they can be brought into Earth-crossing orbits. The HED conveyor belt appears to be an efficient way of getting material from 4 Vesta to the Earth — indeed, based upon the CRE age, some HED meteorites were only launched into space just 3 to 5 million years ago.[2]

13.3   Spin and sunlight

It seems remarkable, the stuff of moonshine and nonsense, but it is possible to move a massive asteroid by the power of sunlight.

---

[2] Alternatively, this could be the time at which a large Vestoid parent body suffered a catastrophic break-up event, thereby allowing otherwise (interior) shielded material to be exposed to cosmic rays.

The process takes time, tens to hundreds of thousands of years, but slow and steady as it goes, the continuous heating of an asteroid's surface by sunlight and the reradiation of this heat at its surface can change its orbit. Depending upon the rate at which an asteroid is spinning, what it is made of (stone or iron), and the sense of the spin (prograde or retrograde), the sunlight effects can either shrink or expand the orbit. Three physical processes are at play, the Poynting–Robertson (PR) effect, the Yarkovsky effect, and the YORP effect, and the key element in each case is how the asteroid reradiates back into space the heat energy it has absorbed from the Sun. The PR effect causes small centimeter-sized meteoroids to spiral inwards towards the Sun; the Yarkovsky effect changes the orbital semi-major axis of meter to kilometer-sized objects, and comes about due to uneven surface heating. The YORP effect is a process that causes meteoroid spin up and is due to shape and/or albedo asymmetry.

The Poynting–Robertson (PR) effect, first described by John Poynting and Howard Robinson in the early 1900s, always acts to reduce the size of a meteoroid's orbit — that is it works to reduce the orbital semi-major axis over time, causing the meteoroid to slowly spiral inwards towards the Sun. The PR effect is a relativistic correction that comes about through the way in which a meteoroid is heated by the Sun's radiation and then reradiates its heat energy back into space. Specifically, a meteoroid in a heliocentric orbit that reradiates solar energy isotopically in its own frame of reference, emits more momentum in the forward direction as seen in the Sun's reference frame. This comes about because the frequency of the photons emitted in the forward direction are relativistically increased or blueshifted and those emitted in the backwards direction are relativistically reduced or redshifted. In effect, the direction from which the Sun's radiation appears to emanate, according to the meteoroid, is shifted slightly ahead of its direction of motion, and this results in a retardation effect, decreasing the size of the orbit (Figure 13.9).

To order of magnitude, the retardation force due to the PR effect can be written as $F_{PR} = \beta(V/c)F_{\mathrm{grav}}$, where V is the velocity of the meteoroid, c is the speed of light, F_{grav} is the gravitational force

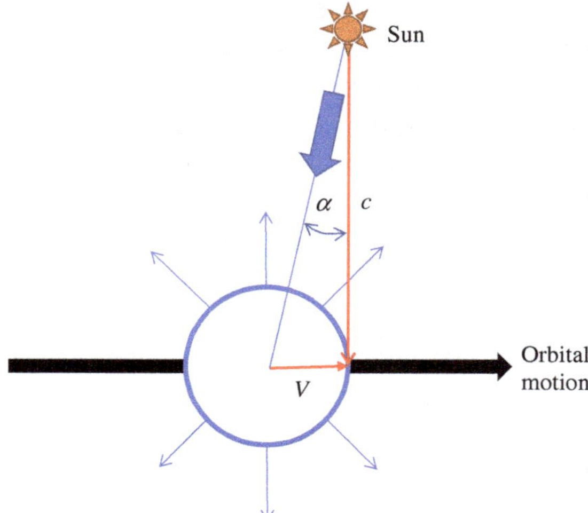

Figure 13.9.   Illustration for the origin of the Poynting–Robertson effect. Due to the motion of the meteoroid, with speed V, a solar aberration effect comes about, with the angle of aberration being given by $\tan \alpha = V/c$, where c is the speed of light. The thin blue arrows indicate the isotropic reradiation of solar energy by the meteoroid, while the large blue arrow indicates the apparent direction of solar radiation in the reference frame of the meteoroid.

acting upon the meteoroid, and $\beta = F_{\mathrm{rad}}/F_{\mathrm{grav}}$, where F_{rad} is the radiation pressure force. Clearly, F_{PR} is going to be characteristically small since the orbital velocity V is typically going to be much smaller than the speed of light — but, remember, the process operates continuously over long intervals of time. Now, the momentum of a photon is given as $P = E/c$, where E is its energy, and the force due to a photon is the time differential of the momentum: $F_{\mathrm{rad}} = dP/dt$, and accordingly, $F_{\mathrm{rad}} = (dE/dt)AQ/c$, where A is the surface area over which photons are being absorbed by a meteoroid, Q is the efficiency with which photons are absorbed, and dE/dt is the energy flux. For the Sun, the energy flux at heliocentric distance R is $L/(4\pi R^2)$, where L is the Sun's luminosity. We have therefore that

$$F_{\mathrm{rad}} = \frac{L}{4\pi 2}\frac{AQ}{c} \tag{13.9}$$

With the radiation pressure term now described, the value of $0 \leq \beta \leq 1$ is accordingly determined for a given sized meteoroid. The spiral-in timescale required for the PR effect to collapse an orbit is to order of magnitude $T_{PR} = P/[\beta(V/c)]$. Normalizing to a meteoroid at Earth's orbit, where $P = 1$ year and $V = 30\,\mathrm{km}/s$, the characteristic PR orbital decay time, for an initial orbital radius R (expressed in AU), will be

$$T_{PR} \text{ (years)} \approx \frac{400}{\beta} R^2 \qquad (13.10)$$

The orbit decay time indicated by Equation (13.10) is rapid — certainly very rapid compared to the age of the solar system, and it typically amounts to a few tens of thousands of years. Since the value of β decreases as the inverse of the meteoroid's radius, the timescale for orbital decay increases with meteoroid size, and this limits the PR effect to smaller meteoroids. For asteroids, the PR effect is extremely small so additional processes are required to change their orbits, and this is where the Yarkovsky effect comes into play.

The Yarkovsky effect, named after Polish engineer Ivan Yarkovsky, takes into account the fact that all asteroids are spinning, and this results in the region of maximum surface temperature being offset with respect to the direction of the Sun. The amount of offset will depend upon the thermal properties at the asteroid's surface along with its spin rate. The faster an asteroid spins so the more evenly will its surface be heated, and the smaller will the Yarkovsky effect be. The displacement effect is illustrated in Figure 13.10. Incoming radiation is always received by the asteroid from the direction of the Sun (taken in the diagram to be the local noon location), but since the asteroid is spinning the region where its surface attains it highest temperature (and where the greatest amount of energy is reradiated into space) is offset from the noon direction, and this results in a recoil force that either increases (in the case of prograde spin) or decreases (in the case of retrograde spin) the size of the orbit. What we have called the Yarkovsky effect is generally known as the diurnal Yarkovsky effect. There is an additional and analogous seasonal Yarkovsky effect which is

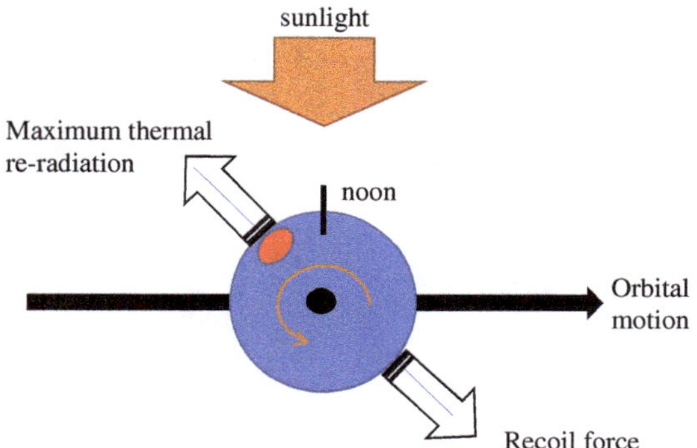

Figure 13.10.   Illustration for the origin of the Yarkovsky effect. In this case the asteroid (blue circle) rotates in a prograde fashion and the hottest region on the asteroid's surface (where thermal reemission will be at its maximum — shown as the red spot) is situated in the post-noon direction, producing a recoil effect in the post-midnight direction. The recoil force (as illustrated) will tend to increase the size of the asteroid's orbit.

produced by the temperature differences that preferentially arises on the alternate spring/summer and autumn/winter hemispheres of an asteroid in its orbit about the Sun. This seasonal effect always acts to reduce the size of an asteroid's orbit.

The rate at which the orbital semi-major axis changes due to the Yarkovsky effect varies according to the distance the asteroid is from the Sun as well as its spin period and spin direction. Even though the recoil force from the Yarkovsky effect is small, it has been measured for a good number of asteroids. The semi-major axis of asteroids 433 Eros and 101955 Bennu, for example, are decreasing at a rate of 0.6 and 12.4 astronomical units per million years respectively at the present time. In contrast, the semi-major axis of asteroids 3361 Orpheus and 4197 Morpheus are increasing at a rate of 6.3 and 30.6 astronomical units per million years. Given these values, the timescale for a significant change in the orbit of an asteroid is of order a few hundred thousand years.

The YORP (Yarkovsky–O'keefe–Radzievskii–Paddack) effect does not as such change the orbit of an asteroid, but it is responsible for changing an asteroid's spin rate. Once again it is an effect produced by the manner in which an asteroid reradiates heat energy

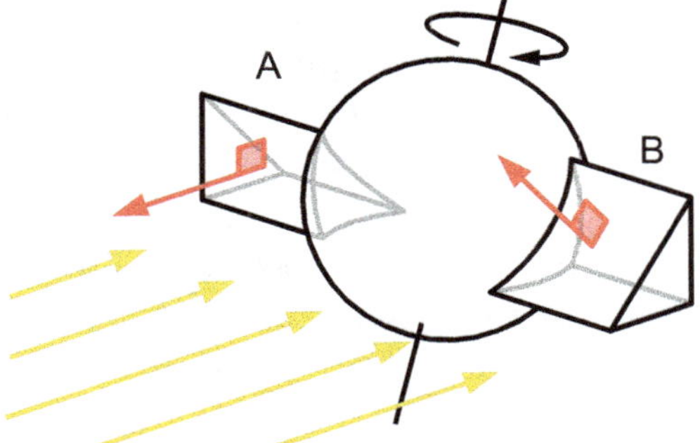

Figure 13.11. Illustration for the origin of the YORP effect. In this idealized picture a spherical asteroid has two wedge-like fins (A and B), and these reradiate energy and scatter sunlight (yellow arrows) in slightly different directions (as shown by the red arrows), resulting in a torque being applied about the spin axis. Image courtesy of Wikipedia Commons.

back into space, and it builds upon the idea that asteroids are not perfect spheres, but present lumpy, asymmetric profiles to the oncoming solar radiation. In this situation, the reradiated energy (and scattered sunlight) moves along non-parallel paths, the direction running according to where the radiation is emitted or reflected from (Figure 13.11). That the outgoing radiation is not isotropic or moving in some common direction results in a torque being developed around the asteroid's spin axis. Then, according to the actual spin sense (prograde or retrograde) and the extent of the asymmetry of the asteroid, the spin rate can either increase or decrease.

The first asteroid in which the YORP effect was measured was discovered on August $3^{\rm rd}$, 2000, and is now, appropriately, designated 54509 YORP. This asteroid, which measures some 150 by 128 by 93 meters in size, currently spins on its axis once every 12.174 minutes (or 730.44 seconds). The measured YORP effect, however, indicates that each day the asteroid rotates through an additional $d\phi/dt = 2 \times 10^{-4}$ degrees per day — after 10,000 days, therefore, it will have spun through an additional 2 degrees on its rotation axis compared to the situation if no YORP effect was in play. The change in the rotation period over a time interval T is given as $360/P = 360/P_0 - (d\phi/dt)T$, where P_0 is the spin period at the start time and $d\phi/dt$ is the additional angular rotation in degrees per day, every day. The time interval T required for the spin period of 54509 YORP to decrease by 1 second (that is $P = 729.44$ seconds) is accordingly about 800 years. For the spin period to decrease by one-half, that is $P = 6.087$ minutes, the required time interval T is of order 583,000 years. Direct observations of asteroid 101955 Bennu further indicate that its current rotation rate of 4.3 hours is decreasing by about 1 second ever 135 years — that is, its angular spin rate is increasing by some 2.64×10^{-6} degrees per day, every day at the present time.

Chapter 14

Space Invaders on Ice

That the Antarctic ice sheets provide for the best meteorite hunting grounds seems a little extreme and a little startling. Of all the places to travel to in order to find extraterrestrial rocks — why Antarctica? The answer is straightforward (in hind sight), but relies on the fact that it is extremely cold and dry, hence meteorite weathering and decay is very slow, and it is a large land mass, indeed, Antarctica constitutes 9% of the Earth's total land area. Just as important as having a large surface (collecting) area for meteorites, Antarctica also supports a glaciological process that conveniently results in meteorite accumulation zones — these zones being recognized as blue ice fields (Figure 14.1). In Antarctica, meteorites, it turns out, are ripe for the picking, standing out from the background of snow and ice as distinct, dark rocks, protruding like so many stygian jewels set against a sea of white.

The first Antarctic meteorite was found by Francis Bickerton on December 5th, 1912 during a sledging expedition to map part of the Cape Denison coast and interior. The mapping expedition was part of the 1911–1914 Douglas Mawson[1] Australasian Antarctic Expedition, and the meteorite was a healthy, 1-kg, L5 ordinary chondrite. Now housed in the South Australian Museum in Adelaide, the Adelie Land meteorite sits as a reminder to the *heroic days*

---

[1](Sir) Douglas Mawson was an Australian geologist famed for his pioneering mapping and exploration expeditions to Antarctica. He was one of the first people to climb Mount Erebus, and to trek to the location of the South Magnetic Pole.

Figure 14.1.   A blue ice field in the Miller Mountain region of Antarctica with an exposed meteorite in the foreground. Image courtesy of Wikipedia.

of Antarctic exploration. The meteorite story only unfolds slowly[2] from 1912, with just three other finds being made over the next 50 years. Then, in 1969 something unexpected and remarkable happened. Quite by chance and unrelated to their primary program, Japanese glaciologist, in that Moon-Landing year, discovered 9 distinct meteorites in a relatively small 5×10-km area of the Yamato Mountain Range. This was unprecedented, but it took several more years before the full implications of these finds were realized.

The person to make the imaginative leap was William Cassidy (University of Pittsburgh), and indeed, Cassidy recalls in his memoirs of Antarctic exploration [1] that it was during a talk on these meteorites at the 1973 Meteoritical Society Meeting in Davos,

---

[2]This, of course, is a somewhat unfair statement since Antarctica, even to this day, is visited by very few people. Indeed, Antarctica itself was not seen by any human eye until 1820, and the first confirmed landing on the continent was not made until 1895. It was additionally not until 1911 before Amundson, and a few weeks later the ill-fated Scott expedition, first reached the South Geographic Pole.

Switzerland, that induced an *idée fixe* that meteorites are concentrated in Antarctica. Cassidy needed to be stubborn, seeing his granting proposal for exploration rejected several times, approval was granted by the US National Science Foundation in 1976. Thus, was born the Antarctic Search for Meteorites (ANSMET) program, which has run every year since (with the exception of the 1989/90 season). Cassidy recalls with unconcealed excitement [1] the remarkable start of the program, with 2 meteorites being found within the first 20 minutes of searching on the very first day. Unfortunately, after this tremendous running start, no more finds were made for 6 weeks — but that's meteorite hunting. As of 2020, ANSMET has collected over 22,000 meteorites and the grand total of officially recognized Antarctic meteorites collected by all international expeditions exceeds 44,623 specimens.[3]

Why are there so many meteorites on the blue ice fields of Antarctica. This question puzzled Cassidy and colleague Ian Whillans, and it seems that the answer is a story written in the topology of the Antarctic mountains [1, 2]. Cassidy and coworkers realized that a number of quite specific conditions for meteorite concentration needed to be in place. First of all, there must be a general sense of ice flow from the Antarctic interior outward towards the coast. Then, rather than the ice flow simply calving into the sea, and dropping its precious meteorite cargo into the ocean, there must be some geographical feature that acts as a barrier to slow and trap the ice (Figure 14.2). Once trapped, however, the ice must then undergo active ablation and not accumulate a too deep layer of snow. The snow is removed from standing regions by a strong katabatic wind that blow from the continent's interior, and continuous sunshine during the long summer days accounts for the ice ablation. In this manner ice is removed and its meteorite cargo is revealed at the surface — that surface being the dynamical equilibrium zone between the ice loss mechanisms and ice accumulation. Given these conditions a standing, blue ice field will form, and meteorites collected from the

---

[3]This number is given according to the Meteoritical Bulletin Database (April 2020).

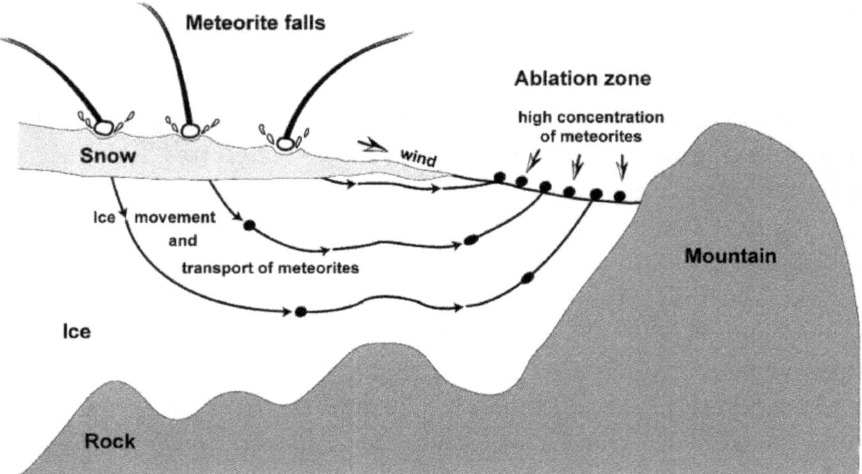

Figure 14.2.   Schematic diagram for the meteorite accumulation process within blue ice fields of Antarctica. Image courtesy of A. Bischoff [3].

vast interior of Antarctica will be continuously brought into the zone, continuing to accumulate as time goes by.

The Antarctic meteorite accumulation model outlined by Whillans and Cassidy provides a good overall picture, there are (as ever) complications that have yet to be fully understood. One issue is related to the ages of blue ice fields. Some measure of this can be gauged by the determination of meteorite terrestrial ages, and these mostly vary from 10,000 to 300,000 years, although some appear to be millions of years old. This suggests that the ice fields are no older than a few hundred thousand years, but this is a remarkably short age when compared to the 20 million years during which Antarctica has been covered in snow and ice. Indeed, Bill Cassidy has wisely pointed out that while the surface density of meteorites in the Allan Hills ice field is amongst the highest in the world, it is also remarkably low given the long history of the continent. The answer to this apparent riddle is climate variation. In this respect it may be that the interior ice fields collect meteorites during exceptionally cold epochs, and then effectively flush these meteorites out during relatively warmer times.

One additional, not fully resolved, issue with the meteorite accumulation model developed by Whillans and Cassidy is that concerning the mystery of the missing irons. Quite literally, the question is, where are they? The statistics illustrate the problem: as of January 2017, the ANSMET totals were 20,353 meteorites, of which 19,652 where chondrites and 701 were achondrites. Of the achondrites just 119 were irons, and this indicates a total percentage of irons as 0.6%. The find statistics from meteorite collected elsewhere in the world, however, indicate that irons should make up about 4.5% of the total. For some reason the iron meteorites are significantly underrepresented among the Antarctic finds. There is no reason to suppose that irons don't fall in Antarctica as often as elsewhere in the world, so there must be some mechanism preferentially affecting the transport of irons to the standing zones. The answer to the puzzle may lie within the physical properties of irons, and specifically with their significantly lower (relative to stony meteorites) thermal properties.

This idea has been studied in detail by a team of geologists, mathematicians, and atmospheric scientists at the University of Manchester in England [4]. They argue that iron meteorites located just a few centimeters below an ice surface can become sufficiently warmed by sunlight that they can melt their surrounding ice and accordingly slowly sink downwards. Indeed, the team argues that the iron meteorites sink into the ice faster than the ablation process can otherwise reveal them. In this way, the irons are still accumulating in the standing zones, but they are located below the ice surface, perhaps a few tens of centimeters down. Chondrites, being largely composed of stone, do not heat as quickly as irons, and accordingly they are revealed as the ice ablates away at the surface of the standing zone.

Well, it's a good idea, and it can be modeled on a computer, but is it really what is going on? To find out Katherine Joy (University of Manchester) led a team to Antarctica in late 2018 to search for the missing, subsurface irons with a custom-built metal detector towed behind a snowmobile. This survey, in spite of a long and distinguished history of Antarctic exploration, was the very first British expedition to search for meteorites in Antarctica. The first season of exploration turned up some 40 meteorites, but, as of yet, no clear evidence has emerged with respect to there being a significant number of iron meteorites lurking below the surface ice — but these are still early days, and the meteorite hunting season in Antarctica is both harsh and short.

14.1 Manitoba on ice

The Antarctic standing zones have been very productive in yielding up meteorites, but they are not a unique feature to the southernmost continent — similar such ice fields exist in Greenland. Not only this, researchers are finding that sometimes the search should be taken to where the ice was, rather than where it currently is. One case in point is the southeastern border region of Manitoba, Canada [5].

While Manitoba is certainly cold, wind-blown, and frozen during the winter months, some 12,000 years ago, during the last glaciation epoch, it was deep under the Laurentian ice sheet. There is no specific topology in Manitoba that would have resulted in the formation of stagnant ice fields, but over the southeast corner of the province two distinct ice sheets encountered each other at a near right-angle (Figure 14.3). One lobe was moving approximately southwestward from the region of what is now Hudson Bay, from what is known as the Keewaitin dispersal center, while the other, the so-called Koochiching Lobe, was moving southeastward. These two flows meet in the region that is close to the present-day Manitoba–Ontario border, and this has turned out to be one of the richest regions for finding meteorites in Canada. Since both ice flows would have collected meteorites over a vast area, the boundary along which they slided and collided would have acted as an accumulation zone, meteorites effectively piling up in the crush. Once the glaciers retreated, about 10,000 years ago, the meteorites were left behind awaiting discovery. And, that discovery began in 1998 under the keen eye of Canadian Derek Erstelle [6]. Not realizing at first that he had found a meteorite, the lump of iron was used as doorstop on his patio. Other meteorite finds in the same area were collected in 2002

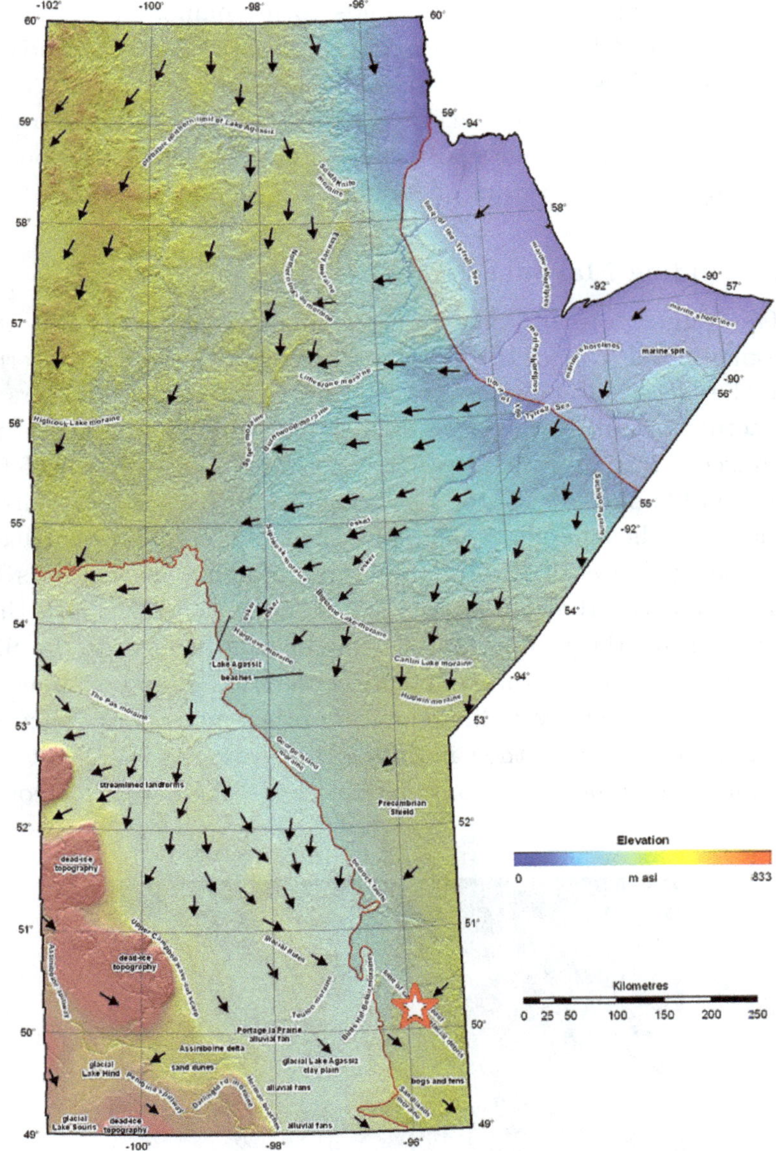

Figure 14.3. Map of Manitoba showing the direction of glacial flows (dark arrows). The approximate location of the Pinawa, Lone Island Lake, and Bernic Lake iron meteorite finds are shown by the large red star (in the southeastern corner of the province). Background map courtesy of the Provincial Government of Manitoba.

and 2005. Having three distinct finds to his name, Erstelle is the Canadian record holder for meteorite discoveries, and his finds are now recognized as the Bernic Lake, Lone Island Lake, and Pinawa meteorites. These are most likely just the first three finds from what could be an extensive lag deposit, deposited some 11 to 10 thousand years ago along the corridor of colliding ice sheets.

14.2 Finding Martians

H. G. Wells wrote of Martians invading the Earth in his 1898 dystopian novel *War of the Worlds*. His story was largely concerned with the decay of human society, but in terms of invaders he was actually writing 83 years behind the times. The first, now recognized, Martian invader was the Chassigny meteorite that fell in France on October 3rd, 1815. The second, now recognized, Martian invader was the Shergotty meteorite that fell in India on August 25th, 1865 (Figure 14.4). And the third Martian invader was the Nakhla meteorite that fell in Egypt on June 28th, 1911. These three falls constitute the prototypes of what became known as the SNC meteorites. As soon as these meteorites were found and examined it was clear that they were different from other meteorites. They did not show chondrules in their interiors, they were not irons, but they were similar in appearance to basaltic rocks — these were meteorites

Figure 14.4. A 179-g slice of the Shergotty meteorite. Image courtesy of the Smithsonian Institution.

made of molten rock that once oozed from ancient volcanoes. The big question was, of course, where in the solar system were these volcanoes located.

That the SNC group of meteorites (currently some 254 distinct finds and 5 falls have been identified) are basaltic is clear enough to the trained geological eye, but what makes them really stand out compared to other meteorites (specifically the eucrite achondrites) is their oxygen isotope ratios and their crystallization ages. Such data first became available in the 1970s to 80s. As seen in Section 1.2, oxygen has two key isotopes, and the variation in the ratios of ^{17}O and ^{18}O to ^{16}O can be used to distinguish between distinct groups of bodies within the solar system. For the SNC meteorites the oxygen isotope values plot in a group that is separate from the Earth and Moon and from all other meteorite groups. This result alone has important implications for interpreting the SNC meteorites. It means that they cannot be some strange terrestrial or Moon-derived basalt, and that they are not derived from the asteroid reservoir responsible for all other known meteorite groups.

The story became even more interesting in the late 1970s when the first crystallization ages became available, and the SNC meteorites were shockingly young, with formation ages between a few hundred million to about one billion years. This was in stark contrast to the 4.5-billion-year formation age associated with all other meteorite groups. This result informed researchers that wherever it was that SNC meteorites came from, then that object must have been volcanically active within the past one billion years. This observation rules out planet Mercury, since it has been geologically dormant for much longer than the SNC meteorite formation age. Venus is still volcanically active to this day, but successfully launching material from its surface is extremely unlikely (recall Chapter 6), and especially so since the SNC meteorites have not undergone extensive shock alteration. Of the terrestrial planets, this leaves Mars. Certainly, there is evidence to indicate that Mars has been volcanically active up until at least the last few tens of millions of years, and additionally, Mars has a low escape velocity, just 5 km/s, and a relatively thin atmosphere. As the 1970s closed and

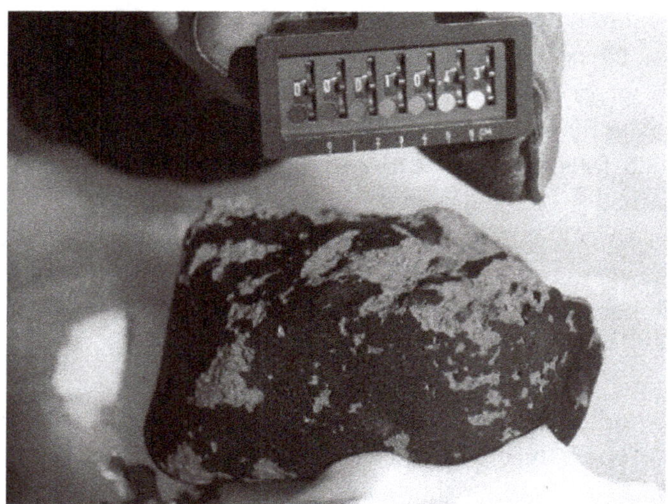

Figure 14.5. Discovery picture of the (later classified) shergottite meteorite EET79001.

the 1980s opened up, so researchers became increasingly convinced that the SNC meteorites must be from Mars, but what clinched the deal was the chemical analysis of a breakthrough ANSMET find [1].

The first SNC meteorite find of the ANSMET program, ALH77005, was collected in the Alan Hills area during the 1977/78 field season, but its importance was not recognized until a few years after it had been cataloged. During the 1979/80 field season, however, another SNC meteorite, EET79001 (Figure 14.5), was discovered by searchers exploring a region known as Elephant Moraine. It was this latter meteorite that sealed the Martian connection. Both ALH77005 and EET79001 are shergottites, but researchers at the Johnson Space Flight Center in Houston found that EET79001 had been highly shocked during the impact event that launched it from the Martian surface. The interior of the meteorite accordingly contained many small blebs of glass. As the shock wave moved through the parent Martian material it liquified small pockets of its interior, which were subsequently cooled very rapidly once the material found itself in interplanetary space, turning the melt to glass. By slowly heating separated-out glass bled from EET79001, researchers at the

Johnson Space Center in 1984 were able to measure the composition and abundances of evolved gasses, and the results were an exact match with the Martian atmosphere as measured by the *Viking* landers as they descended to the surface of Mars in 1976. The SNC meteorites were thus definitively linked to Mars. Since the discovery of ALH77005 and EET79001, an additional 29 Martian meteorites have been found (as of 2020) on the Antarctic ice fields, and the catalogued total of such meteorites collected from around the world is 259.

During the 1984 collecting season, another Martian meteorite, ALH84001, was discovered, and analysis of this object once again brought out a new result in the ever-developing SNC saga. This meteorite turned out to be distinct from the recognized SNC types, and it had a derived formation age of 4.1 billion years — it is a piece of ancient crust. Not only was ALH84001 a new type of Martian meteorite, it also solidified at a time when water flowed on Mars. This meteorite caught great media attention in 1996 when it was suggested that samples extracted from its interior contained fossilized microbes — this claim, however, has since been refuted by many, but not by all researchers (see Section 7.11).

While the crystallization ages of SNC meteorites span the range from a few hundred million to 4.1 billion years, the cosmic-ray-exposure ages are characteristically a few millions of years. ALH84001 took about 14.5 million years to find its way to Earth, while EET79001 took just 730,000 years to find its Antarctic home. Interestingly, however, there appears to be a restricted (even discrete) range of cosmic-ray-exposure ages for the SNCs, suggesting that only a small number, perhaps 10, Mars impact events are required to account for the entire Martian meteorite contingency on Earth. All of the Nakhlites appear to have been ejected from Mars, for example, some 11 million years ago, while Chassigny appears to have been launched on its way 11.6 million years ago. The majority of shergottites appear to have been launched in 4 cratering events that occurred 1.1, 2.2, 2.7, and 3.1 million years ago. Given the relatively small number of ejection events required to account for Martian meteorites, so relatively large impact craters are implicated as the launch locations. Using remote compositional sensing of the Martian surface, one 58.5-km diameter crater, called Mojave crater, has been implicated as the source region of at least some of the shergottites. A subbranch of the great scar that cuts across the equator of Mars, *Valles Marineras*, known as *Eos Chasma*, has also been tagged by Vicky Hamilton (University of Hawaii) as a possible launch region for ALH84001.

14.3 Finding the Moon

The first meteorite from the Moon was recognized less than 50 years ago, although, just like the SNC meteorites, they must have been falling on Earth for billions of years. The trick to recognizing anything new is to look at it with the right *eyes*, and in the case of the lunar meteorites it was the Apollo Moon landings, in the early 1970s, that established the right backdrop for *seeing* things as they really are. The very first lunar meteorite, Yamato 791197, was found in the Yamato Nanataks region of Antarctica by Japanese field researchers in 1979. For all this, however, and in spite of its original classification as an anorthite brecciate, it was not identified as a lunar meteorite until 1984. The honor for the first *bona fide* recognition of a lunar meteorite goes to ANSMET find ALH81005 (Figure 14.6).

From the outset, researchers knew that this find was something altogether different, and the field notes written up at the time of its discovery on January 17th, 1982 read "#1422 — strange meteorite. Thin, tan-green fusion crust, ~50%, with possible ablation features. Interior is dark grey with numerous white to grey breccia (!). Somewhat equidimensional at ~3cm" [1]. After transportation from Antarctica to Houston, researchers at the Johnson Space Center immediately saw that there was something special about ALH81005 — it was clearly a piece of the Moon. The great advantage, of course, that the scientists at the Johnson Space Center had over everyone else, is that they had been curating and analyzing the Moon rocks from the Apollo program for over a decade. To them the lunar connection was obvious — in every detail, the meteorite looked like a lunar breccia. From the Johnson Space Center, small samples of

Figure 14.6. ALH81005 — found in the Allan Hills mountain range in Antarctica, this was the first lunar meteorite to be identified on Earth. Image courtesy of NASA.

ALH81005 were passed along to researchers at different institutions and universities and the monumental news of the discovery was made at the 1983 meeting of the Lunar and Planetary Science Conference in Houston. A total of 42 lunar meteorites have now been collected from the Antarctic ice fields, with a total of 408 lunar meteorite finds being officially catalogued as of 2020. The majority of the finds outside of Antarctica have been from those warmer dessert regions of northwest Africa, Oman, and Saudi Arabia.

Lunar meteorites appear to be mostly derived from the old highland terrains, with component clasts having ages between about 2.5 and 4.0 billion years. The latter age closely corresponds to the 3.9-billion-year age deduced for the Moon rocks returned during the Apollo program. The fact that the oldest clast components in lunar meteorite breccias are not significantly older than the Apollo Moon rocks, and that the meteorites are derived randomly from the entire surface of the Moon, has been taken as evidence that the Late Heavy Bombardment (LHB) really did happen (Figure 14.7). Not every researcher agrees that the LHB is real [7], but argue that it is an artifact of collecting bias — that is, the Apollo samples

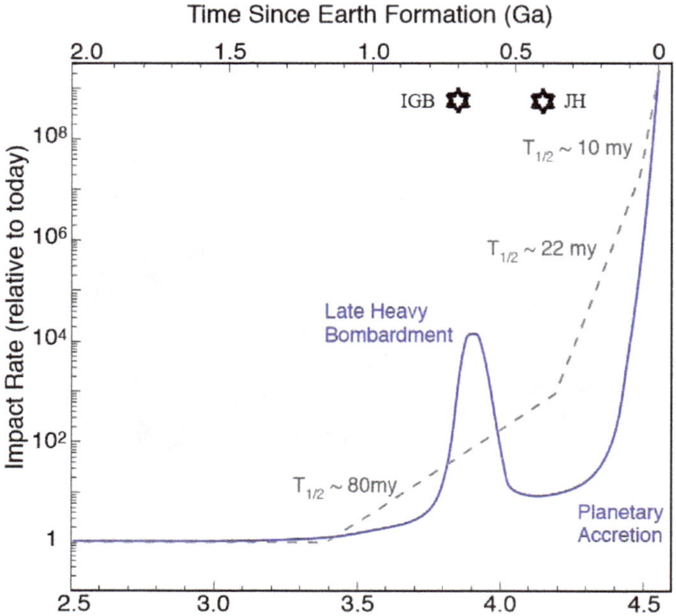

Figure 14.7. Estimated variation in the Earth–Moon impact rate as a function of time T. Two interpretations of the data are shown in the diagram. The smooth curve allows for the LHB, while the dashed lines correspond to an alternative interpretation of the lunar cratering data (in which no LHB occurs) with the impact rate decaying exponentially with $\exp(T/T_{1/2})$ [7]. The two stars in the upper righthand corner indicate the ages associated with the Isua Greenstone Belt in Greenland (IGB) and the Jack Hills (JH) zircons. Background image from cococubed.com.

were all collected from the near-side hemisphere of the Moon, and from locations that were often close enough to each other that cross-contamination of impact ejecta could have occurred.

Evidence for the occurrence of the LHB has been derived from a study of rocks from the Isua Greenstone Belt region in the southwestern region of Greenland — these Arctic rocks are amongst the oldest known on Earth, dating to about 3.8 billion years in age (Figure 14.7). Importantly, however, measurements of the iridium abundance in rock specimens from Isua have been found to be elevated, some 150 parts per trillion (ppt), compared to average crustal rocks which have an iridium abundance of about 20 ppt.

The relevance of this elevated abundance has been investigated by Uffe Jorgensen (Niels Bohr Institute, Denmark) and coworkers [8], and they argue that the elevated level is the result of terrestrial impacts towards the end of the LHB, and, furthermore, they suggest that the elevated levels are consistent with the LHB impactors being predominantly cometary, rather than asteroidal, in nature. Their argument begins by adopting three starting conditions: (1) The impact speed of comets is characteristically larger ($\sim$25 km/s) than that asteroids ($\sim$15 km/s); (2) The escape velocity for the Earth and Moon are 11.2 km/s and 2.4 km/s respectively; (3) Both comets and asteroids contain enhanced iridium abundances compared to that of Earth's crustal rocks.

With these starting conditions in place, Jorgensen *et al.* note that, given its low escape velocity, lunar material is mostly expelled into space during an impact, with more material being expelled during a cometary impact than an asteroidal one, because of their higher encounter speeds. Detailed calculations for Earth impacts, however, revealed that about 10 percent of the ejecta material is retained during an asteroid impact, compared to about 50 percent of the material being retained in a cometary impact. This retention difference should accordingly leave different iridium enhancements on the Moon and the Earth depending upon the nature of the impactors. That is: if the LHB impactors were asteroids then both the Earth and the Moon should show modest iridium enhancements, with detailed modeling indicating that the Earth's enhancement should not exceed 1.8 times that on the Moon. In contrast, if the LHB impactors are predominantly comets, then the Earth should show a large iridium increase, but not the Moon.

The raw numbers now complete the narrative. The Isua rocks have an iridium abundance of 150 ppt, while Moon rocks show an iridium abundance of about 10 ppt. On this basis, Jorgensen and coworkers argue that the LHB was caused by cometary impacts. This result is consistent with the implications from the Nice Model (recall Chapter 5) that invokes the outward migration of Uranus and Neptune into a disk of cometary (Kuiper Belt) objects as the cause of the LHB. Jorgensen *et al.* additionally determine from their calculations that during the LHB the Earth suffered of order 4000 impacts comparable to and larger than the Chicxulub impact, and that of order one million kilograms of extraterrestrial material fell on each square meter of the Earth's surface. These same cometary impacts can readily account for the formation of Earth's oceans (recall Chapter 8).

References

[1] W. Cassidy. *Meteorites, Ice and Antarctica — A Personal Account.* Cambridge University Press (2003).

[2] I. Whillans and W. Cassidy. "Catch a falling star: Meteorites on old ice." *Science,* **222**, 55–57 (1983).

[3] A. Bischoff. "Meteorite classification and the definition of new chondrite classes as a result of successful meteorite search in hot and cold deserts." *Planetary and Space Science,* **49**, 769–776 (2001).

[4] G. Evatt *et al.* "A potential hidden layer of meteorites below the ice surface of Antarctica." *Nature Communications,* **7**, 10679 (2016).

[5] A. Hildebrand, M. Beech, S. Kissin, and G. Quade. "A possible meteorite lag deposit after continental glaciation in southeastern Manitoba." Paper presented at the 69th Annual Meteoritical Society Meeting, Zurich, Switzerland (2006).

[6] C. Kulyk. "Meteorite magnet — is Manitoba harbouring an ice-age meteorite dumping ground?" *Sky News,* September/October, p. 8 (2005).

[7] W. Hartmann, G. Ryder, L. Dones, and D. Grinspoon. "The time-dependent intense bombardment of the primordial Earth–Moon system." In *Origin of the Earth and Moon* (R. Canup and K. Righter, Eds.) University of Arizona Press (2000), pp. 393–512.

[8] U. Jorgensen *et al.* "The Earth–Moon system during the late heavy bombardment period — geochemical support for impacts dominated by comets." *Icarus,* **204**, 368–380 (2009).

Chapter 15

Terrestrial Meteorites

Meteorites have been found on Earth, but is it possible to find Earth-derived meteorites on other solar system bodies? The answer to this question is yes, and this affirmation has direct consequences for the spread of life across the greater solar system and potentially to worlds well beyond.

In the same way that Martian and lunar meteorites are launched into space by asteroid and comet impact events, so too will terrestrial material be lofted into space during large crater forming epochs. The processes, as previously described in Chapter 11, are difficult to model and are described according to complex physical interactions (Figure 15.1). In this chapter we are primarily concerned with the melt zone, ejecta curtain, and ejecta trajectories that accompany the excavation of material. Effectively, the question to be answered is, what happens to the target material that is blasted away from the impact point [1].

There is no simple, analytic theory to describe crater formation and target material ejection, but there are simplified formulae, based upon numerous detailed computational models, that describe what occurs in general. A superb summary and description of such equations has been given by Jay Melosh (Purdue University) [2], and here we extract the ones relevant to the ejecta material. Accordingly, the first formula of interest is the speed V_{eject} with which material is ejected from the target substrate, as a function of distance Δ from the impact point. If the impactor has radius a and impact speed V_{imp}, then

$$V_{\text{eject}} = V_{\text{imp}} \left(\frac{a}{r}\right)^{2.87}, \text{ where } r = \sqrt{\Delta^2 + a^2} \qquad (15.1)$$

343

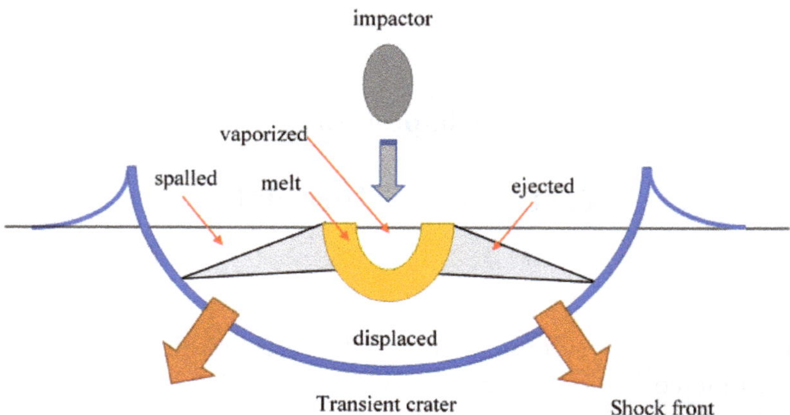

Figure 15.1.   Schematic diagram of the contact, compression, and excavation phases of a crater forming event. The impact results in the vaporization of ground material and the impactor, the melting of ground material (yellow region), the spallation ejection of ground material (grey area), and the fragmentation of surrounding substrate material due to the advancement of a strong shock wave.

Accordingly, as one would expect, the ejection velocity decreases the further the launch point is from the contact location. From this formula we progress to the amount of material ejected, m_{eject}, with a speed greater than V — this is expressed according to the ratio formed with the impactor mass, m_{imp}:

$$f = \frac{m_{\text{eject}}}{m_{\text{imp}}} = 0.75 \frac{P_{\max}}{\rho_{\text{target}}\, c\, V_{\text{imp}}} \left[\left(\frac{V_{\text{imp}}}{2V} \right)^{5/3} - 1 \right] \tag{15.2}$$

where $P_{\max}$ is the maximum shock pressure that the ejecta is subject to, ρ_{target} is the density of the target material, and c is the speed of sound in the target material. Furthermore, the characteristic boulder size D_B within the ejecta spray is given by the formula

$$D_B = \frac{2\,T\,a}{\rho_{\text{target}}\, V_{\text{eject}}^{2/3}\, V_{\text{imp}}^{4/3}} \tag{15.3}$$

where T is the tensile strength of the target material. Combining these formulae with those describing the mass of the impactor in terms of its radius and density, and that for the crater size as given in Chapter 11 [specifically Equation (11.2)], the ejecta history associated with a specific impact and crater can be unraveled [2]. Of specific interest in this chapter is that some material can be ejected from a crater with a speed greater than the Earth's escape velocity: $V_{\text{eject}} > V_{\text{esc}} = 11.2\,\text{km/s}$ — this material can leave the Earth environment, without falling back, and thereby begin a journey that could see it either hit another planet or a moon, and indeed, potentially leave the solar system altogether.

The Chicxulub crater, with a diameter of some 150 km and age of 65 million years, is the largest crater to have formed in the past 550 million years — a timespan that covers the great blossoming of life on Earth since the Cambrian explosion to the present day. Adopting the characteristics of a stony asteroid impactor of size 17 km across and impact speed of 25 km/s, then of order 5.5×10^{12} kg of material will be ejected into space (with $V > V_{\text{esc}}$) from this one event, and the characteristic boulder size of the ejecta will be 1.8 meters across. A more detailed study [2] has identified a total of 96 impact craters with ages less than 550 million years and sizes more than 5 km across, and the formation of these craters has potentially launched of order 10^{13} kg of terrestrial material into interplanetary space. The Chicxulub event accounts for more than half of the total amount of terrestrial material distributed into the solar system in the past 550 million years — importantly, this material will likely contain terrestrial microbes, and this has direct consequences for the transfer of life between planets and their moons via lithopanspermia (see below). There is a certain irony, if not yin–yang symmetry, introduced by the idea that the Chicxulub impact, which on Earth is associated with the mass extinction of life, might have seeded life elsewhere in the solar system.

Once ejected from Earth, with a speed greater than its escape velocity, the first possible target that any terrestrial ejecta might encounter is the Moon, and remarkably, a small terrestrial meteorite (Figure 15.2) was found in a sample of Moon rock (#14321) brought

Figure 15.2.   The terrestrial meteorite (felsite clast) is arrowed in this image of Moon breccia 14321 brought back to Earth by the Apollo 14 astronauts. Image courtesy of NASA.

back during the Apollo 14 mission in 1971 [3]. Detailed study of the terrestrial material inclusion suggests that it is at least 3.9 billion years old, suggestive of the possibility that it was launched during the Late Heavy Bombardment. That additional exploration of the Moon will identify terrestrial meteorites in the future seems clear [4], and there is no reason to suppose that there are not terrestrial meteorites to be found on Mars as well.

15.1   Tektites

Obviously, not all of the material excavated from an impact crater will achieve escape speed, and the greater amount of material will fall back to Earth. Some of this fall-back material, however, could well land many hundreds to even thousands of kilometers away from the impact location, and spend some considerable time moving through the Earth's atmosphere. In some sense, these are reverse meteorites — they start at the Earth's surface, become ejected into the Earth's atmosphere in a molten condition, and then fall

Figure 15.3. A flanged tektite showing the effect of molten front-surface material being shaped into a rounded, heat-shield-like, aerodynamic structure. Image courtesy of Wikipedia.

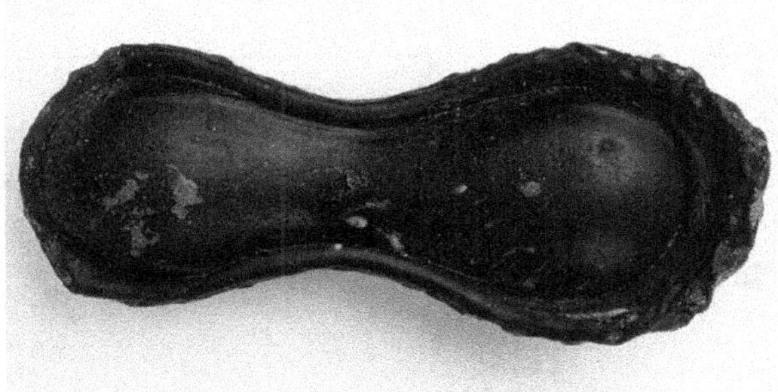

Figure 15.4. A bar-bell tektite showing the effects of melting and extension due to rapid rotation. Image courtesy of the Natural History Museum, London.

back to the ground in a chemically and structurally altered state. Such objects, in fact, abound and are called tektites (a name derived from the Greek word for *molten*). Tektites take on various characteristic forms, from near perfect spheres, to spheres with flanges (Figure 15.3), tear-drop shapes and bar bells (Figure 15.4). All of these forms come about through the heating of material at the point of impact, as well as in the Earth's atmosphere, and rotation.

Detailed analysis indicates that tektites have a similar composition to that of terrestrial sedimentary rocks, although the material has been heated to form a homogenous (volatile-free) glass.

Four extensive strewn fields of tektites are now generally recognized — these are the Australasian strewn field, the Central European strewn field, the Ivory Coast strewn field and the North American strewn field (see Figure 15.5). The North American, European, and Ivory Coast strewn fields have associated impact structures — the Chesapeake Bay impact crater, the Nördlinger Ries crater, and the Lake Bosumtwi crater respectively — but no source crater for the extensive Australasian strewn field has, as yet, been identified. Using the associated crater age estimates as a guide, the North American, European, and Ivory Coast strewn fields are thought to be of order 35.5, 15.1, and 1.07 million years old respectively. While there is no known source crater for the Australasian tektites, this strewn field is estimated to be of order 0.8 million years old.

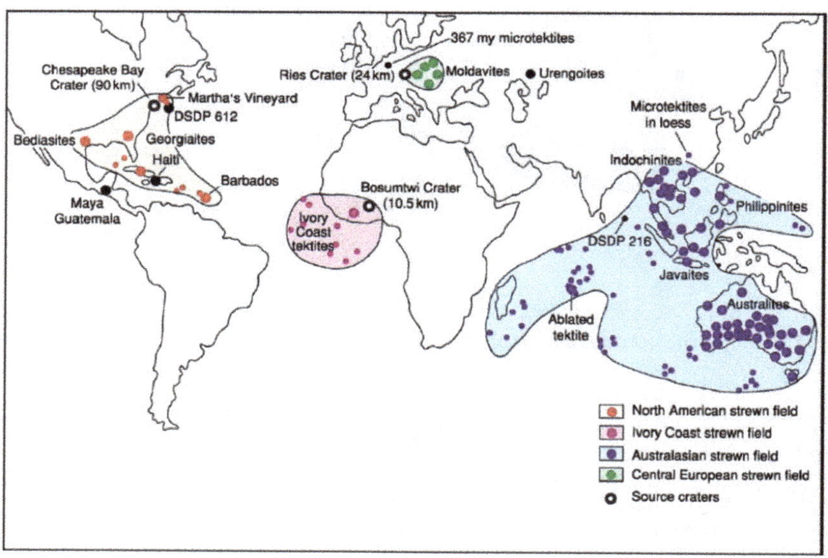

Figure 15.5. Distribution map of the major tektite strewn fields. Where known, the source craters are indicated by an open circle. Image adapted from http://earthsci.org/space/space/tektites/tektites.html.

15.2 Lithopanspermia

It has long been argued that life may have first evolved on planet Mars, at a time when it had a denser atmosphere, a warmer climate, and extensive oceans. At this same time, in the young solar system, the Earth was still in a state generally hostile towards the nurturing of life. Life, under this hypothesis, was transferred to Earth from Mars by the action of a meteorite conveyor belt. Jay Melosh has estimated that at the present epoch, and it may well have been different in the distant past, something like 500 kg of Martian material (in the form of meteorites) is delivered to Earth each year. This same conveyor belt can, in principle, also distribute material to any other planet or moon in the solar system [4]. The conveyor belt of ejected materials, especially the terrestrial component, is capable of carrying a living cargo of microbes, making it an important lithopanspermia vector.

The idea of spreading life by panspermia,[1] not only throughout the solar system but through the greater galaxy as well, has a long history, and it has taken many forms. Some researchers have proposed that interstellar space is actually populated by microbes, and that these, rather than interstellar dust grains, are primarily responsible for the reddening of starlight.[2] Others have suggested that some microbes can be wafted between the stars along tenuous gas currents driven by stellar winds, and they possibly ride along the blast waves produced by supernovae. Yet other researchers have suggested that there is a deliberate panspermia action in play, with advanced civilizations literally seeding the galaxy with microbial lifeforms. The idea of lithopanspermia simply extends, and perhaps refines the idea of panspermia, to the transportation of microbes through space aboard rocky substrates. The rock is important since

---

[1]Panspermia literally means "life everywhere."

[2]Interstellar reddening of starlight is generally attributed to interstellar dust grains that have a characteristic size comparable to the wavelength of blue light. The greater the column length of dust through which a given star is seen, therefore, results in more of its blue light being scattered away from the line of sight, and accordingly the redder the star will look.

its interior provides locations where microbes will be shielded from the worst effects of space travel and cosmic ray exposure.

The whole process of lithopanspermia is driven by a whole series of low probability processes, but the very existence of life itself tells us not to down play the possibility of the extremely unlikely physically coming about. For all this, the numbers are not especially encouraging with respect to spreading life between star system. If we imagine N_0 rock fragments being ejected from the solar system, with a characteristic speed of U, then after T years they will have travelled a distance $D = UT$ from the Sun. If the cross-section area for an impact encounter with another star system is σ, and given that the density of stars in the solar neighbourhood is of order $\rho^* = 0.1$ stars per cubic parsec, then the characteristic time interval for at least one collision to occur is $1/N_0 U \sigma \rho^*$. Taking U to be, say, 10 km/s, $\sigma = 10^8$ km^2 (which is about the cross-section area of the Earth), then the characteristic time before an impact will be $10^{25}/N_0$ years — this is vastly older than the age of the universe. Even if the cross-section area resulting in an impact is increased to say the orbit of Neptune about the Sun (giving $\sigma = 6 \times 10^{19}$ km^2), and thereafter allowing for orbital perturbations to drive an ultimate encounter with a terrestrial planet, so the typical impact time is still some $10^{13}/N_0$ years. The remaining parameter that might save the day is N_0, but this is unlikely to be a particularly large number — even if being very generous and set $N_0 \sim 10^4$ (thinking of the escaping ejecta as being millimeter-sized pebbles) then the characteristic encounter time is still going to be of order a billion years. While this time interval is, perhaps, realizable, the problem is that it is much longer than the expected survival time of any pebble within interstellar space (as a result of collisions with, and erosion by interstellar dust grains), and the survival time of any hibernating microbes.

While it seems, therefore, that solar-system-directed lithopanspermia is not going to be an effective way of spreading life to the rest of the galaxy, the situation may well have been very different in the distant past. During the Late Heavy Bombardment (some 4 billion years ago), for example, large quantities of solar system material would have been ejected into interstellar space. Also, at that time, the Sun would have still been relatively close to its stellar siblings — those other stars that had formed and were forming within its natal interstellar nursery. At that time the exchange of life-bearing rocks between planetary systems was much more likely, and especially so since the signs of microbial life have been detected on Earth as far back as 3.5 to 3.8 billion years — these include the microbe alterations described in the Apex cherts located in the Australian Pilbara Craton, and the Nuvvuagittuq Belt rocks in Quebec, Canada. Indeed, life is everywhere on and in Earth, being discovered in its upper atmosphere and at depths of at least 6 km below its surface. Indeed, life is dominated by bacterial lifeforms — the exact entities required to make lithopanspermia work. One recent example of this extravagance of bacterial life being found in a place on Earth, where it might otherwise be thought that none should exist, was described by Yohey Suzuki (University of Tokyo) and coworkers in the April 2020 issue of *Communications in Biology*. Incredibly, Suzuki *et al.* found single-celled bacterial life living within the tiny cracks inside volcanic rock that had been extracted from a core driven 125 m into the seabed in a region 5.7 km below the water surface in the South Pacific. Not only were these bacteria thriving, there were as many as 10 billion of them per cubic centimeter of rock.

Thinking more parochially than interstellar space, the exchange of material between planets and moons within the solar system certainly allows for species invasion, and potentially the evolution of bi-planetary biospheres. The issue, as we shall see, is not so much in the physics behind the idea of transfer, but in the universal manner in which evolutionary biology operates. On Earth, when expanding into a new environment, an invading organism typically requires multiple attempts before it can gain a permanent foothold. Lithopanspermia, however, is generally an all or nothing process, with any passenger organisms needing to survive first time, or not at all. One evolutionary strategy that has evolved on Earth to counter unpredictable environmental conditions is that of bet-hedging in which genetic diversity is held high within a colony of identical organisms [6]. In this manner, even under extreme environmental variation some members of the colony should survive. Interestingly, bet-hedging is common among many bacteria species. Indeed, several species of bacteria, such as *Deinocossus radiodurans* and *Bacillus subtilis*, have already shown the ability to survive the conditions encountered in space, and can survive under simulated present-day Martian conditions. Such bacteria are good candidates for interplanetary lithopanspermia.

Indeed, calculations conducted by Curt Mileikowsky (Royal Institute of Technology, Stockholm, Sweden) and coworkers [7] indicate that a fraction of order one millionth (10^{-6}) of a bacterial population should survive the vagaries of interplanetary travel for of order 10^6 years, provided that they are protected by about $1\,\mathrm{m}$ of rock (this rock layer acts to protect the bacteria from cosmic rays, solar radiation, and the harsh vacuum of space). If the bacteria are shielded by 2 to $3\,\mathrm{m}$ of rock, then about 10^{-6} of the original population should survive a journey of 25 million years in space. It seems generally clear that bacteria can survive both the launch, journey, and landing stages invoked in the lithopanspermia hypothesis. Likewise, survival of the remnant bacterial colony, upon landing in an entirely hostile environment (at least initially), is also entirely possible.

The efficiency of the transfer process of material between two planets X and Y can be readily determined, and will depend upon N_X, the number of boulders ejected from planet X, and the properties of the donor planet Y. If N_Y is the number of boulders ejected from planet X that impact planet Y, then

$$N_Y = N_X \left(\frac{\sigma_Y}{4\,\pi\,D_{XY}^2} \right) \left(\frac{T}{\tau_{XY}} \right) \qquad (15.4)$$

where σ_Y is the cross-section area of planet Y for capture, D_{XY} is the characteristic distance between planet X and Y, T is the time interval over which material is actively ejected from planet X, and τ_{XY} is the characteristic transit time for material to move between planet X and Y. That is, $\tau_{XY} = D_{XY}/V$, where V is the typical velocity of the ejected material. The cross-section area for capture is cast in terms of the Hill sphere radius,[3] and given by

$$\sigma_Y = \pi a_Y^2 \left(\frac{M_Y}{3\, M_{\mathrm{Sun}}} \right)^{2/3} \tag{15.5}$$

where a_Y is the orbital radius of planet Y and M_Y is the mass of planet Y compared to the Sun.

Of interest now is the cumulative fraction $P(X \to Y)$ of rocks ejected from planet X that impact upon planet Y. Taking characteristic values of $T = 3.5$ billion years, $\tau_{XY} = 10$ million years,[4] then for the Earth/Mars system, $P(\mathrm{Earth} \to \mathrm{Mars}) \approx 0.002$, whereas $P(\mathrm{Mars} \to \mathrm{Earth}) \approx 0.05$. These fractions are quite favorable for the interchange of spore-bearing rocks between Earth and Mars, although the transfer efficiency of material from Mars to Earth is some 25 times higher than that from Earth to Mars — this is primarily because of the larger mass and capture radius of the Earth in relation to those for Mars. Not only do these characteristic numbers indicate that lithopanspermia is a mechanism by which life-bearing rocks might be exchanged between planets, but in principle allow for back contamination as well — that is, life-bearing rocks could be exchanged back and forth between planets X and Y allowing for the development of twin-biospheres. Life may accordingly have first evolved on Earth and been transferred, via lithopanspermia, to Mars, and then been brought back to Earth again. The process might also work in reverse, with life evolving on Mars and then being carried to Earth — and, this back and forth process could have proceeded multiple times.

The likelihood of material exchange between planets is maximized when their separation is small, and Manasvi Lingam and Abraham Loeb at Harvard University [7] have suggested that one planetary system where exchange may be rampant is that in orbit about the star TRAPPIST-1. This particular star, an M-spectral

---

[3]This is the radius about a planet within which it will dominate the gravitational motion of a satellite.

[4]This transfer age is based upon the characteristic cosmic-ray-exposure age of Martian meteorites.

type dwarf star with a mass of about $1/10^{\text{th}}$ that of the Sun, has a retinue of seven Earth-sized planets, three of which orbit within the region of the habitability zone[5] and which have orbital separations of order 0.01 AU. Discovered in 2015, this planetary system is some 12 parsecs away from the Sun, and is not a candidate for Earth-based lithopanspermia, but, if life originated on any one of the planets within the system's habitability zone, then it will rapidly spread to the others. This system provides one potential location where the exchange of life between planets is possibly rampant, and possibly, in the not-so-distant future, something that might be testable. The test of this hypothesis would proceed via infrared reflectance spectroscopy, whereby biotic markers for each planet might be analyzed.

References

[1] B. M. French. *Traces of Catastrophe — A Handbook of Shock-Metamorphic Effects in Terrestrial Meteorite Impact Structures.* LPI Contribution No. 954, Houston (1998).

[2] M. Beech, I. Coulson, and M. Comte. "Lithopanspermia — the terrestrial input during the past 550 million years." *American Journal of Astronomy and Astrophysics,* **6**(3), 81–90 (2018).

[3] M. Beech, M. Comte, and I. Coulson. "The production of terrestrial meteorites — Moon accretion and lithopanspermia." *American Journal of Astronomy and Astrophysics,* **7**(1), 1–9 (2019).

[4] A. Crawford *et al.* "On the survivability and detectability of terrestrial meteorites on the Moon." *Astrobiology,* **8**, 241–251 (2008).

[5] M. Reys-Ruiz *et al.* "Dynamics of escaping Earth ejecta and their collision probability with different solar system bodies." *Icarus,* **220**, 777–786 (2012).

[6] I. von Hegner. "Interplanetary transmission of life in an evolutionary context." *International Journal of Astrobiology,* **19**, 335–348 (2020).

[7] C. Mileikowsky *et al.* "Natural transfer of microbes in space; 1. From Mars to Earth and Earth to Mars." *Icarus,* **145**, 391–427, (2000).

[8] M. Lingam and A. Loeb. "Enhanced interplanetary panspermia in the TRAPPIST-1 system." *Proceedings of the National Academy of Science,* **114**(26), 6689–6693 (2017).

---

[5]This zone recall is the region about the parent star where a planet with an atmosphere can support liquid water at its surface.

Chapter 16

Meteorites on Mars

Technically, the term meteorite is reserved, by definition, to mean a natural object (meteoroid) that has survived passage through the Earth's atmosphere to produce a rock or iron mass on the ground. But this definition can be readily stretched to include meteoroids surviving passage through any planetary (and one moon[1]) atmosphere. Certainly, fireballs and impact plumes have been observed in the atmosphere of Jupiter (Figure 16.1), but since this planet has no formal surface, there will be no surviving meteorite.[2] Impacts onto the Moon's surface have also been detected (Figure 16.2) but since the Moon has no atmosphere all impacts will accordingly be at cosmic speeds and the outcome will be a crater-producing event,[3] with little

---

[1]This moon being Titan. Larger, in fact, than planet Mercury, Titan, in orbit about Saturn, is the only moon in the solar system with a substantive atmosphere.

[2]By far the most dramatic impact event observed on Jupiter is that of comet Shoemaker-Levy-9 (D/1993 F2) in July 1994. The comet nucleus split into numerous fragments prior to hitting the Jovian upper atmosphere. The resultant impacts were the first direct observations of cometary impacts into a planetary body. The energy released in the 21 impacts that were recorded is estimated to have been the equivalent of some 300 gigatons of TNT explosive. The impact scars imprinted upon Jupiter's atmosphere were visible for many months, and the equivalent impacts on Earth would have totally decimated the biosphere and resulted in catastrophic climate change and extinction events. The estimated size of the parent comet is about 2 km across, and the impact speed into Jupiter's atmosphere was some 60 km/s.

[3]There is a long history of seeing *flashes* upon the Moon's surface. Some of these optical events were no doubt caused by meteoroid impacts. The first instrumental recordings of meteor impact flashes upon the Moon were made in response to the recent spate of Leonid meteor storms in the late 1990s and early 2000s [1].

Figure 16.1.   A fireball impact flash (bright spot to the right) in Jupiter's upper atmosphere. The event was captured by Anthony Wesley on June 3rd, 2010 from his observatory at Broken Hill (Australia). The impactor was estimated to be about 10 m across. Image courtesy of ESO.

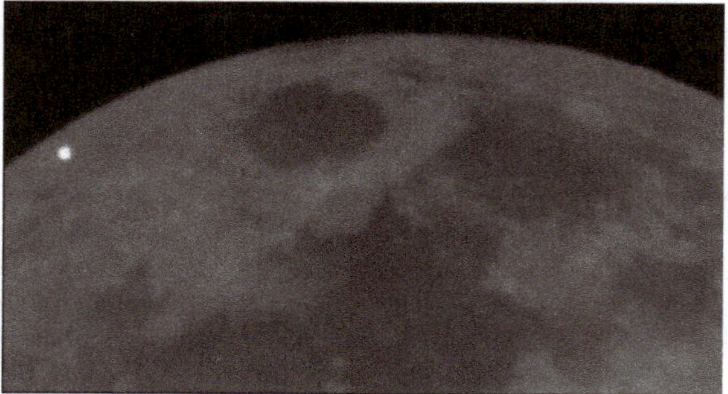

Figure 16.2.   Impact flash (bright spot to the upper left) resulting from a meteoroid hitting the Moon's surface on January 4th, 2008. It is not an ablation fireball that is being seen, but rather reflected sunlight from the impact plume of dust thrown up as a crater is formed. Image courtesy of NASA.

of the impactor surviving (but recall Section 11.1). No specific fireball or impact event has ever been observed in the atmosphere of Venus, but since its atmosphere is so dense, only very larger impactors can make it to the ground (recall Chapter 6) — that there are some meteorites on the Cytherean surface, however, is highly likely, but they will have to bide some time yet before being identified.[4]

16.1 Meteorite finds

The first meteorite on Mars was discovered by the *Opportunity* rover in January 2005. Indeed, it was fortuitously located close to where its protective heat shield landed. This meteorite (Figure 16.3), now officially known as Meridiani Planum (after the find location), shows distinct surface regmaglypts and analysis by *Opportunity's* instrumentation indicates that it is an IAB iron. The meteorite is remarkable for showing no impact crater or plunge pit, and very little rusting or weathering effects. It is not known when Meridiani Planum actually fell, but the lack of weathering is consistent with that expected in a dry, cold environment, and accordingly it could have a residency time on Mars measured in millions of years.

Other meteorites were soon found by *Opportunity* and its companion rover *Spirit*. Remarkably, on May 25th, 2014, the *Curiosity* rover stumbled upon the first Martian strewn field (Figure 16.4). The larger of the iron fragments (officially named Aeolis Palus 002) is some 2 meters in length, and shows distinct evidence of erosion, suggesting a lengthy Martian residency age. To date, six officially named meteorites have been found on Mars, but many more potential meteorites have been encountered and imaged. Finding meteorites

---

[4]Venus is perhaps the least well observed of the larger terrestrial planets. The dense CO_2-rich atmosphere that hides its surface from direct (optical) view has produced a dramatic greenhouse heating effect, resulting in a surface temperature of 464° centigrade. The surface topology has been mapped, however, at radar wavelengths (by NASA's *Magellan* spacecraft in the early 1990s) and the surface is largely covered by smooth basaltic plains. The first surface images were obtained by the Soviet *Venera* 9 and 10 landers in 1975. These were followed by more images from the *Venera* 13 and 14 landers in 1982. A number of Earth-intercepting meteoroid streams make a close approach to the orbit of Venus, and several periodic comets additionally make close approaches to its orbit. There are, therefore, meteor showers on Venus, but they will be difficult to observe from the Earth. It is distinctly possible that bright fireballs occurring in the Cytherean atmosphere could be observed from the Earth, but no such event has ever been reported — largely because no one is actually looking for such events.

Figure 16.3.   The Meridiani Planum iron meteorite as found by the *Opportunity* rover on Mars. Note the distinctive surface regmaglypts and the lack of surface rusting. Image courtesy of NASA.

Figure 16.4.   The Aeolis Palus strewn field of meteorites as found by the *Curiosity* rover in May 2014. Image courtesy of NASA.

on the surface of Mars is perhaps not overly surprising, but what is surprising, and indeed, remains a mystery is why all the finds so far have been iron meteorites. On Earth, and there is no reason to believe that it should be any different on Mars, only about 6% of meteorite falls are irons, the vast majority of falls being stony chondrites. There must, accordingly, be stony meteorites on the surface of Mars, some possibly still sporting dark fusion crust. At this stage it is the case, however, that the rovers are picking out the visually more obvious iron meteorites — just as on Earth, irons on Mars will constitute some 5% of all finds. The stony meteorites are there, but they are simply less easily recognized.

While no stony meteorites have as yet been identified on Mars, it is the case that the stony precursors of Martian meteorites (as found on Earth) have been identified. The classic example of this relates to what has been called Bounce Rock — so named as the *Opportunity* rover hit the rock during its bouncing decent to a final landing location. This football-sized rock (Figure 16.5) was dusted, drilled, and studied in detail by the instruments aboard the *Opportunity* rover and found to be an excellent match for the shergottite meteorites. This identity provides for the verification of the idea of material exchange between planets. The reverse process of material exchange will also apply and there are presumably terrestrial meteorites on Mars, but until Mars is much better explored by human beings then their presence will remain, for the present, a hidden treasure.

16.2   Blueberries

The apparently straightforward detection of large numbers of iron meteorites on Mars's surface bodes well for future Mars exploration by humans. Indeed, Geoffrey Landis (NASA John Glenn Research Center) has argued [2] that such meteorites will be a highly valuable resource of nickel, iron, cobalt, and various platinum-group elements. In terms of steel making, Mars is better resourced than Earth in the sense that not only are the iron meteorites ripe for *picking*, but its surface is rich in hematite (Fe_2O_3), conveniently stored

Figure 16.5.   Bounce Rock after drilling and analysis by the instruments aboard the *Opportunity* rover. Image courtesy of NASA.

and arranged in centimeter-sized concretions (commonly called *blueberries*, Figure 16.6). As with iron ore on Earth, the iron can be extracted from hematite via the carbothermic reduction process. In this reduction process carbon monoxide (CO), which is abundantly available on Mars,[5] is used to extract the iron through the reaction: $Fe_2O_3 + 3CO \Rightarrow 2Fe + 3CO_2$. Once the metallic iron has been extracted then steel can be manufactured through the controlled addition of carbon. The carbon can be extracted by the Sabatier reaction in which hydrogen (H) and carbon dioxide (CO_2) react in the presence of a nickel catalyst to produce methane (CH_4) and water (H_2O). The carbon can then be extracted from the methane by pyrolysis. The Sabatier reaction also has the added benefit of producing water. Landis has additionally noted that on Mars carbonyl processing might prove important with respect to the purification of iron and nickel. In this case carbon monoxide is used

---

[5]The Martian atmosphere is 95.32% carbon dioxide.

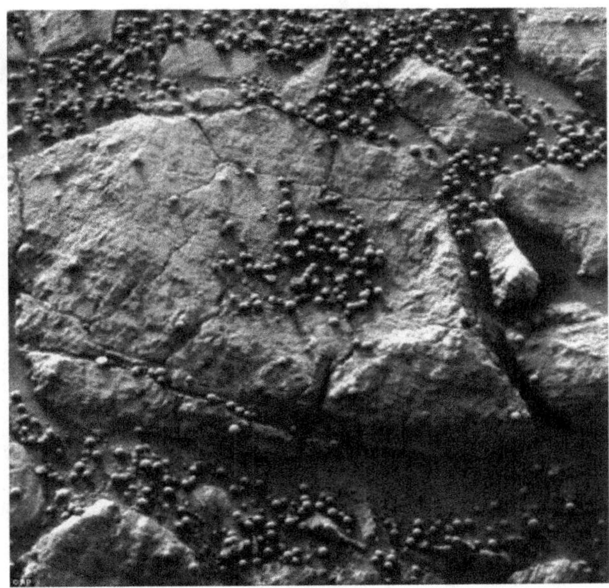

Figure 16.6. Blueberries (hematite concretions) on the surface of Mars. The millimeter to centimeter-sized concretions are about 50% hematite by weight. Image courtesy of NASA.

to form iron and nickel carbonyl gas. This gas can then be distilled, or made to undergo vapor deposition, to build up manufactured surfaces directly. While on Earth have been fashioned into jewels and daggers, on Mars they will be used to make houses — this is surely a positive step forward.

16.3   Meteor storm

Not only have meteorites been found on the surface of Mars, but a meteor storm has been recorded in its atmosphere. This event took place on October 19th, 2014 and was related to the close passage of comet C/2013 A1 (Siding Spring). The comet was discovered by observers at Siding Spring Observatory in Australia on January 3rd, 2013 and while it was not a spectacular comet from Earth, it passed (on the day of the meteor storm) within 150,000 kilometers of Mars. Indeed, at the time of its closest approach the comet was imaged by

Figure 16.7. A 10-second exposure on the *Opportunity* rover's panoramic camera captured this *fuzzy* image of comet Siding Spring (circled) on October 19th, 2014. Image courtesy of NASA.

the camera aboard the *Opportunity* rover — the first ever image of a comet taken from a planet other than Earth (Figure 16.7).

No comet in recorded history has passed as close to Earth as comet Siding Spring did to Mars. The closet Earth approach by a comet was that on July 1st, 1770, by comet D/1770 L1 (Lexell). This comet, at its closest, was located 0.0146 AU from Earth (about six times the distance to the Moon). Comet Siding Spring was just 0.0008 AU (about 1/3rd of the distance between the Earth and the Moon) from Mars at its closest approach. No actual meteors were imaged in the Martian atmosphere, but instruments aboard the Mars Reconnaissance Orbiter (MRO), the Mars Atmosphere and Volatile Evolution (MAVEN) spacecraft, and the Mars Express spacecraft, all detected a dramatic enhancement in the number of ions and electrons in the Martian ionosphere (Figure 16.8) — the ions would have been produced by meteoroid ablation in the Martian atmosphere. Instruments aboard the MAVEN spacecraft specifically detected the presence of metal ions, including those of sodium, magnesium, and

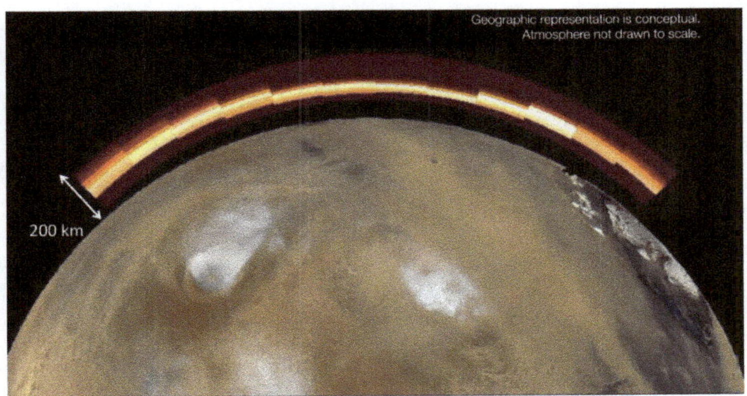

Figure 16.8. This graphic illustrates the strong ultraviolet emission detected in the ionosphere of Mars by the MAVEN spacecraft following the close approach of comet Sinding Spring on October 19th, 2014 [3]. Image courtesy of NASA.

iron, and these too would have been produced through the ablation of meteoroids within Mars's atmosphere [3].

References

[1] Brian Cudnik. *Lunar Meteoroid Impacts and How to Observe Them.* Springer, New York (2009).

[2] G. Landis. "Meteorite steel as a construction resource on Mars." *Acta Astronautica,* **64**, 183–187 (2009).

[3] N. Schneider *et al.* "MAVERN IUVS observations of the aftermath of the comet Siding Spring meteor shower on Mars." *Geophysical Research Letters,* **42**(12), 4755–4761 (2015).

Chapter 17

Interstellar Interlopers

17.1 Visitors in the solar system

It is easy to forget that the solar system is an integral part of a vast galactic structure, itself embedded within an even more expansive universe. For all this *space*, however, the solar system is not entirely cut off from events that have occurred elsewhere, in other planetary systems situated around other stars in other distant parts of the Milky Way galaxy. There is every reason to expect that there are such objects as interstellar meteoroids, but the question is, how might we recognize such objects when and if we chance to see them. Observationally, the key feature that will identify an interstellar interloper will be the eccentricity of any deduced orbit. As outlined in Chapter 3, if the eccentricity $e > 1$ then the orbit has the form of a hyperbola, and this would indicate an origin from outside of the solar system. The question that follows this, of course, is how did the material observed end up in interstellar space to begin with.

There are a number of mechanisms that will result in the ejection of material from a planetary system. In the solar system, and other planetary systems by analogy, this could be via gravitational perturbations induced by the planets, or by gravitational scattering at the time of close stellar encounters. This latter mechanism being important in our solar system for the ejection of cometary nuclei into interstellar space from the Oort cloud. In addition, very small meteoroids can be driven out of a planetary system through radiation pressure — in this manner the parent star pushes material directly into the interstellar medium.

In Chapter 3 the equation of motion for a meteoroid moving under a central gravitational force was briefly introduced. This equation is written as $F(r) = m\frac{d^2r}{dt^2}$, where m is the mass of the orbiting body, and $F(r) = -GMm/r^2$ is the gravitational force between the orbiting body and the Sun (having mass M), and where G is the gravitational constant. For small meteoroids, however, the Sun's radiation pressure[1] generates an outward (that is away from the Sun) radial force. In effect, therefore, the radiation pressure acts to reduce the attractive gravitational force and this can be accommodated into the equation of motion by writing

$$F(r) = -G\frac{Mm}{r^2}(1 - \beta) \tag{17.1}$$

where the β term accommodates the effects of radiation pressure. Written in this manner, the force term in the equation of motion will become positive once $\beta > 1$, and this will result in the ejection of the (otherwise) orbiting body from the solar system.

The β term is expressed in terms of the ratio between the radiation pressure acting upon a meteoroid of mass m, radius δ, and density ρ, and the gravitational force, at a heliocentric distance r. Accordingly,

$$\beta = \frac{(L/4\pi r^2 c)\,A}{G\,M\,m/r^2} = \left(\frac{3\,L}{4\,\pi\,G\,M\,c}\right)\left(\frac{1}{\rho\,\delta}\right) \tag{17.2}$$

where L is the Sun's luminosity, c is the speed of light, and A is the cross-section area of the meteoroid. Notice that the r^2 distance term cancels out in Equation (17.2), and accordingly, the value of β is essentially controlled by the density and size of the meteoroid. The first bracketed term on the right-hand side of Equation (17.2) evaluates to a constant value of 0.0023. Setting the density of the meteoroid to be that typical of stony material, say $\rho = 2000\,\text{kg/m}^3$,

---

[1]This pressure is exerted upon the surface of an object due to an interaction involving the exchange of momentum with the Sun's radiation field. To order of magnitude the solar radiation pressure is given by the ratio $P_{\text{rad}} = I/c$, where I is the solar flux — corresponding to some specified heliocentric distance — and c is the speed of light.

so the $\beta = 1$ condition is achieved for all meteoroids smaller than $\delta = 2.3 \times 10^{-6}$ meters across (that is with masses smaller than $\sim 10^{-14}$ kg).

Such β-meteoroids, on their way out of the solar system traveling along hyperbolic orbits, have been detected by a number of spacecraft-borne dust-detector experiments. Instruments carried aboard the *Ulysses* spacecraft, for example, determined a β-meteoroid flux of 1.5×10^{-4} per m^2 per second. This material represents a continual rain of microscopic particles pouring into interstellar space, and indeed, by analogy, it is to be expected that other planetary systems will additionally eject microscopic dust grains into interstellar space.

One of the important design characteristics of the *Ulysses* mission was to take the spacecraft out of the plain of the ecliptic, thereby allowing its instruments to study the magnetic field structures emanating from the Sun's polar regions. And, indeed, at times when the spacecraft was well above the ecliptic plane, the onboard dust detector recorded collisions with interstellar dust grains (IDGs).[2] These grains are true interstellar interlopers and they must have entered the solar system from the vastness of the interstellar space. That the *Ulysses* dust detector was truly recording IDG impacts was based upon the fact the spacecraft was not in the ecliptic, the known pointing direction of the spacecraft, and the high speed (between 20 and 50 km/s) of the impacting particles.

The characteristic size of the IDGs detected by the *Ulysses* dust detector was 3×10^{-5} meters (0.3 microns), giving a characteristic mass of order 10^{-10} kg. These IDGs represent a small but vital component of the interstellar medium (ISM) and their origins will be discussed in more detail below. Only about 1 percent of the mass of the ISM is in the form of dust grains (the remaining 99 percent is mostly composed of hydrogen atoms), but it is the dust grains that

---

[2]The *Ulysses* detections were made over a sixteen-year time interval starting in 1993. Dust-detector instruments aboard the *Galileo* spacecraft at Jupiter and the *Cassini* spacecraft at Saturn have also recorded impacts from interstellar dust grains.

Figure 17.1.   The Horsehead Nebula in Orion. While not presenting a typical view of the interstellar medium, this classic nebula shows the effect of bringing together large quantities of interstellar gas and dust in order to make a star formation column. The obscuration caused by the cloud is due to the dust grains scattering and absorbing background starlight. Image courtesy of NASA.

act as important platforms for the growth of interstellar molecules (Figure 17.1). The latter are important for our very existence, since it is within the cold depths of giant molecular clouds (and some of these can be many parsecs across and contain millions of solar masses of matter) that new stars and planetary systems form. The dust component to the interstellar medium is also responsible for the dimming and reddening of starlight — this is due to the fact that the interstellar grains are similar in size to the wavelength of blue light ($\sim 2 \times 10^{-7}$ m). Indeed, careful study of this reddening effect[3] indicates that the characteristic mass of interstellar dust grains is of

---

[3]The reddening is due to the blue light component of starlight being scattered via a process known as Rayleigh scattering — this is the very same scattering process that makes the sky appear blue on Earth.

order 10^{-16} kg, and this leads to an apparent anomaly — where are the low-mass IDGs?

The data available from spacecraft observations suggest that some mechanism is systematically filtering out the lower-mass interstellar dust grains, denying them entry into the solar system. The most likely mechanism behind this filtering effect is the build up of an electrical charge upon grain surfaces. This charge build-up is important since once established, the grain will begin to *feel* the presence of the solar system's interplanetary magnetic field (IMF) — specifically, the grains will experience the Lorentz force. This force will act to restrict the grain dynamics along magnetic field lines and/or deflect a grain's path away from the direction that they would otherwise have if no charge-magnetic-field interaction was taking place. The contribution of the Lorentz force upon grain motion will vary according to the IMF strength and direction, as well as the particle charge-to-mass ratio (Q/m). Modeling of the *Ulysses* dataset revealed that the observations are best explained if the charge-to-mass ratio is $Q/m \approx 1$ coulomb per kilogram.

There are many mechanisms that can result in the charging of dust grains while in interstellar space, and while crossing into the solar system where they have to traverse the heliopause — this is the complex, shifting boundary at about 125 AU from the Sun where the outward flow of the solar wind is effectively stopped by the interstellar medium.[4] Numerical simulations concerning the dynamics of charged dust particles indicate that the larger the charge-to-mass ratio so the greater the deflection from the erstwhile path under gravity alone. Accordingly, for a given surface charge, smaller-mass IDGs are going to be systematically and preferentially deflected away from the inner solar system — and this accounts for the spacecraft instruments only *seeing* the more massive IDGs. At 10^{-16} kg the spacecraft data indicate a flux of about 3×10^{-5} IDGs per square meter per second in the inner solar system.

---

[4]It is the (non-static) boundary at which the pressure from the solar wind and the pressure from the surrounding interstellar medium through which the solar system is moving balance.

17.2   Visitors on Earth

In addition to spacecraft-borne instrument detections, interstellar dust has also been discovered ablating, as meteors, in the Earth's atmosphere. The grains are still very small, being perhaps 10 to 100 microns in size, and the meteors are not visible to the naked eye. Indeed, these interstellar meteors are only detectable with radar arrays. Again, speed is the key identifying feature of interstellar meteors, and the various radar surveys that have been conducted indicate that they are very rare. A 2005 radar-survey report by Robert Weryk and Peter Brown (University of Western Ontario, Canada) deduced that of some 1.5 million meteoroid orbits determined over a 2.5-year time interval, only 10 meteors were strong candidates for having an interstellar origin. While the detection of microscopic dust grains arriving from interstellar space is truly remarkable and reminds us that the solar system is not isolated from the rest of the galaxy,[5] the question arises as to the detection of much larger interstellar interlopers. Indeed, are there such entities as interstellar meteorites, asteroids, and comets? The answer to this is assuredly yes, but the detection of such objects is something only just within the reach of present-day survey technology.

Only one meteorite has ever been touted to be interstellar in origin, and that is the Pultusk meteorite which fell on January 30th, 1868 (Figure 17.2). The evidence for an interstellar origin is based upon an analysis of the eyewitness accounts by Johann Galle,[6] at the time of the fall, and later by physicist Ernst Opik in 1933. It is extremely unlikely, however, given the unavoidable uncertainties inherent to all eyewitness observations, that the meteorite really did have a hyperbolic orbit and therefore an interstellar origin. It is also the case that Pultusk appears to be, at least chemically, a typical H5 ordinary chondrite meteorite.

---

[5]In addition to interstellar microscopic dust grains, there are also cosmic rays, produced in distant supernovae explosions, that interact with and tie the solar system, and the Earth directly, to the rest of the galaxy.

[6]It was Johann Gale who, following the predictions made by Urbain LeVerrier, first swept up the planet Neptune with the Berlin Observatory telescope in 1846.

Figure 17.2.   Front page of the Polish illustrated magazine *Klosy* (*Ear's of Corn*), showing an artist's rendition of the Pultusk fireball.

There are strong selection effects acting against any interstellar meteorites ever being found on Earth's surface, and this is because of the high encounter velocity that such objects must necessarily have. As described in Chapter 6, the higher the velocity with which an object enters into the atmosphere, so the greater the ablation and mass loss that must follow, and accordingly, the most likely end result of an interstellar meteoroid encountering the Earth's atmosphere is complete destruction. This being said, the associated fireball could be detected and the interstellar origin betrayed through an analysis of its trajectory. Indeed, one such candidate for an interstellar fireball has been described in a research paper by Amir Siraj and Abraham Loeb (Harvard University) published in the *Astrophysical Journal* in April 2019. This particular fireball was detected by sensors carried aboard U.S. Department of Defence (DoD) satellites designed to look for missile launch and re-entry events.

The fireball was captured on the January 8th, 2014 entering Earth's atmosphere over Manus Island in the South Pacific. The associated meteoroid suffered catastrophic brake up at an altitude of 29.6 kilometers, and the energy associated with the event was of order 0.1 kilotons of TNT equivalent energy (that's about 1/10th of the energy liberated by the first atomic bomb). Siraj and Loeb note that the satellite data indicate a pre-impact (heliocentric) velocity for the fireball's parent meteoroid of about 60 km/s — and this implies a hyperbolic orbit.[7] The DoD satellite data that have been released contain only a partial record of fireball events detected between 1988 to 2019, and accordingly the available data suggest that the interstellar meteoroid detection rate is 0.03 per year, there being just one such event recorded in 31.5 years of data collection.[8] Siraj and Loeb suggest that another two fireball events might have been interstellar, but the evidence for these is much less convincing.

That the Earth encountered one interstellar meteoroid, with an estimated diameter of 0.9 m, in a survey interval of 30 years implies that interstellar space must be fairly teeming with such bodies. The number density n_{IM} of interstellar meteoroids can be estimated by dividing the yearly detection rate D_R by the speed of the interstellar meteoroid V multiplied by the cross-section area of the Earth A_E — accordingly,

$$n_{\mathrm{IM}} = \frac{D_R}{V\,A_E} = \frac{0.03}{13\,\times\,6\,\times\,10^{-9}} = 3.85\,\times\,10^5\ (\mathrm{AU}^3) \qquad (17.3)$$

---

[7]The solar system escape velocity at 1 AU is 42.2 km/s.

[8]In all, a total of 799 fireball events were recorded by the DoD spacecraft in the 31.5-year time interval (see Figure 24.2), suggesting that the detection of an interstellar fireball is essentially a one-in-a-thousand event.

The available data, taken at face value, suggests, therefore, that there might be of order 385,000 1-m diameter meteoroids per cubic AU of interstellar space in the solar neighborhood. If we assume that these interstellar meteoroids are derived from those stars within a radius of 10 parsecs of the solar system, then the 400 or so stars in this volume of space[9] must have ejected into interstellar space about 3.5×10^{22} meteoroids a piece, in order to produce the apparent interstellar fireball flux of 0.03 per year at the Earth.

The mass of these interstellar meteoroids (each taken to be about 1 m across) is of order 4×10^{25} kg (or about 6 Earth masses) per star. There are several remarkable consequences that follow from such estimates (on the basis that they actually stand up to future scrutiny): firstly, on the basis that the influx rate is constant over long intervals of time, then some 130 million 1-m sized interstellar meteoroids have been intercepted by the Earth since it formed, and this is a powerful argument for the seeding of life on Earth via lithopanspermia (recall Section 15.2). Secondly, the rate of interstellar fireball detection is sufficiently high that there is the very definite hope of one day being able to study such objects spectroscopically as they ablate in the Earth's atmosphere.

In late 2019, Siraj and Loeb extended their earlier analysis of interstellar meteoroids by suggesting that they might be studied via lunar impacts. Indeed, Siraj and Loeb suggest that a telescope, with a 2-m diameter, or larger, objective, set in orbit about the Moon at an altitude of about 100 km should be able to detect about one interstellar meteoroid, with a size of order 1 m, impacting the lunar surface per year.[10] The non-interstellar origin of the impactor, its speed and direction, would be determined by tracking changes in the meteoroid's position (and its shadow on the Moon's surface) as a function of time. The optical flash upon impact and the resultant crater diameter would additionally be measured in order to determine

---

[9]There are about 0.1 stars per cubic parsec of interstellar space in the solar neighborhood.

[10]This estimate is based upon the rather optimistic flux of 0.1 1-m sized interstellar meteoroids hitting the Earth's atmosphere per year.

the meteoroid mass and density. The instrumental challenge and the data analysis associated with such a mission, however, would be formidable — but just because something is hard to do is no reason not to try. Funding, however, for a lunar-orbiting, meteoroid-impact-recording telescope has not, as yet, been forthcoming.

17.3 Pre-solar grains

While the odds of sampling or even detecting interstellar material, in the centimeter to meter size range, interacting with the Earth's atmosphere or the lunar surface are very much against a successful outcome, it is the case that interstellar grains have been found within the matrix material of carbonaceous chondrites (Figure 17.3). These grains, sometimes called pre-solar grains (and sometimes, more evocatively, *stardust*), were a part of the interstellar gas cloud that collapsed to produce the Sun and the solar system — they constitute original material for us, but they were formed elsewhere in the galaxy. These alien grains are very small, typically having sizes of order a few micrometers, and while they have travelled a very long way through

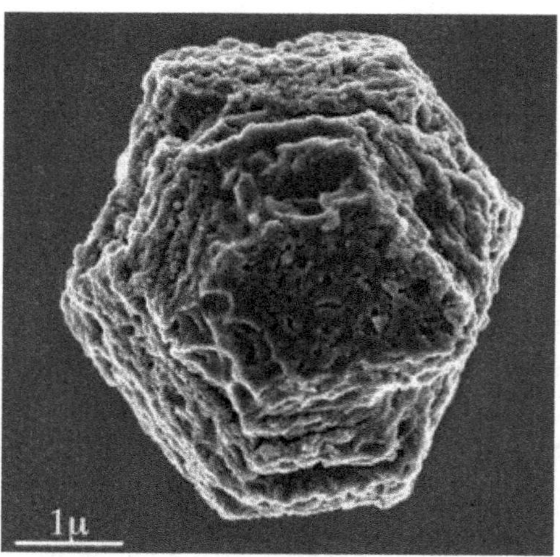

Figure 17.3. A silicon carbide pre-solar dust grain, about 5×10^{-6} meters across, extracted from the matrix material associated with the Murchison meteorite. Image courtesy of WUSTL.

the galaxy, and have been around for longer than the Sun has existed, we do not know from where in the galaxy they specifically originated. For all this, however, we do know how these orphaned pre-solar grains come about (as will be seen below).

The existence of pre-solar grains is betrayed by their chemistry, specifically in the form of isotope ratios. The important isotope ratios that are measured are those of $^{17}O/^{16}O$, $^{18}O/^{16}O$, $^{12}C/^{13}C$, and $^{14}N/^{15}N$, since variations in these ratios betray different environments of formation (recall Section 1.2). Figure 17.4 shows one such diagram in which the measured $^{17}O/^{16}O$ and $^{18}O/^{16}O$ ratios indicate the existence of pre-solar grains in various meteorite samples. The key reference point is that given by the intercept of the

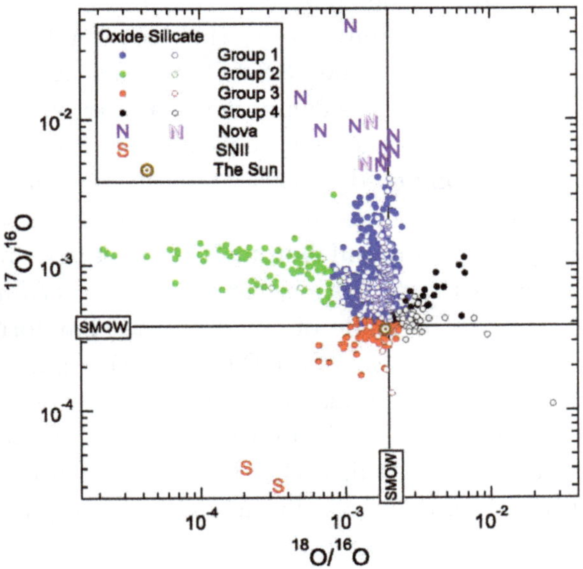

Figure 17.4. Oxygen isotope ratios for pre-solar oxide and silicate grains. SMOW stands for Standard Mean Ocean Water, and solar system material (as measured in Earth rocks, lunar samples) will generally cluster around the intercept point delineated by the large blue circle. Four main groups of pre-solar grains are highlighted (in the four ellipse regions). These groups are populated by grains derived from either massive star supernova (SN) explosions or in the final (asymptotic giant branch — AGB — and planetary nebula) evolutionary phase of low-mass stars. Image adapted from Reference 1.

two lines labeled SMOW, which gives the Standard Mean Ocean Water value for the two ratios as determined on Earth (as well as from lunar rock samples). All the dust grains formed during the condensation sequence phase within the solar nebula disk, prior to planet formation, are expected to plot very close to the SMOW intercept point in the diagram. This intercept point is effectively the value for the $^{17}O/^{16}O$ and $^{18}O/^{16}O$ ratios that define solar system formation and membership — anything else, plotting away from the intercept point, must have formed somewhere else in the galaxy.

Having recognized that there are such entities as alien dust grains within meteorites, the question arises as to where the grains came from in the first place. Indeed, the question now arises as to the origin of the dust component to the interstellar medium. To start this story at the beginning, it is required to go back to the first hour of the origin of the universe — the proverbial Big Bang. At this creation moment, set some 13.8 billion years ago, the universe came into existence and it was composed of a near pure distribution of hydrogen and helium. Hydrogen dominated, and some 75% of the early universe was composed of this simplest of all atoms. Of the remaining 25%, most of it was in the form of helium, and there was a smattering of beryllium and boron. In essence the universe was born sterile, since little interesting chemistry, and certainly not the chemistry of life, can be made of the elements that formed during the primordial nucleosynthesis of the Big Bang. Fortunately, however, hydrogen and helium gas clouds are good at making stars (initially, mostly very massive stars) and importantly, stars are very good at making new elements. Indeed, within the hot, pressure-cooker cores of stars new atoms are forged through nuclear synthesis. By this process, the hydrogen and helium are gradually converted into other atomic elements, all the way along the periodic table up to iron.

Having synthesised iron, however, there is no more that a star can do in a quiescent state, and the next step, to generate those atomic elements all the way up to uranium, requires that a star explodes as a supernova. In this moment of destruction, a vast amount of energy is liberated and a frenzy of atomic synthesis and neutron capture events take place. The newly generated atoms and the shredded remnants of a star are blasted back into the surrounding interstellar medium. The supernova end phase is not the end fate of all single stars, however. Only those stars initially more massive than about 8 times that of the Sun die through supernova disruption. Lower-mass stars, in contrast, go out with a whimper rather than a bang. Low-mass stars end their days as bloated red giants. The Sun will eventually swell to some 250 times its present size, consuming as it grows planets Mercury and Venus, and possibly Earth. And then, having obtained giant proportions, a star seems to have second thoughts, developing at its surface a strong outflow of material — this is the so-called asymptotic giant branch (AGB) phase. This intense stellar wind gradually strips away half or more of the star's original mass, and the remnant core evolves into a diminutive white dwarf. It is the strong stellar wind that interests us here, since this outflow carries nuclear-processed material back into the surrounding interstellar medium.

All stars, whether they end their days in a dazzling supernova, or quietly shed their outer envelope in the production of a white dwarf, return processed material to the interstellar medium (Figure 17.5). Slowly, indomitably, eon after eon, the hydrogen and helium mix that started at the dawn of the universe is being converted, through star formation, stellar nucleosynthesis, and star death, into heavier elements — indeed, all those atoms that constitute the periodic table of elements. Each old generation of stars feeds newly processed elements into the interstellar medium, and these are then incorporated into the next generation of stars. Gradually, generation after star generation, the interstellar medium has become richer in the abundance of those elements other than hydrogen and helium. Not only are new elements synthesized in supernovae and red giant envelopes, but so too are complex molecules — silicate and graphite dust grains, nano-diamonds, and various oxides. These latter components make up the dust component of the interstellar medium, and they are the pre-solar, alien grains found within meteorites.

Returning to Figure 17.4, the various data points labeled with an S or an N indicate dust grains that formed in the past supernova destruction of massive stars. The other data points (labeled Groups 1, 2 and 3) are thought to indicate grains that formed around

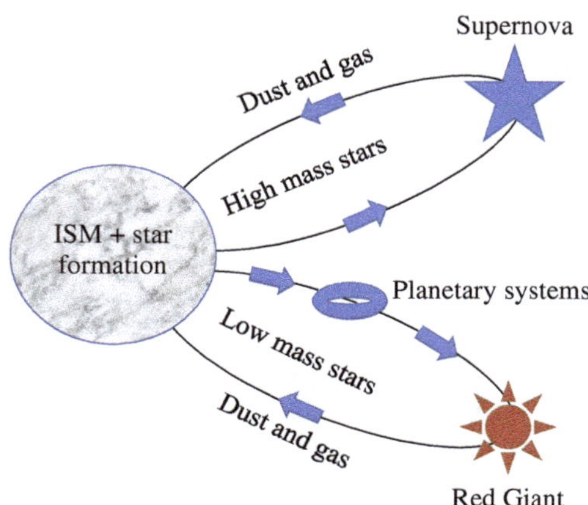

Figure 17.5. A schematic diagram illustrating the recycling and chemical transformation of the interstellar medium (ISM) by successive generations of star formation, stellar evolution, and star death. High-mass stars are those stars initially more massive than about 8 times that of the Sun, and their end phase is associated with supernova disruption. Low-mass stars end their days as bloated red giant stars prior to their eventually evolution into diminutive white dwarf remnants. Each end phase, however, implants newly synthesized material into the interstellar medium, thereby enriching the composition of the star generations that will follow.

old, large, cool, low-mass stars. It has been estimated that perhaps 90 percent of the pre-solar oxide grains originate in old, low-mass, red giant stars, with the remaining 10 percent originating in what are called core-collapse supernovae (SN II). About 98 percent of silicon carbide pre-solar grains are thought to be produced in the extended envelopes of red giant stars, with just 2 percent being contributed by supernovae. The graphite pre-solar grains are more evenly produced between supernova and red giants, with the supernova contribution being perhaps slight greater at 60 percent of the total. The nano-diamonds that have been found within numerous meteorites are produced by several mechanisms, of which supernova production is but one process. Some (small) fraction of the nano-diamonds are accordingly thought to be pre-solar in origin, but the majority are probably the result of processes operating in the early solar system (collisional shocks and/or the differentiation of large bodies).

A recent study of the Murchison carbonaceous chondrite meteorite by Philipp Heck (The Field Museum, Chicago) and coworkers has delved deeply into the past of its constituent grains. Indeed, they set about crushing and chemically dissolving small fragments of the meteorite (turning it initially into a paste that apparently had a "smell like rotten peanut butter") and extracting from this mixture, through an acid wash, some 40 pre-solar silicon carbide grains. Having determined the extrasolar nature of these grains via isotopic analysis, Heck and coworkers then set about estimating their cosmic-ray-exposure ages. To do this they analysed the cosmogenic radionuclide neon-21 (recall Chapter 4). Accordingly, they found that the extrasolar grains had an age range that placed them between a few million to a few billion years older than the solar system (which, remember, formed about 4.56 billion years ago). Incredibly, some of the pre-solar grains extracted from the Murchison meteorite represent the oldest material known to reside on Earth today (Figure 17.6). Arguing, in fact, that the pre-solar grains that they studied were most likely produced in the condensation flows associated with old asymptotic giant branch (AGB) stars, Heck *et al.* suggest that they are derived from stars which physically

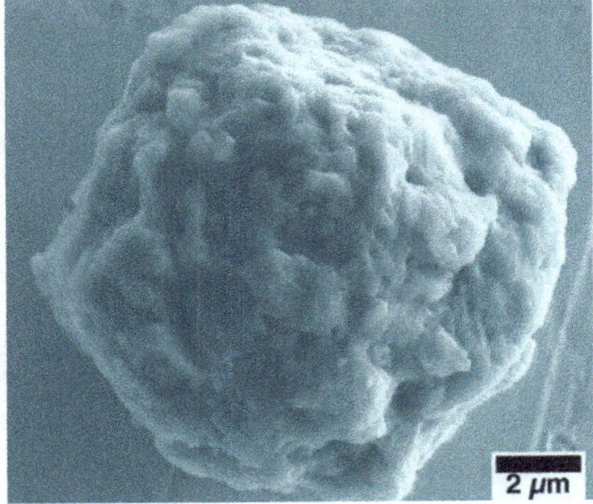

Figure 17.6.   One of the miniscule pre-solar silicon carbide grains extracted from the Murchison meteorite. This particular grain is estimated to be 6.2 billion years old. Image credit: Philipp Heck *et al.* [2].

formed about 7 billion years ago (when the universe was only half its present age).

17.4   Oumuamua

The solar system is lightly bathed in a particulate sea of interstellar dust grains (as recorded by space-borne detectors and radar systems on Earth), and meteorites are impregnated with pre-solar grains. Interstellar meteoroids no doubt exist, and one or two may have even been detected ablating in Earth's atmosphere, but the first large interstellar object to be identified was I1/2017 U1 Oumuamua — the I1 indicates that it is the first confirmed member of the class of interstellar objects. Oumuamua, which approximately translates from the Hawaiian to mean *first distant messenger*, was discovered by Robert Weryk on October 19th, 2017. Using the PANN-STARRS telescope system, this interstellar interloper was nearly missed and it was caught on its way out of the solar system (Figure 17.7), having rounded the Sun some 40 days prior to its detection. The special nature of Oumuamua was immediately recognized when its orbit

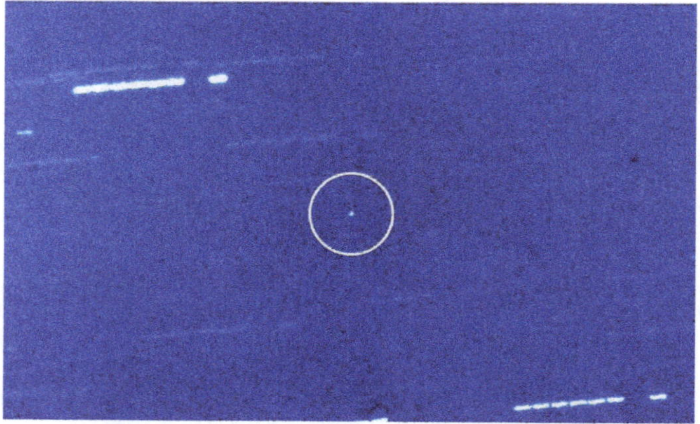

Figure 17.7.   William Herschel Telescope image of I1/2017 U1 Oumuamua (circled). The dashed lines are star trails produced as a result of the telescope tracking motion. Although point-like in this image, Oumuamua has proved to have a decidedly strange (and unexpected) shape. Image courtesy of A. Fitzsimmons, QUB/Isaac Newton Group, La Palma.

was found to be clearly hyperbolic with an eccentricity of $e = 1.2$. While the detection of interstellar objects was generally deemed by astronomers to be inevitable, it is the case that Oumuamua has posed more questions for researchers than it has potentially answered.

Among the first questions that astronomers wanted to answer in the wake of Oumuamua's discovery were, what is its shape, and is it an asteroid-like body made of rock and/or nickel-iron, or is it a cometary nucleus composed predominantly of water ice. These seem like very basic questions, but in spite of intensive effort by many observers over many months, the answers are still unclear. The shape question is perhaps the least complicated to resolve, although its resolution has thrown out a mystery.

The shape of a solar system object (and that of any interstellar interloper) can be gauged according to how much sunlight it reflects. Indeed, the more sunlight an object reflects towards a telescope, so the brighter it will appear, and in general, the light intensity I is taken as being proportional to the cross-section area A in the line of sight to the object: $I = kA$, where k is a constant relating to the surface reflectance, color, and roughness. For a sphere the cross-section area is $A = \pi R^2$, where R is the radius, and in this manner, providing the entire surface has the same reflectance characteristics, so the measured intensity of reflected sunlight will be constant irrespective of the object's rate of rotation. The resultant light curve, that is the diagram plotting brightness against time, will in this case be a flat line. The light curve deduced for Oumuamua is shown in Figure 17.8, and it is most definitely not a straight line.

The brightness of Oumuamua shows a looping pattern that repeats every 7 to 8 hours (see below), and this is indicative of it having an elongated shape. When seen long-side on, it presents a large surface area to the line of sight and it appears bright; a quarter of a rotation period later, however, it is seen end on, reflects less light and therefore appears fainter. Another quarter rotation period on and the larger area, long-side is seen, and so on. The simplest model to account for the light curve shown by Oumuamua is that of a cylinder, of long-side L and end-on radius R. In this manner the

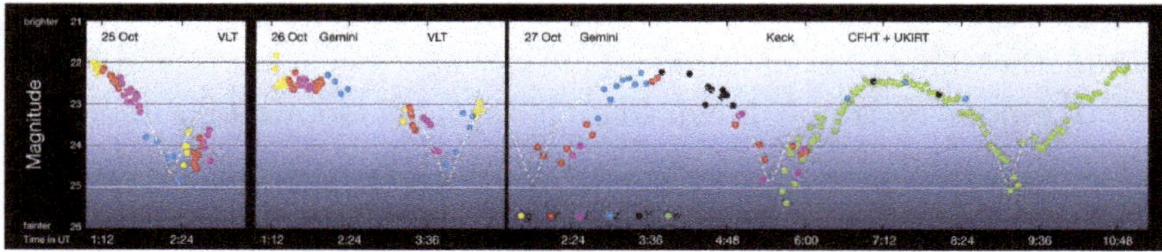

Figure 17.8.   The light curve (brightness) variations measured for Oumuamua between October 25th and 27th, 2017. Time is measured along the horizontal axis, while the brightness of Oumuamua (measured in magnitudes) is indicated on the vertical axis. Image courtesy of ESO.

maximum and minimum intensity ratio is expressed as

$$\frac{I_{\max}}{I_{\min}} = \frac{2RL}{\pi R^2} = \frac{2}{\pi}\left(\frac{L}{R}\right) \tag{17.4}$$

Measurements, with multiple telescope systems, of $I_{\max}$ and $I_{\min}$ revealed that Oumuamua is highly elongated, even cigar-shaped (Figure 17.9), with its longer side length (L) being somewhere between 5 and 10 times larger than its diameter ($2R$). This elongated profile is mysterious in the sense that nothing of a similar physical nature has been found in the solar system. Remarkably, the first interstellar object to be detected looks nothing like any asteroid or cometary nucleus that has been seen before.

The shape deduced for Oumuamua's profile (Figure 17.9) represents a mystery that requires explanation — although, to date, no fully convincing model has been forthcoming. According to how much light is actually reflected from its surface (that is its albedo), the various studies that have been published indicate that Oumuamua is a cylinder-like object about 250 meters long and about 35 meters wide. The problem with such a profile, however, is how and when did this shape come about. Are we seeing Oumuamua as it originally formed, prior to being ejected from some distant planetary system, or are we seeing a profile that has come about over the many eons that it has been moving through interstellar space? Indeed, both scenarios have been explored but no clear consensus has emerged between competing research groups.

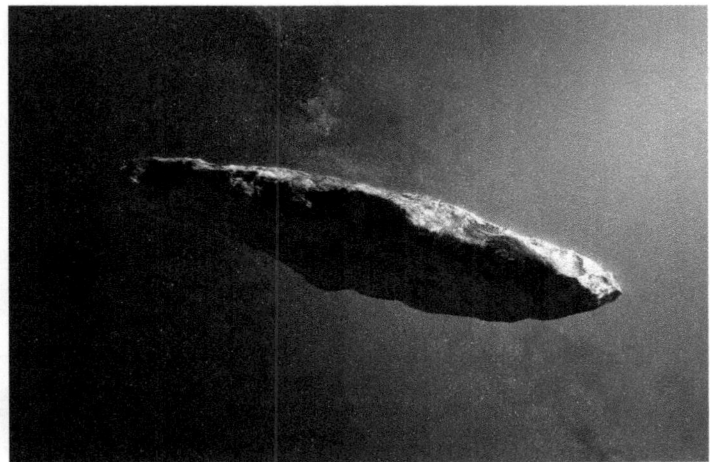

Figure 17.9.   Artist impression of what Oumuamua potentially looks like. It has a highly elongated profile and a dark, low reflectance surface. Image courtesy of ESO.

Not only is the shape of Oumuamua strange with respect to known solar system asteroids and cometary nuclei, its rotation sense is also odd. Rather than rotating about some fixed principle axis, Oumuamua appears to be in a tumbling mode — it is this effect that accounts for the apparent imprecise determination of the rotation period. This tumbling motion is important, since it suggests that Oumuamua was set in this mode by a past collision, and it is to be assumed that that collision was the last one prior to being ejected into interstellar space. There are mechanisms that will gradually damp out tumbling motions, but they operate on very long timescales (of order of billions of years), and accordingly, the fact that Oumuamua is tumbling suggests that it can't be more than a few billions of years old. Since the speed and direction of motion are known, attempts have been made to trace Oumuamua back to its parent star system, but again no consensus between various studies has been reached, and it may be the case that it is not associated with any of the present nearby stars and/or star formation regions.

Perhaps the biggest unresolved mystery concerning Oumuamua relates to its physical nature — is it a comet or is it an asteroid?

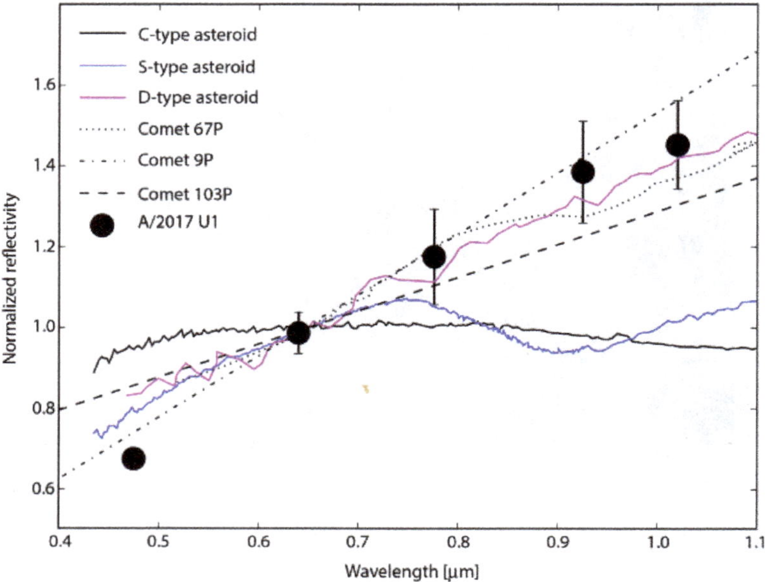

Figure 17.10. Reflectance spectra of interstellar object I1/Oumuamua (A/2017 U1 — large dots) and representative spectra of C-, S-, and D-type asteroids (solid lines). Also shown (dotted and dashed lines) are the spectra of three comets: comet 67P/Churyumov-Gerasimenko, comet 9P/Tempel-1, and comet 103P/Hartley-2.

Reflectance spectroscopy (recall Section 13.1) provides a few clues, but has not fully answered the question. Figure 17.10 shows a series of infrared spectral measurements made of Oumuamua, and compares them against the spectra for C-, S-, and D-type asteroids along with three short-period comets. The data points for Oumuamua are a good fit to the spectra produced by D-type asteroids and cometary nuclei. This indicates that Oumuamua is not a typical stony (S-type) or carbonaceous (C-type) asteroid as found within our solar system. Indeed, the spectra of Oumuamua suggests it is quite possibly an extinct comet, its surface having been effectively de-iced and chemically modified so as to appear like a D-type asteroid. An additional hint that Oumuamua is, in fact, a comet-like body, was revealed by the detection of what are called non-gravitational forces. These forces relate to outgassing and jetting effects that arise from the localized sublimation of surface ices,[11] and have long been implicated in the modification of periodic comet orbits within the solar system. As ever, however, the situation with Oumuamua is somewhat unclear. While an additional, very small, non-gravitational effect was detected with respect to the motion of Oumuamua, no indication of actual outgassing was observed. This set of circumstances is unprecedented, and indeed, Shmuel Bialy and Abraham Loeb (Harvard University) have suggested that this might indicate an even more remarkable possibility for the origin of Oumuamua. Perhaps, they suggest, it is a light sail.

Bialy and Loeb investigated the small deviation of Oumuamua from that of a purely gravitational hyperbolic orbit in a research paper that appeared in the November 12th, 2018 issue of *Astrophysical Journal Letters*. Noting the observational constraint that there was no indication of the observed acceleration being due to the sublimation of surface ices, Bialy and Loeb suggested that direct solar radiation pressure might be the mechanism at play. Here, we are back (in some sense) to the β-meteoroids discussed at the beginning of the chapter. The pressure P due to the Sun's radiation acting upon an object at a heliocentric distance r is

$$P = \Gamma \frac{L/4\pi\, r^2}{c} \tag{17.5}$$

where Γ is a coefficient related to the reflectivity of the object, c is the speed of light, and L is the Sun's luminosity. This solar radiation pressure will result in an acceleration a of order

$$a = P\left(\frac{A}{m}\right) = \Gamma\left(\frac{L}{4\pi\, r^2 c}\right)\left(\frac{A}{m}\right) \tag{17.6}$$

where A is the surface area presented to the Sun's radiation field and m is the mass of the object.

---

[11]The name non-gravitational is somewhat of a misnomer, but the idea is that there are additional forces, beyond that of gravitational attraction, at play in determining the orbital dynamics.

For a perfect reflector (that is a good mirror) the constant $\Gamma = 2$, and accordingly Equation (17.6) reduces to

$$a = 9.14 \times 10^{-6} \left(\frac{r}{\text{AU}}\right)^{-2} \left(\frac{m/A}{\text{kg/m}^2}\right)^{-1} \qquad (17.7)$$

where the heliocentric distance is now expressed in astronomical units. This result can be compared to the actual observational constraints that indicated an anomalous acceleration of

$$a = 4.92 \times 10^{-6} \left(\frac{r}{\text{AU}}\right)^{-2} \qquad (17.8)$$

Accordingly, the mass-to-area ratio required to produce the observed anomalous motion of Oumuamua is $m/A \approx 1.86\,\text{kg/m}^2$. If we take the maximum area of Oumuamua to be that of a cylinder of length $L = 250\,\text{m}$ and end-diameter $2R = 35\,\text{m}$ (as suggested by the light-curve observations — Figure 17.8), so the implied mass of Oumuamua is 16,000 kg, which is an absurdly low mass.

Keeping the same cylindrical geometry, but assuming Oumuamua is composed of water ice with a density of $900\,\text{kg/m}^3$, so its mass would be $2 \times 10^8\,\text{kg}$ (giving a mass-to-area ratio of 2.3×10^4). Its mass (and mass-to-area ratio) would be even greater if it is made of rock and/or nickel-iron. One way to reconcile the numbers, according to Bialy and Loeb, however, is to accept the observations at face value, and this implies that Oumuamua is really a very thin, highly reflective, sheet-like object — that is, it something like a light sail[12] (Figure 17.11). The remarkable implication that follows from this statement, of course, is that Oumuamua could potentially be a light sail built by an extraterrestrial civilization. Could Oumuamua be, Bialy and Loeb question, the first detection of alien technology — a scout ship deliberately sent into interstellar space, or perhaps some defunct space debris that escaped into interplanetary space? Indeed,

---

[12]The Japanese Space Agency successfully used a small light sail in its 2010 IKAROS (Interplanetary Kite-craft Accelerated by Radiation of the Sun) mission to study the planet Venus. The light sail was $14 \times 14\,\text{m}$ with a total mass of 315 kg (giving a mass-to-area ratio of 1.6). The more recently launched (2019) Light Sail 2 has a sail area of $32\,\text{m}^2$ and mass of 5 kg (giving a mass-to-area ratio of 0.16).

Figure 17.11. Image from Light Sail 2, showing Baha California, Mexico, and part of the light sail itself. Image courtesy of The Planetary Society.

Oumuamua was monitored by the Allen Telescope Array operated by the SETI Institute at Mountain View, California. No outgoing radio emissions were (sadly?) detected, however.

17.5 Borisov

After Oumuamua the next interstellar object to pass through the inner solar system was comet C/2019 Q4 (Borisov). This latter discovery was again a great surprise — the surprise this time, however, being that it occurred within just two years of the very first interstellar object being detected — the count of known interstellar objects passing through the solar system jumping from zero throughout recorded history, to 2 objects within 2 years. Discovered by Gennady Borisov on August 30th, 2019 from the Crimean Astrophysical Observatory in the Ukraine, this latest interstellar interloper passed through perihelion on December 10th, 2019. Since this object, unlike Oumuamua, sports a very definite tail feature (Figure 17.12), it can be readily classified as a comet, and in this

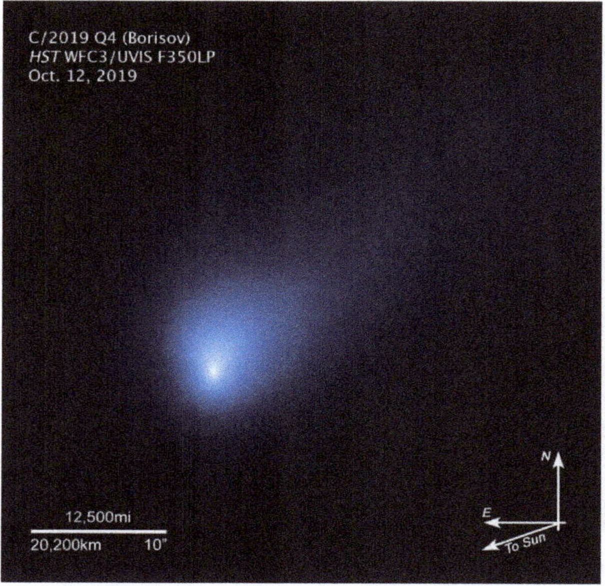

Figure 17.12.   Image of interstellar comet C/2019 Q4 (Borisov). Taken on October 12th, 2019, this early-detection image shows the faint nebulosity associated with the comet's nucleus. Image courtesy of NASA/ESA HST.

case the nucleus, estimated to be about half a kilometer across, must have formed in the outer, cooler reaches of some distant planetary system before being ejected into interstellar space. Indeed, the orbital eccentricity of C/2019 Q4 (Borisov) = 2I/Borisov is deduced to be $e = 3.36$, clearly indicating a hyperbolic orbit (recall Chapter 3), and betraying the fact that it cannot be an object that formed within our own solar system and is now moving in towards the Sun from the Oort cloud.

For all this, however, infrared spectra of 2I/Borisov indicates a close similarity to that betrayed by cometary nuclei found within the solar system — the nucleus containing the expected H_2O, CO_2, and CO ices. One great oddity, however, and one that probably relates to the host star about which comet 2I/Borisov formed, is that the observed ratio of CO/H_2O is substantially higher than that observed for any other comet within the solar system. One interpretation of this result suggests that the 2I/Borisov formed around a low mass, low luminosity, low temperature, M-spectral-type star. These stars are the most common type of star within the galaxy,[13] and also the coolest. The retention of so much CO in the nucleus of 2I/Borisov indicates that its nucleus has never been greatly warmed, and this observation suggests a formation location within the accretion disk around a newly forming, low temperature star.

As a final surprise for observers, as 2I/Borisov began its exit from the solar system, in March 2020, it underwent a series of sudden brightening events. Observations obtained with the Hubble Space Telescope indicated that its nucleus had split into at least two components. Such fragmentation is not an uncommon ending for solar-system-born comets, but this is perhaps a more remarkable event since, henceforth, the two nuclei will drift further and further apart as they head back into interstellar space — what was one interstellar comet is now two.

While 1I/2017 U1 Oumuamua and 2I/Borisov are, as deduced from their orbital paths, most definitely interstellar objects, there are a few additional objects within the solar system that suggest they might have had an origin from outside of the solar system — that is, that they have been captured from interstellar space, or that they were captured at a very early stage of the solar system's history when it was still within its natal cluster. Perhaps the most likely such captured interstellar object is asteroid 514107 Ka'epaoka'awela (the name being Hawaiian for *the tricky Jupiter*). Unique amongst the asteroids, this object orbits the Sun in a retrograde, 1:1 resonance with planet Jupiter. In this manner it has a similar orbit to that of Jupiter, but distinctively orbits the Sun in the opposite sense to the planets and all other asteroids. It is far from clear how 514107 Ka'epaoka'awela might have been captured into the orbit that it has,

---

[13] About 78 percent of all stars are classified as being of an M spectral type, and they are roughly 10 times more common in the galaxy than Sun-like stars.

but numerical studies indicate that the orbit is remarkably stable over many millions, if not billions of years — accordingly, 514107 may have been captured from a star–planet system forming in the same interstellar cloud as the Sun and planets. Rather than being a sibling of the Sun it may well be, in fact, a distant cousin.

References

[1] A. M. Davis. "Stardust in meteorites." *Proceedings of the National Academy of Sciences of the United States of America,* **108** (48), 19142–19146 (2011).

[2] P. R. Heck *et al.* "Lifetimes of interstellar dust from cosmic ray exposure ages of presolar silicon carbide." *Proceedings of the National Academy of Sciences of the United States of America,* **117** (4), 1884–1889 (2020).

Chapter 18

Hammer Stones

Destiny is a difficult idea to pin down. Are things destined to be, or does the entirely improbable happen by brute random chance? Imagine the situation: it is a hot and humid afternoon, and a quiet sleep on the sofa is just what the hard-worked body calls for. Sleep soon follows and the mind is at ease. Suddenly, the brain awakes and the body starts — something is wrong. What's that noise? A great crashing and splintering can be heard and out of the blue, in the comfort and confines of your own home, a rock punches through the ceiling, careens off a piece of furniture, and smacks you hard and painfully in the side. This may all sound improbable, but it is exactly what happened to Ann Hodges, in Sylacauga, Alabama in the early afternoon on November 30th in 1954 (Figure 18.1).

The rock that smashed its way through house roof, floor, and ceiling and eventually struck Mrs. Hodges was a meteorite — an ordinary H4 meteorite, weighing in at 3.86 kilograms. But what are the odds of such an event taking place? They must be, and indeed are, astronomical. Incredibly, based upon its cosmic-ray-exposure age, the parent body to the two recovered Sylacauga meteorites (the other, having a mass of 1.68 kg, landed in a field) set out on its journey, being released from its parent asteroid, some 37.7 million years ago. What are the odds of such a story — this meteorite-producing body took flight before human beings had even evolved. Indeed, it would have been launched in the Priabonian Age of the Eocene, and our nearest ancestor would have been the primate *Aegytopithecus*, which thrived in what is now central Egypt. This ancient ancestor lived before the evolutionary split between Old- and New-World monkeys

391

Figure 18.1.   The 3.86-kg Sylacauga meteorite that struck Ann Hodges in November 1954.

occurred, and we, as Homo sapiens, a side branch of the Old-World line, only appeared about 2 million years before the present.

Was it pre-ordained that this ancient asteroid shard was to injure Ann Hodges, or was it just her bad luck? If it was destiny, then the event brought Mrs. Hodges no joy and she grew to resent and shun any attention — but one can only imagine the media pressure that she must have experienced: Meet Ann Hodges — the woman who was hit by a meteorite. Ann Hodges was 34 years old when the Sylacauga meteorite horribly bruised her side, but she survived; it was not until some 18 years later, in 1972, that she died of liver failure, at the age of 52. Ann Hodges's name has gone down in history, but the meteorite was no win fall. Indeed, the Hodges and the landlord of the house they rented vied for ownership of the meteorite and a legal battle ensued. Eventually, however, the meteorite found its way to the Alabama Museum, where it can be seen to this very day — a famous relic, and testament to the fact that unbelievably small odds can, in spite of everything, come true.

Astronomers have a special name for meteorites that hit someone or something — they are hammer stones, and these meteorite events are the ones that tend to be remembered over the otherwise forgetful arc of history.

18.1   Hit and miss

History, in fact, abounds with near-miss and supposed meteorite hits upon people and buildings. Most are probably only partially true — stories of events, indeed, quite possibly real meteorite fall events, that have become exaggerated and twisted through repeated retellings. One such tale of a close encounter concerns that of Drake's Cannonball. Sir Francis Drake was most definitely a real person, and his marriage to Elizabeth Sydenham most definitely took place in 1585. What is less commonly known, however, is that this marriage would not have taken place but for a strange occurrence that took place just before the wedding day. It transpires that while Francis and Elizabeth had agreed to be wed, when Drake was away at sea, on his famous world circumnavigation voyage in the *Golden Hinde*, Sir George Sydenham (Elizabeth's father), sensing a more profitable arrangement and arguing at the time that Drake was probably dead, convinced his daughter to marry another wealthier suitor. However, this underhand arrangement came to a dramatic end, it is said, when, as the marriage couple stood before the alter, an iron meteorite crashed through the roof of the church. Convinced that this was a warning sign and ill omen, Elizabeth refused to complete the marriage vows. That same day, it is also recounted, Drake returned to Plymouth, and the couple were soon, thereafter, reunited and wed. The iron mass, some 36 cm across, has been kept at Combe Sydenham Hall ever since; it is still on view to this day, and it is appropriately called Drake's Cannonball. Indeed, there is no indication that Drake's Cannonball is an actual meteorite, and the story of an iron mass falling through the church may or may not be true, but it all makes for a grand yarn, and it taps into the mystery, magic, and wonderment that follow upon a close encounter with a meteorite fall.

Other historic fall accounts are even more far fetched (that is, distorted in the telling) than Drake's Cannonball. There are tales of stones falling from the sky, 'like rain,' killing more than 10,000 people in Shansi, China in 1490; a monk being killed by a falling stone in Lombardy, Italy in 1511; and two sailors being killed onboard ship by falling stones while at sea in 1648. In 1790, it is said, a meteorite destroyed a farmhouse, killing the farmer and his cattle, in Barbotan, France. In 1810, it is said, a great stone fell in Shahbad, India, destroying five villages and killing several people. An interesting, but still far from convincing, case for a man being killed and another paralyzed during a fall of meteorites on August 27th, 1888, in Sulaymanigah, Iraq has recently been presented in the March 2020 issue of *Meteoritics and Space Science*. In 1896, a great meteor was seen in the early morning from Madrid, Spain, the sighting being accompanied by loud detonations that shattered windows and shook buildings. While this last account begins to sound very much like a real event, accurately described (see the Chelyabinsk account in Section 21.1), the earlier accounts are less likely to be accurate. Certainly, it can be assumed that something really did happen at the time of the various historic reports, but the retelling of circumstances, often over many human lifetimes, has resulted in associated distortions and exaggerations.

18.2   What are the odds?

For all this, however, meteorites have, do, and will hit buildings and even people — and they may have even killed animals and possibly hapless bystanders. The problem for the historian is separating myth from reality — that is, the problem is one of signal to noise. While the historical accounts are *noisy* with *narrativium*,[1] in the modern era the probability of a person, a building, or a village being hit by a meteorite can be calculated directly. The hit probability P_{hit} is readily calculated from the product of the flux F of meteorites,

---

[1]Narrativium was the term invented by Terry Pratchett to describe the essential ingredient that must be placed within a story to make it work.

larger than a given size, falling to Earth per year per square meter, and the target area A_{target} exposed to the flux. The probability of being hit within a given time interval T is accordingly,

$$P_{\text{hit}} = F \times A_{\text{target}} \times T \qquad (18.1)$$

While the flux of impacting bodies is dictated by the dynamics and evolutionary history of the main-belt asteroid region, it is something that can be determined by observations and numerical modeling. For a given flux, however, Equation (18.1) indicates that the probability of some specific object being hit increases in accordance with the time that it is exposed to the sky, the longer the exposure time the better (as it were), and with respect to the size of the target, the bigger the target area the better.

Determining the actual flux, that is the number per unit area per unit time, of meteorites (and asteroids) hitting the Earth's surface is a non-trivial calculation and is based upon specific fireball observations and population modelling. All survey methods have their own specific limitations and biases, but Figure 18.2 is a summary plot of the annual global flux of meteorites and asteroids as deduced by Phil Bland and Natalia Artemieva in a 2006 publication [1]. A logarithmic scale is used in the diagram and covers the mass range from 1 kilogram to 10^{22} kilograms — the former object being just 10 cm across and the latter being some 1853 kilometers across (assuming a density of $3000\,\text{kg/m}^3$). Now, impacts at the upper mass end of the scale are never going to happen — the solar system is not old enough, even at 4.56 billion years in age, for any such impact to reach a probability of $P_{\text{hit}} = 1$, and additionally, there are no actual objects as large as the upper mass limit within the solar system's reservoir of potential Earth impactors (see Section 12.1).

At the lower end of the scale, however, the global flux of 1-kilogram meteorites is of order 100,000 per year. Most of these meteorites, however, will never be found since the Earth's surface is 2/3rds covered by ocean. Even so, something like 33,000 meteorites larger than 1 kg should fall on land — the question of discovery, or of hitting something or somebody, is now strongly biased upon exactly which bit of land the impactor falls. If a meteorite falls in

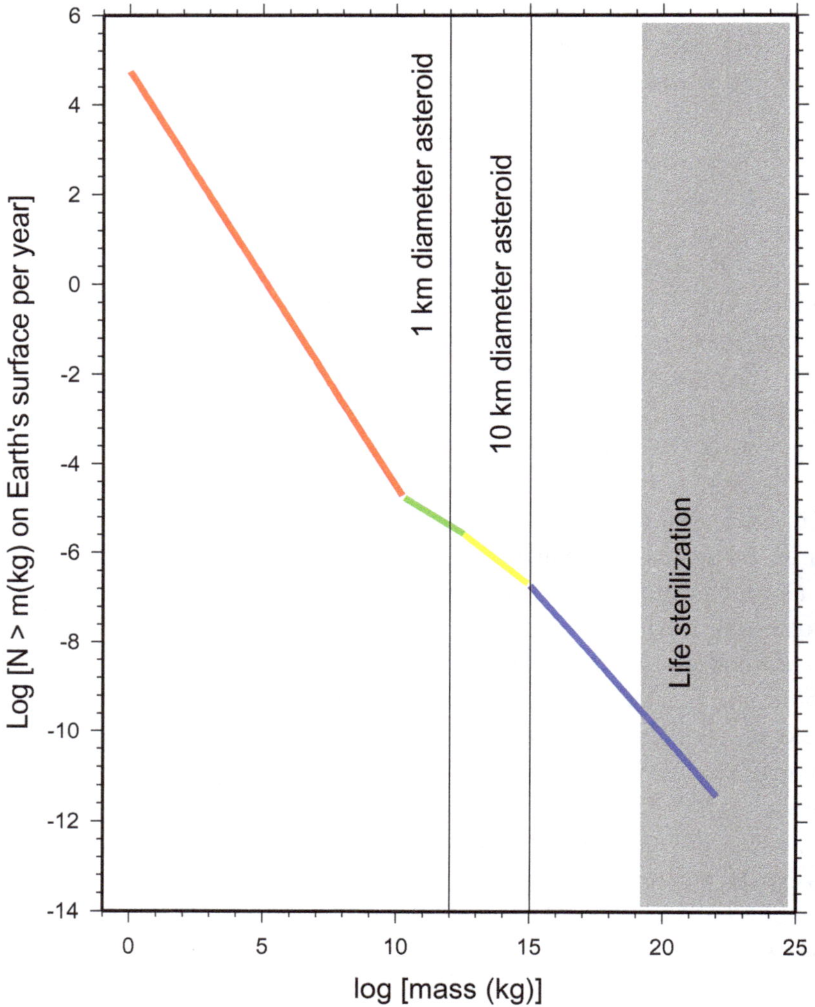

Figure 18.2. The annual global flux of impactors larger than a specific mass as deduced from fireball survey data and solar system crater counts.

the Arctic, Antarctic, the Amazon jungle, or in the Sahara dessert, regions of very low human population density, then it will most likely go unobserved and undetected (as a fall). This being said, there are physical mechanisms that make some remote and uninhabited locations good for meteorite recovery as finds (see Chapter 14).

One of the distinctive features evident in Figure 18.2 is that the number of impactors larger than a given mass hitting the Earth's surface per year is a strongly decreasing function of the impactor mass. This simply means that impacts with large (that is asteroid-sized) objects are rare. The vertical line labeled "1-km diameter asteroid" (an object with a mass of order 10^{12} kg) in Figure 18.2 cuts the regression line at about 3×10^{-6}, indicating that a hit from such an object is likely to occur somewhere on Earth once every 300,000 years. At 10 km across (corresponding to a mass of 10^{15} kg), impacts are likely to occur at intervals of order several tens of million years. One such impact with a 10-km sized asteroid was that which took place 65 million years ago and contributed to the demise of the dinosaurs. Other similar sized impacts must have taken place since the dinosaur were driven to extinction, but the geological and fossil signatures relating to these events have not, to date, been forthcoming. The far right (shaded) region in Figure 18.2, labeled "life sterilization," indicates those impacts that could potentially result in the extinction of all life on Earth. The limiting impactor mass for the sterilization of all life on Earth has been determined by David Sloan (Oxford University) and coworkers [2], and is estimated to be 10^{19} kg (corresponding to a diameter of about 185 km). This mass limit was calculated according to the energy requirements needed to perturb the Earth's biosphere to such a degree that the humble tardigrade (Figure 18.3) could no longer survive.

Tardigrades are remarkable animals that live almost anywhere, and we are surrounded by them. Small by human standards, they are just a millimeter or so in size. The tardigrades are an example of a great survivor in the game of life. They have flourished for at least the past 500 million years, and various studies and experiments indicate that they can survive the rigors of non-protected spaceflight, near complete desiccation, high radiation dosages, and even explosive shock-wave compression that would kill almost anything else. As a great survivor on Earth, the tardigrade is also a good candidate for being the solar system's contribution to lithopanspermia (see Chapter 15). Indeed, the humble tardigrade has already made it to the Moon, all be it in a desiccated form scattered amidst the wreckage

Figure 18.3.   Electron microscope image of a tardigrade. Such *Slow Walkers* (or *Water Bears*) are amongst the most resilient of animals known.

of the *Beresheet* lander.[2] The tardigrades were encased in resin, and formed part of an analog data archive cache assembled by the Arch Mission Foundation. The lunar colony is certainly not viable, but the all-important tardigrade DNA has been preserved — and in the game of life, it's the DNA that counts.

In order to use the data shown in Figure 18.2 and calculate impact probabilities, through Equation (18.1), the area A_target needs to be expressed in terms of the Earth's surface area: $A_\text{Earth} = 5.101 \times 10^{14}\,\text{m}^2$. It is this factor that dramatically reduces the probability of even very small mass meteorites hitting something like a human being or a car. Accordingly, taking the flux of 1-kg meteorites hitting dry land to be 33,000 per year, then the probability of an adult human being, with a surface area of about $1.0\,\text{m}^2$, suffering a hit is $P_\text{hit} = 33,000 \times 1/(5.101 \times 10^{14}) = 6.5 \times 10^{-11}$ per year. The global human population is (as of 2020) set at 7.7 billion and accordingly something like one person should be hit by a 1-kg meteorite or larger every 2 years ($P_\text{hit} = 0.5$). This is clearly an overestimate and does not include adjustment for hit-probability-reducing factors such as people are not uniformly distributed over

---

[2]The first lunar lander mission attempted by a private company, *Beresheet* unfortunately crashed in Mare Serenitatis on April 11th, 2019.

the land area of the Earth, children have a much smaller surface area than an adult [A_{target} (child) $\sim \frac{1}{2} A_{\text{target}}$ (adult)], and that a high fraction of people live in buildings which will typically protect them from a hit (although not in the case of Ann Hodges). The impact probability just determined, that someone, somewhere, should be hit by a meteorite every few year, is perhaps surprisingly high, given that only three documented accounts[3] of such strikes have taken place in the 66 plus years since the 1954 fall in Sylacauga,[4] but it does underscore the point that historically (although the global population was much smaller) we should certainly expect to find occasions when humans were truly hit by a meteorite, and that fatalities should, indeed, have occurred.

While much of the Earth's surface is inhospitable to both the recovery of meteorites and the long-term survival of such materials, some regions are exceptional in their ability to preserve. These places are those with an inherently dry climate and include Antarctica (a cold desert) and the Sahara and the Atacama Desert (which are hot deserts). The temperature of the region is not so much the issue with respect to meteorite preservation; the key point is that a region must have very little rainfall. Indeed, in terms of meteorite survival, once they have landed on Earth, the key factor is liquid water. Meteorite weathering is often a rapid process and within just a few short years of exposure to wind, rain, freezing, and thawing, a meteorite will either be totally degraded (this is especially so for carbonaceous chondrites) or highly rusted and chemically altered. A meteorite weathering scale that varies from W0 to W6 is often used to account for the terrestrial modification, with W0 essentially corresponding to a newly fallen meteorite and indicates that no oxidation (rusting) is observable. From W1 to W4 the metal

---

[3] A young boy was reportedly hit by a small fragment of the L5-6 Mbale, Uganda, chondrite on August 14th, 1992, and several villagers were reportedly hit by small fragments of the Kendrapara H4-5 chondrite meteorite fall in India on September 27th, 2003. Neither of these accounts, however, can be verified by supporting evidence or documentation.

[4] These statistics, at face value, suggest that one person per ten-year time interval might be struck by a falling meteorite.

component shows minor to complete oxidation; from W5 to W6 the stony (silicate) component shows stronger and stronger replacement by clay minerals and oxides.

In very dry regions, where rain rarely (or never) falls, the weathering of meteorites is slowed and meteorites can accumulate over thousands, even millions of years as distinct and recognizable objects. Under these circumstances the flux of meteorites can be estimated by careful survey and collection work over a specified area of land, and then by measuring the terrestrial residency age of the meteorites found (recall Chapter 4) to set an actual timescale. Just such a survey was recently completed by Alexis Drouard (Aix-Marseillie University, France) and coworkers in 2019 [3]. Drouard and collaborators studied a small region of the ultra-dry Atacama Desert (Figure 18.4) from which some 388 meteorites have been recovered. Of this sample of meteorites, 54 were randomly selected and their terrestrial residency ages were determined by measuring ^{36}Cl cosmogenic radionuclide

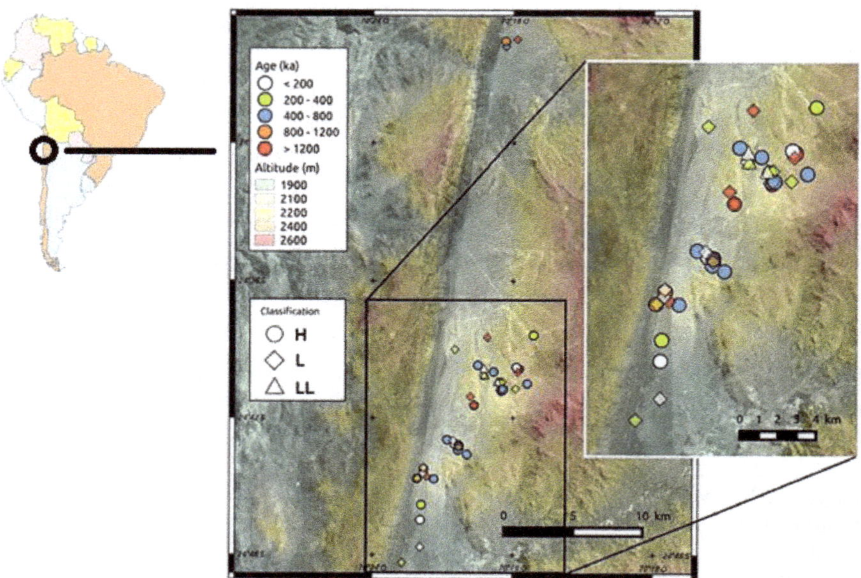

Figure 18.4.   Map of the El Médano and Caleta el Cobre areas of the Atacama Desert showing the locations of 54 meteorite finds, along with their class (H, L, or LL) and age. Image adapted from [3].

concentrations. Terrestrial residency ages from a few thousands of years to over 2 million years were determined — the mean terrestrial residency age was found to be 710,000 years, which attests to the long-term stability of the dry climate in the region being studied.

Combining the terrestrial residency ages with the statistics concerning the search area and the total number of meteorites discovered, Drouard *et al.* concluded that average meteorite flux over the past 2 million years is 2.22×10^{-4} meteorites more massive than $0.01\,\mathrm{kg}$ per km^2 per year. While determined from data collected in the Atacama Desert, this flux, given the random nature of meteorite falls, applies to any location on Earth.

18.3 Buildings, cars, and other objects

While it has been seen that meteorite hits upon human beings are rare, meteorite hits upon buildings and cars are more commonly reported. This is the case since buildings and cars will have larger surface areas than human beings, and because buildings are permanently exposed to the elements rather than human beings who live, sleep, and work inside of buildings. Table 18.1 provides a table of known meteorite hits upon cars. Remarkably, while the first Model-T car rolled out of the Ford production plant in 1908, it was just 30 years before the first reported strike was documented following the fall of the Benld meteorite (Figure 18.5)

If the typical surface area of a car roof is taken to be $2\,\mathrm{m}^2$, the expected number of car strikes around the world per year, at the present time, can be calculated. There are some 1.2 billion cars in the world at the present time, and this corresponds to a total exposed area of $2.4 \times 10^9\,\mathrm{m}^2$ (if these cars were all put together in one giant parking lot, they would cover an area the size of Luxembourg[5]). Using the meteorite flux estimate of Drouard *et al.*, the probability of a car being hit by a meteorite with a mass greater than 10 grams,

---

[5]Only one meteorite fall has ever been recovered from Luxembourg — our super-sized parking lot — and this fell in the village of Tintigny in February 1971. This particular meteorite was a 210-gram eucrite achondrite, and it was a hammer stone, being recovered after smashing its way through the roof of a building.

Table 18.1.   Confirmed meteorite strikes upon cars since 1938. The first two columns indicate the date and location, while the third and forth indicate the meteorite mass and type. The last column indicates where the car was at the time of the strike.

Date	Location	Mass (kg)	Type	Car
September 25th, 2009	Grimsby, Ontario	0.46	H5	Parked car
March 26th, 2003	Park Forest, IL	—	L5	Parked car
September 1st, 1997	Worden, MI	1.6	L5	Car in garage
February 18th, 1995	Neagari, Japan	0.42	L6	Parked car
October 9th, 1992	Peekskill, NY	12.0	H6	Parked car
January 31st, 1977	Louisville, KY	—	L6	Parked car
December 24th, 1965	Barwell, UK	—	L5	Parked car
December 10th, 1950	St. Louis, MO	1.0	H4	Moving car
September 29th, 1938	Benld, IL	1.8	H6	Car in shed

Figure 18.5.   The Benld (hammer stone) meteorite fall of September 29th, 1938 resulted in damage to Mr. Edward McCain's 1928 Pontiac coupe.

somewhere in the world, is $P_{\text{hit}} = 2.2 \times 10^{-4} \times 2400 \approx 0.5$ — remarkable, the probability of a car being hit by a 10-gram meteorite, somewhere in the world, is about the same as that for a person being struck by a 1-kg meteorite. The fact that more car hits have been reported in the last 70 years (8 in total) than hits upon human beings (3 in total) is largely (but not exclusively) due to the fact that much of the time human beings are protected inside of buildings whereas cars are not.

In modern times no human fatality, due to a meteorite hit, has been recorded, but this does not mean that the risk is so small that it can be ignored. Indeed, there are potential meteorite hits that could have a significant (dare one say, high impact) effect upon human confidence and society. Such possibilities could be a meteorite hitting an aeroplane while in flight or the destruction of a communications satellite in space. Satellites have certainly been hit by small-mass meteoroids (see below), and a meteorite hit has been investigated in the case of TWA flight 800.

Indeed, the tragic downing of TWA flight 800 is the third worst airline disaster in American history, with all 230 passengers onboard being killed in the incident on July 17th, 1996. It must be said at the outset that a meteorite hit on the Boing 747 aircraft is one of the least likely explanations for the disaster, but at the time of the incident many eyewitness accounts reported a streak of light being seen in the sky prior to the explosion taking place. Investigators did rule out a terrorist bomb or missile attack as the cause of the explosion, and concluded, after an intensive investigation, that the most probable case was the ignition of fuel vapor. Geologist William Cassidy (University of Pittsburgh) was asked to investigate the probability of the TWA 800 explosion being the result of a meteorite hit [4], and while Cassidy applied an analysis relating to the hammer stone hits on houses, the calculation can also proceed via Equation (18.1).

The area of a Boing 747 is $A_{747} \approx 500\,\text{m}^2$, and let us assume that at the altitude in question (about 4 km altitude) a meteorite will be traveling at about 100 meters per second — this is significantly slower than its cosmic velocity due to the meteoroid's passage through the atmosphere. Now, an M67 hand grenade releases about 5×10^5 joules worth of explosive energy, and we shall assume that a similar such explosion is enough to cause catastrophic aircraft damage. The mass of a meteorite with the same kinetic energy as an M67 hand grenade is 4 kg. The graph provided by Bland and Artemieva (Figure 18.2) indicates that at 4-kg mass, there are something like 32,000 meteorite falls per year around the globe. The TWA flight 800 catastrophe occurred 12 minutes after take off, and we can now determine from Equation (18.1) the probability of the aircraft being hit by a 4-kg meteorite within this timeframe. Accordingly, $P_{\mathrm{hit}} = [32,000/(5.101 \times 10^{14} \times^3 .156 \times 10^7)] \times 500 \times (12 \times 60) \approx 10^{-12}$, which is comfortingly very small and essentially negligible. Indeed, for P_{hit} to approach say 0.01 (1 percent) a single 747 aircraft would need to clock some 320,000 years of flight time.

At any one instant of the day it has been estimated that there are something like 10,000 aeroplanes in the sky around the world, so even with this enhanced number of aircraft, the time interval required for the aircraft impact probability to approach 1 percent is some 32 years — or 3,200 years for the impact probability to approach certainty (100 percent). These numbers are generally comforting and death by being caught in an aircraft destroyed by a meteorite strike is certainly not something worth worrying about. This comfort, however, is only won statistically, and indeed, there is absolutely nothing to say that the very next flight that you are on won't be the first to be hit by a meteorite.

The cost of building and launching a communications satellite can run into the many hundreds of millions of dollars range. The Olympus-1 communications satellite (Figure 18.6) launched by the European Space Agency (ESA) in 1989 was no exception to this rule, and while it cost some 850 million dollars to assemble and launch, its mission was ended by a single impact from a meteoroid with a mass of about 5×10^{-5} grams. This remarkably small mass for producing serious satellite damage comes about because satellites are hit by meteoroids having their cosmic speed. In the case of the Olympus end-of-mission anomaly it was the time of the event that provided a distinct clue as to the speed of the impactor. Indeed, at

Figure 18.6.   Technicians work on the Olumpus-1 spacecraft. When launched on July 12th, 1989 it was the world's largest telecomunications satellite. Image courtesy of ESA.

23:32 UT on August 11th, 1993 an anomalous momentum impulse set the Olumpus-1 satellite into an uncontrolled spin mode. The impulse that set the spacecraft spinning was a meteoroid hit upon one of its large solar panels, and the clue as to the speed of the impactor is that the event occurred as the annual Perseid meteor shower was approaching maximum activity.

The key point about Perseid meteoroids is that they have a high cosmic speed of about 60 km/s at Earth's orbit — they are speeding bullets. Additionally, in the year in question the parent comet to the Perseid stream, comet 109P/Swift-Tuttle, had recently returned to perihelion and the shower was undergoing enhanced outburst activity. It was a near perfect storm for the Olympus-1 satellite — it presented a large area to the stream and the Perseids were passing through Earth's orbit thick and fast. Given the high cosmic speed of Perseid meteoroids even a very small mass impactor can deliver a lot of impact energy and momentum transfer. A Perseid meteoroid with a mass of just 5×10^{-5} grams (an object a mere 3 millimeters across) is enough to account for the recorded spacecraft spin anomaly, and at such a mass the kinetic energy of motion is equivalent to about $1/5^{\text{th}}$ that of an M67 hand grenade.

The Olympus-1 mission-ending anomaly is not the only instance in which an active satellite has been struck by a meteoroid. Indeed, spin anomalies have been observed on several other spacecraft (e.g., the Russian Ekspress AM11 communications satellite in 2006). Increasingly, however, it is the case that satellites have to attend with collisions from other spacecraft and space debris as well as meteoroids — e.g., the Cosmos 2251 and Iridium 33 collision in 2009, and the French Cerise microsatellite impact event in 1996.

The threat associated with meteorite and meteoroid impact events stretches beyond the mere physical. Continuing the ancient theme of fireballs being the portend of doom, in the modern era, military and nuclear test-ban treaty analysts have to distinguish between natural fireball events and enemy action. Monitoring of atmospheric and underground nuclear testing is fundamental to preserving modern peace treaties, and one of the ways in which this monitoring is achieved is through the detection of infrasound anomalies. These long-wavelength sound waves can be analysed to yield energy thresholds and are discussed in more detail in Chapter 21. For now, the point is that a meteorite-producing event can mimic the infrasound signal produced by a nuclear explosion, and the challenge for the analyst is to distinguish between the two phenomena.

Additionally, the fireball associated with the entry of a large meteoroid into Earth's atmosphere mimics that of a missile launch, and satellite monitoring for such transient events could be mistaken for enemy action. Since the mid-1970s the U.S. Department of Defence (DoD) has used satellites to continuously monitor the entire Earth for potential missile launch and re-entry events, and over this time many hundreds of fireball and small-asteroid explosions in the atmosphere have been detected. These satellites work at optical and infrared wavelengths and can pinpoint precise entry locations, and determine velocity and energy-related characteristics. While most of the data gathered by the monitoring satellites is not available for public (or academic) analysis, some data have been released and form the CNEOS catalog (https://cneos.jpl.nasa.gov/). For our discussion, however, the point is that analysis of satellite trigger events is made in real time and accordingly the timing and location of an event are important. For example, the DoD satellites detected an atmospheric explosion, with an estimated impact energy of 5.2 kilotons of TNT equivalent energy, on October 1st, 1990 at 03:51:47 UT. The event was actually centered over the Pacific Ocean, north of Papua New Guinea, but it occurred during the machinations of the Dessert Shield War in Iraq. If the entry point of the asteroid had chanced to be centered over Bagdad, then such a detection might, in the heat of the moment, have been interpreted as the detonation of a nuclear bomb.

References

[1] P. A. Bland *et al.* "The flux of meteorites to the Earth over the past 50,000 years." *Monthly Notices of the Royal Astronomical Society*, **283**, 551 (1996).

[2] D. Sloan, R. A. Batista, and A. Loeb. "The resilience of life to astrophysical events." *Scientific Reports*, **7**, 5419 (2017).

[3] A. Drouard *et al.* "The meteorite flux of the past 2 m.y. record in the Atacama Desert." *Geology*, **47**, 673–676 (2019).

[4] W. A. Cassidy. "Estimated frequency of a meteorite striking an aircraft." https://www.dartmouth.edu/~chance/teaching_aids/books_articles/Cassidy.pdf (article accessed June 2020).

Chapter 19

Draco Volans

"It's a bird... It's a plane... It's Superman" is the title of the 1966 Broadway musical composed by Charles Strousse, and its sense of confusion and otherworldliness encapsulates how many eyewitnesses respond to the sight of a bright fireball — the erstwhile *Draco Volans* or *flying dragon*. Indeed, there is something unreal and magical about the sight of a very bright meteor — it is an unexpected, shocking, rarely experienced event, and the typical human being has no readily accessible framework within which to interpret what it is that is being seen. The fireball experience is not just something beyond the norm, it is an experience largely outside of human experience, and descriptive words accordingly tend to fail the individual observer (Figure 19.1). One eyewitness to the 1908 Tunguska event (see Section 21.2) commented that, "I saw the sky open ... to the ground and fire pour out" — indeed, an apocalyptic description of an extraordinary event, but a description that carries tremendous imagery [1].

19.1 Singular objects of veneration

It is not an uncommon situation, throughout human history, that when something is big, bright, powerful, and generally misunderstood, that it is turned into a god or some object of veneration. Such are the workings of Pascal's wager: even if you don't believe such things are gods or portends, better to treat them as such — just in case. Indeed, the two most widely spread religions in the world, Christianity and Islam, may both have founding meteoritical aspects.

409

Figure 19.1. The Peekskill fireball as it careened across the sky from six different viewing angles. A truly remarkable sight and one that many eyewitnesses described as being *unworldly*, like a UFO, and/or feared that it was an exploding aeroplane, or an incoming missile.

It has long been suggested that the black stone housed in the eastern corner of the Kaaba at Mecca is a meteorite. According to Islamic tradition, it is believed that the stone was placed by the prophet Mohamed into the Kaaba in 605 BCE, five years before his first revelation. Indeed, the tradition holds that Adam and Eve were inspired to build the first altar and temple after seeing the stone fall from heaven (that is, Jannah — *the paradise gardens*).

This is truly beautiful imagery and incorporation. Many of the miraculous phenomena that are described in the Christian Bible have additionally been interpreted in terms of astronomical phenomena. William Hartmann (Planetary Science Institute, Tucson) has recently argued, for example, that the events leading to the conversion of Saul of Tarsus (later Saint Paul) on the road to Damascus may well have been the result of observing a bolide similar in appearance to the Chelyabinsk fireball or the Tunguska conflagration. The story of Saint Paul conversion is one of the pivotal events in the early history of the Christian Church, and it is described in the *Acts of the Apostles*. The accounts indicate that at midday a bright light, brighter than the Sun, burst from the heavens, producing thunderous sounds. People on the road fell to the ground and Saul was temporarily blinded and heard a divine voice telling him to change his ways. The experience of epiphanic sounds, 'burning eyes,' and 'prickly skin' were all phenomena reported by eyewitnesses to the Chelyabinsk fireball, and indeed, many described what they experienced in terms of a religious-like euphoria.

The sudden, unexpected appearance of bright, thundering fireballs and wayward meteorites falling from the sky, gain relevance and mystery from the very fact that, to a human being, they are singular events. Once all the excitement and surprise have passed, no similar such event is seen again. Repeat phenomena imply order, whereas one-off events imply chaos or special, even divine intervention. For all this, there are no preferred days or times of the year when meteorites are more likely to fall. Figure 19.2 indicates the number of meteorite falls (recorded according to day of the year) over the 400-year time interval from 1610 to 2010. In all, 952 falls are indicated, and over the timespan considered, only 27 days remained free of any meteorite fall association. November 5th turned out to be the most auspicious day for the recording of a meteorite fall, with 8 falls being attributed to that specific day. The greatest cumulative number of falls were recorded during the month of June (a total of 92 falls) and the least number of falls were recorded during the month of March (with a total of 63 falls); only April has accumulated at least one fall for every day of the month. Effectively, Figure 19.2 attests to the fact that,

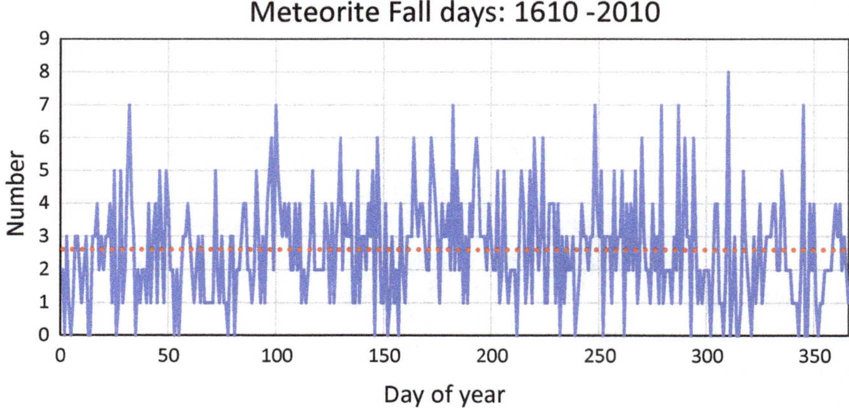

Figure 19.2.   Number of meteorite falls recorded on each day of the year in the time window from 1610 to 2010. The horizontal dotted line indicates an average of 2.5 falls per given day over the time interval considered. Data from https://en.wikipedia.org/wiki/Meteorite_fall.

at least over the past 400 years, there are no strongly preferred epochs (months or days) when meteorite falls are more likely to occur — that is, there are no meteorite showers.

In the same manner that there are no preferred days upon which meteorites fall, there are equally no preferred fall locations. Indeed, there are only two known instances where two distinct meteorites have fallen, on different days in different years, but in the same location. These are Wethersfield in Connecticut and Honolulu on the island of Hawaii. The falls at Wethersfield are the most remarkable, with one meteorite falling on April 8th, 1971, and the second on November 8th, 1982. Both meteorites were L6 ordinary chondrites, and upon landing, both meteorites hit houses. In Hawaii, an L5 ordinary chondrite meteorite fell near Honolulu on September 27th, 1825 and an H5 ordinary chondrite meteorite fell in Palolo Valley (a suburb of Honolulu) on April 24th, 1949. Of course, there are no physical reasons why meteorites cannot fall at the same location, but the odds are very much against such happenings. Additionally, even if meteorites do fall at the same location this does not automatically indicate the existence of a meteorite stream and/or shower. For all

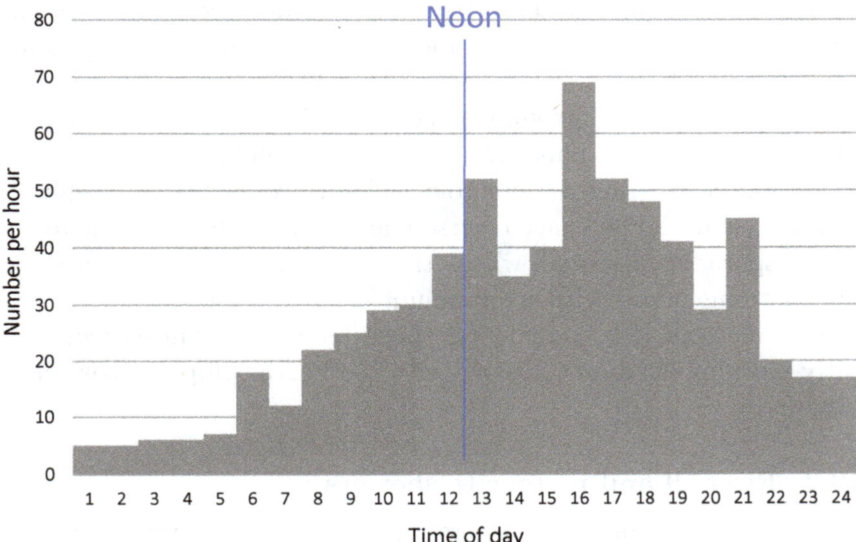

Figure 19.3. Time-of-fall histogram for ordinary chondrite meteorites. It is clear from the plot that more meteorites fall in the afternoon than in the morning hours.

this, however, the idea that there really are meteorite streams has a long and persistent history.

There is one statistic concerning the fall of meteorites that does bear exploration, and this is the observed time of their fall. Figure 19.3 shows the frequency of meteorite falls as a function of the local time of day. The diagram indicates an afternoon excess, with more meteorites falling after noon local time than before. Indeed, the most likely time to witness a meteorite fall is about 4 pm local time. This afternoon excess, with 63% of falls occurring between noon and midnight, is believed to be a real phenomenon, rather than some observational bias (most people being asleep in the early morning hours, for example). The excess is the result of several factors, one of which relates to the manner in which meteoroid orbits evolve to become Earth-crossing, and the other relates to the direction of the apex of the Earth's way. At 6 am local time an observer is orientated in the direction of the Earth's motion about the Sun, where as at 6 pm local time they are orientated 180 degrees away from the direction of the Earth's motion about the Sun. This variation results in the Earth's orbital speed being added to that of a meteoroid in the morning hours, but subtracted from that of the meteoroid in the afternoon hours. This encounter speed difference can be enough to determine whether a meteoroid will survive against atmospheric ablation or not. The exact reverse effect is seen with respect to so-called sporadic meteors,[1] where the greatest numbers are seen in the morning hours with a minimum occurring circa 6 pm. In the case of meteors, the incremental speed increase, resulting from the early-morning head-on collisions with the Earth, simply makes them brighter.

19.2   Fireball and meteorite showers

Unlike cometary meteor showers which are characterized by their periodic, annual displays, the evidence for periodic fireball showers is much less certain — that is, perhaps one should say, the evidence for such showers, at the present time, is not convincing to all practitioners. There are two ways that a shower might be identified, one is to look at the orbital characteristics of fireballs and meteorites and the other is to analyze the times of appearance. Both methods have been applied (more than once and by more than one researcher) and the results are intriguing.

There are still only a relatively few well-determined orbits for meteorites (recall Figure 3.4), but remarkably, of the first 6 deduced 2 have near identical orbital parameters: the Příbram meteorite of April 7th, 1959 and the Neuschwanstein meteorite of April 6th, 2002. In spite of their near identical orbits, however, these two meteorites have very different physical characteristics and cosmic-ray-exposure (CRE) ages. Příbram is an H5 ordinary chondrite meteorite with a CRE age of 12 million years, while Neuschwanstein is a rare EL6 enstatite chondrite meteorite with a CRE age of 48 million years.

---

[1]These are meteors not attached to any specific shower. Characteristically 10 to 15 such visual sporadic meteors are observable every hour of the day, on every day of the year.

On this basis the two meteorites must be derived from two different parent bodies and accordingly are not related or part of a meteorite stream. Adam Pauls and Brett Gladman (University of British Columbia) have numerically studied the orbital dynamics of potential meteorite-producing objects in Earth-crossing streams and find that the decoherence time, such that the fall times of components become random, is of order 10^4 to 10^5 years [2]. These results further argue against a stream association for the Neuschwanstein and Příbram meteorites given the great disparity in their CRE ages. While the numerical simulations of Pauls and Gladman do not rule out the possibility of there being coherent meteorite-producing streams, they do imply that the break-up events responsible for the production of any streams must have occurred in the relatively recent past. All the above being said, however, it is arguably possible (although highly unlikely) that the Neuschwanstein and Příbram meteorites are derived from a coherent meteorite-producing stream associated with the relatively recent break up of a mega brecciated parent asteroid.

The Millman Fireball Archive (MFA) exists partly because of the efforts of a few dedicated researchers, and partly because multitudes of casual observers were sufficiently inspired to report the *hautgoût* of their experiences. The earliest report card in the MFA exemplifies this latter point and relates to the midnight sighting of a simultaneous sound-producing fireball (see Chapter 22) observed from the top of Mount Oscar (in British Columbia, Canada) in August 1927. The primary impetus to begin a Canadian fireball archive, however, followed from the fall of the Bruderheim (Alberta) meteorite on March 4th, 1960, and the archive was initially overseen by Dr. Peter Millman (hence the archive's name) at the Herzberg Institute in Ottawa, Canada. The archive was actively maintained from early 1962 through to the end of 1989, and contains some 3878 report cards relating to the observation of 2131 fireballs [3]. The archive also contains report cards relating to a further 315 fireballs witnessed by observers in the United States. An analysis of the MFA fireball magnitude distribution reveals that most were between magnitude -1 to -5 in brightness (many, however, are simply described as 'bright' or 'very bright'); about 15% of the events were deemed to be brighter than magnitude -10: *Draco Volans*, indeed.

Time of event information is available for 2373 of the fireballs contained in the MFA records. This is less than the full complement of fireballs recorded simply because in a number of cases either no complete time of day information is available, or just a year and month are recorded. For the fireballs with complete time records, the UT time, day, month, and year information have been used to determine a corresponding solar longitude (epoch 2000), and the resultant data set has been 'binned' in 1-degree increments — see Figure 19.4. Solar longitude is used as the time variable in this analysis since it is a more accurate measure (than that of the day of the month) of where the Earth is in its orbit about the Sun. Based upon the tropical year, the zero of solar longitude occurs at the time of the vernal equinox (around March 21st) — the time at which the Sun is located at the point where the ecliptic (defined by the Earth's orbital plane) intersects the celestial equator (defined according to the Earth's obliquity) on the celestial sphere. The solar longitude does not advance at a uniform rate, but takes into account the Earth's variable speed as it moves around its orbit (as required by Kepler's 2nd law — recall Chapter 3).

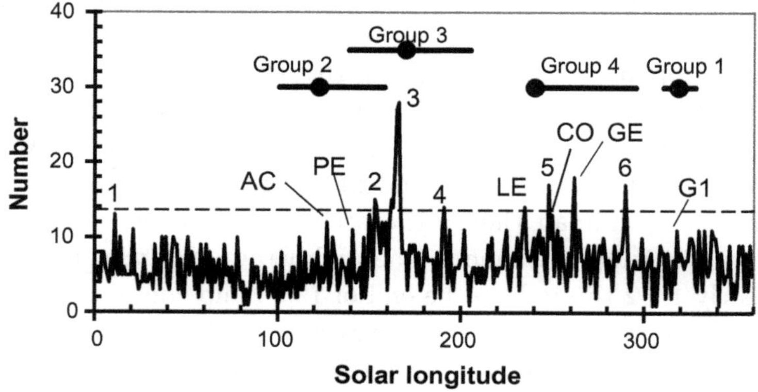

Figure 19.4. Number of fireballs versus solar longitude for the MFA dataset. See Table 19.1 for a discussion of the labeled peaks. The dashed line indicates the two-sigma level above the mean.

Table 19.1. Summary of peaks identified in Figure 19.4. Column 1 relates to the labels shown in the figure. Column 2 provides some brief comments about the possible origins and/or associations for each peak. Column 3 indicates the solar longitude of maximum activity, while column 4 gives the full-width half-maximum (FWHM) value for each identified peak. Column 5 indicates the number of meteors contained in the 'peak bin.' The average number of fireballs per 'one-degree' solar longitude bin is 6.6 ± 3.5.

Label in Figure 19.4	Comments and Possible Associations	$\lambda_\odot$ (deg.) of Maximum	FWHM (deg.)	Number
1	Peace River/Revelstoke (?)	11.0	±0.5	13
AC	α-Capricornid meteor shower	127.0	±0.5	12
PE	Perseid meteor shower	141.0	±0.5	11
2	λ-Aquilid stream	154.0	±3.0	15
3	Group 3 MORP stream	165.0	±4.0	28
4	HC34 meteorite stream (Peekskill)	191.0	±2.0	14
LE	Leonid meteor shower	235.0	±1.0	14
5	Group 4 MORP stream	248.0	±3.0	17
CO	χ-Orionid meteor shower	250.0	±1.0	13
GE	Geminid meteor shower	262.0	±2.0	18
6	Group 4 MORP stream (extreme)	290.0	±2.0	17
G1	Vilna/Group 1 MORP stream	318.0	±1.0	11

A number of features are immediately obvious in Figure 19.4. Firstly, several distinct peaks are delineated and a reasonably clear minimum in activity is evident at around a solar longitude of 95 degrees, close to the time of the summer solstice (June 20th). The features labeled in figure are described (and briefly annotated) in Table 19.1. A number of annual meteor shower maxima are discernible in Figure 19.4, although, and surprisingly, the Perseid meteor shower at $\lambda_\odot = 141$ is not the most prominent peak. The α-Capricornid shower at $\lambda_\odot = 127.0$ is well defined, and so too are the Leonids at $\lambda_\odot = 235.0$, the χ-Orionids at $\lambda_\odot = 250.0$, and the Geminids at $\lambda_\odot = 262.0$. We also identify a number of distinct peaks and relatively strong activity in the time interval $\lambda_\odot = 220.0$ to 230.0 corresponding to the times at which the Southern and Northern Taurids are active. The author along with graduate student Megan Hargrove and Dr. Peter Brown [4] have previously discussed the variable activity of the Taurid shower, as evidenced by the MFA record combined with several additional fireball surveys. Indeed, the Taurid fireball activity appears to be modulated by a 7:2 mean-motion resonance with Jupiter (recall Chapter 3), resulting in enhanced Taurid activity at intervals corresponding to 3, 4, and 7 years.

All of the annual showers known to produce bright fireballs on a regular basis appear to be represented in Figure 19.4. The α-Capricornid shower has a complex radiant structure, but since its initial documentation in the late 19th century, it has become well known for producing slow and bright fireballs. The α-Capricornid stream is commonly associated with comet 45P/Honda-Mrkos-Pajdusakova, although several additional comets and Apollo asteroids may additionally 'feed into' the shower, indicative of the stream and its member components being the product of a break-up event. The χ-Orionid shower is well known for producing infrequent but bright fireballs and it is generally believed to be associated with the Apollo asteroid 2201 Oljato, which has in turn been associated with the 'extended' Taurid complex. Interestingly, the evolutionary status of 2201 Oljato is that of an extinct cometary nucleus (recall Section 12.4). The same extinct cometary nucleus status also applies to the parent body, 3200 Phaethon, of the Geminid meteor shower — again, a shower that is well known for producing slow, bright meteors. Indeed, the Geminid shower (with a peak at $\lambda_\odot = 262.0$) is the most prominent of all the annual meteor showers represented in Figure 19.4. The Leonid meteor shower associated with comet 55P/Tempel-Tuttle is well known for its ability to deliver, in contrast to the Geminid and α-Capricornid showers, very swift and bright fireballs. The fireball-rich Leonid storm of 1965 is clearly evident in the MFA records, and 11 Leonid fireballs are documented for the nights of November 16th and 17th in just that one year.

Six of the prominent peaks identifiable in Figure 19.4 have no apparent commonality with any known and/or distinctive annual meteor showers. These particular peaks possibly relate to weak and/or only intermittently active showers that produce perhaps one or two fireballs every other year. This being said, the peak at $\lambda_\odot = 154$ (August 26th) may correspond to the λ-Aquilid fireball stream, which is active from August 14th to 31st. The peak at $\lambda_\odot = 248$ (November 30th) may be a 'residue' associated with the storm-producing, but now essentially inactive, Andromedid meteor shower, although it is noted that the Andromedids were never recognized as being an especially fireball-rich shower. This peak also falls close to the mean time of occurrence ($\lambda_\odot = 241$) for the Group 4 meteorite stream identified from the MORP camera data gathered by Ian Halliday and coworkers [5].

The Meteorite Observation and Recovery Project (MORP) camera network operated continuously, from across the Prairies of Canada, from 1971 to 1985, and complemented the Prairie Fireball Network, run in the mid-western United States, from 1964 to 1975. Like the MFA, MORP grew out of the interest generated by the Bruderheim meteorite fall, but was run under the direction of Dr. Ian Halliday and supported by the National Research Council of Canada. Well over 700 fireballs were observed with the MORP cameras and one meteorite was recovered: the Innisfree (LL5) meteorite which fell on February 5th, 1977 (Figure 19.5). Halliday and coworkers conducted an analysis of the orbital data derived from the MORP observations and found that about 40% of the fireballs could be linked to known meteor showers. Additionally, however, Halliday *et al.* [5] identified 4 meteorite streams, or groups (see Figure 19.4), within the orbital data set, and suggested that these streams were related to material ejected from near-Earth asteroids, into Earth-crossing orbits, within (perhaps) the last few thousands of years.

The peak at $\lambda_\odot = 165.0 \pm 4$ indicates that a good number of MFA fireballs were reported in the time interval from September 3rd to September 11th. Given the extraordinary dominance of this peak it is worth looking a little more closely at the year to year variation in the fireball numbers in the interval $\lambda_\odot = 161.0$ to 169.0. In most years 2 to 4 fireballs were reported in the time interval, although in 1964, 1965, and 1966, some 7, 8, and 9 fireballs respectively were recorded. The circumstances of September 8th, 1966

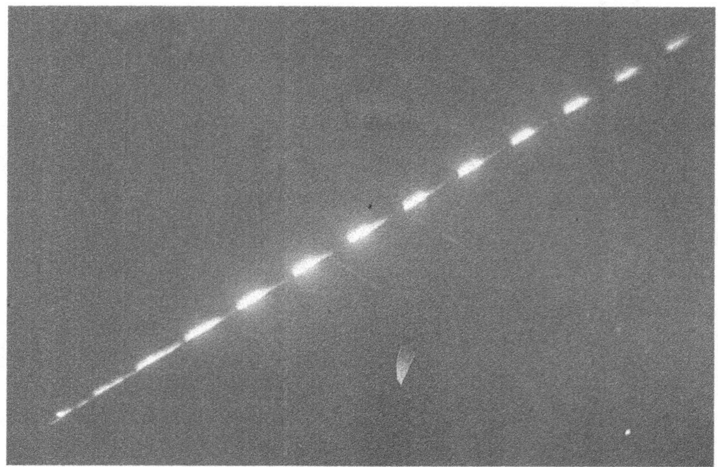

Figure 19.5. The Innisfree meteorite fireball as captured by the Vegreville, Alberta, MORP camera. The fireball trail is broken into dashed segments by a rotating shutter system — this *blinking* allows for the speed and deceleration of the meteoroid to be determined as it moves through the atmosphere. On the other hand, images from multiple camera stations allow for the prediction of the landing location to be made and for the meteoroid orbit to be determined.

seem especially remarkable, in that on that night five distinct fireball reports were logged from observers in Newfoundland, Quebec, and Ontario, as well as data being received from the flight crew of an Air Canada airplane commencing its landing at Winnipeg airport (in Manitoba). The $\lambda_\odot = 165.0$ peak appears to be due to both a 'steady' annual component and to a series of apparent 'outbursts' in the mid-1960s. The mean time of occurrence (September 13th) for the Group 3 stream identified by Halliday *et al.* [5] also falls close to the $\lambda_\odot = 165.0$ peak. It has been suggested that the Group 3 fireball stream may possibly be related to the transitional comet/asteroid object (4015) Wilson-Harrington, and interestingly, Halliday *et al.* [5] indicate that at least one of the fireballs from the Group 3 stream may have been a carbonaceous chondrite. This observation would be consistent with meteoroids being derived from a predominantly devolatized cometary nucleus.

Of the four possible meteorite-producing streams identified by Halliday and coworkers their Group 1 stream is of particular interest since it contains the Innisfree meteorite which fell in Alberta on February 6th, 1977. Group 1 also contains a second MORP fireball, and suspected meteorite-dropping event, that was recorded on February 6th, 1980 with a near identical orbit to that of the Innisfree meteorite. The Vilna, Alberta meteorite also fell on February 6th, 1967 but it is apparently not a Group 1 member, in spite of its time of occurrence, although no good orbit exists for the parent meteoroid to this particular meteorite. The main reason for excluding Vilna from the Group 1 stream is compositional, in that Vilna is an L5 ordinary chondrite, while Innisfree is an LL5 ordinary chondrite. There is no evidence for a particularly strong peak at $\lambda_\odot = 318.0$ in the MFA data (see Figure 19.4) at a time corresponding to the fall of the Innisfree meteorite.

Robert Dodd, S. Wolf, and M. Lipschutz [6] have argued that 5 H-chondrite meteorite-producing streams can be distinguished from the random background of meteorite falls. The streams or clusters, it is argued, are identified according to a commonality of fall times and by a similarity of volatile trace element content and thermal histories. One stream, designated HC34, is deemed to be active from September 27th to October 7th and includes some 17 meteorite members, one of which being the car-hitting Peekskill meteorite (October 9th, 1992). The time interval for HC34 encompasses the distinctive peak in Figure 19.4 at $\lambda_\odot = 191 \pm 2$ (October 4th).

The relatively weak peak in Figure 19.4 at $\lambda_\odot = 11.0$ (March 30th) accommodates the approximate fall times of the Peace River and Revelstoke meteorites. These two meteorites are clearly not petrologically related, however, since the Peace River meteorite is an L6 ordinary chondrite, while Revelstoke is a CI carbonaceous chondrite. There are no orbit determinations for either the Peace River or Revelstoke meteorites, so it is not clear if any 'commonality' (other than date of fall) should be expected between the meteorites. This being said, the peak at $\lambda_\odot = 11.0$ is reasonably distinct and may indicate that an intermittent fireball and/or meteorite-producing shower is active on, or near to, March 31st.

The many selection effects that have shaped the time distribution of fireball reports contained within the MFA are difficult to quantify.

We do not know, for example, how the weather has affected the reporting from one season to the next, or from one year to the next, and neither do we know what specifically motivates people to report their sightings of fireballs — clearly, most people do not report them. All that the archive effectively provides us with is the time of a fireball event, and this by necessity must be a random sample of the true fireball time-of-arrival distribution. Surprisingly the strongest peak in Figure 19.4 is centered at $\lambda_\odot = 165$ degrees (September 7th) rather than at $\lambda_\odot = 141.0$ degrees corresponding to the maximum of the annual Perseid meteor shower. It would be anticipated that a strong Perseid peak should be evident since it is by far the best known and observed of the annual meteor showers, and it does produce many bright meteors and fireballs. Perhaps we are seeing, in the case of the Perseids, a 'desensitized reporting' selection effect. That is, since the Perseid shower is both well known and well advertised the general observer is not greatly surprised if a bright fireball is seen in early to mid-August, and correspondingly does not report the event.

In addition to reporting and weather-related biases, there is an additional natural selection effect that works against the physical identification of a meteorite stream. This latter effect relates to the speed with which material enters the atmosphere. As described in Chapter 6, the faster the entry speed so the greater the amount of material lost through ablation. Accordingly, a fireball stream might be observed (such as in the MFA and MORP data), but no actual meteorites survive to reach the ground (in contrast to the HC34 group). Once the entry speed increases much above 25 km/s then meteoroid survival is highly unlikely, and this largely explains why meteorite falls do not correlated with the peak activity times of meteor showers (this and the fact that meteor shows are generally associated with cometary nuclei). This being said, there is some evidence from fireball observation data that material may have survived atmospheric passage, to deliver material to the ground, from the Geminid and Taurid meteoroid streams. In a recent case study Peter Brown (Western University) and coworkers have argued that meteorites might only be expected from those fireballs which demonstrate terminal heights and velocities smaller than 35 km and 10 km/s respectively (recall also Table 1.1).

The story with respect to meteorite streams is far from finished. Its popularity ebbs and flows according to the data available and the *tastes* of researchers. There is no physical reason to indicate that they cannot exist, at least for time intervals of many hundreds to thousands of years, but the question remains: are any such streams extant or recognizable at the present time.

References

[1] B. E. Schaefer. "Meteors that changed the world." *Sky and Telescope*, February (2005).

[2] A. Pauls and B. Gladman. "Decoherence time scales for meteoroid streams." *Meteoritics and Planetary Science*, **40**, 1241–1256 (2005). K. R. Grazier and M. E. Lipschutz. "A numerical examination of the long-term coherency of meteoroid streams in near-Earth orbit." *Bulletin of the American Astronomical Society*, **32**, 859 (2000).

[3] M. Beech. "The Millman fireball archive." *Journal of the Royal Astronomical Society of Canada*, **97**, 71–77 (2003).

[4] M. Beech, M. Hargrove, and P. Brown. "The running of the bulls: A review of Taurid fireball activity since 1962." *The Observatory*, **124**, 277–284 (2004). D. J. Asher and S. V. M. Clube. "An extraterrestrial Influence during the current glacial-interglacial." *Quarterly Journal of the Royal Astronomical Society*, **34**, 481–511 (1993).

[5] I. Halliday, A. T. Blackwell, and A. A. Griffin. "Evidence for the existence of groups of meteorite-producing asteroidal fragments." *Meteoritics*, **25**, 93–99 (1990).

[6] R. T. Dodd, S. F. Wolf, and M. E. Lipschutz. "An H chondrite stream: Identification and confirmation." *Journal of Geophysical Research*, **98** (E8), 15105–15118 (1993). S. F. Wolf, M.-S. Wang, D. T. Dodd, and M. E. Lipschutz. "Chemical studies of H chondrites, 8: On contemporary meteoroid streams." *Journal of Geophysical Research*, **102** (E4), 9273–9288 (1997).

[7] P. Brown *et al.* "Meteorites from meteor showers: A case study of the Taurids." *Meteoritics and Space Science*, **48**, 270–288 (2013).

Chapter 20

Fireball Processions

Witnessing a single bright fireball is a rare enough sight, and indeed, the majority of people never have the good fortune to experience the thrill of a *Draco Volans*. To witness a whole host of fireballs — a veritable procession of *flying dragons* — well, that is altogether something else, and indeed, very few human generations get to witness such a marvel.

20.1 A grand display

One of the first fireball procession that was reasonably well observed and well reported upon occurred on August 18th, 1783. This grand meteor descended down the backbone of England, being first reported in flight over eastern Scotland, flying over London, across the English Channel, and then onwards to disappear over central Europe — some reports suggest that it was even seen in northern Italy. The ground path of the procession was at least 2000 kilometers. By chance the grand display was observed from the terraces of Windsor Castle and amongst the eyewitnesses was natural philosopher Tiberius Cavallo and artists Paul and Thomas Sandby. Indeed, the Sandby's produced a number of sketches of what they saw and in one image produced what is essentially a time-lapse montage of the fireballs flight and development (Figure 20.1). Indeed, in mid-sequence (picture center) the Sandby's show the fireball having dissolved into a whole host of lesser fireballs all moving along the same atmospheric trajectory. It is this common motion of fragments and trailing bodies, along with the great length of trajectory parallel to the horizon, that distinguishes

Figure 20.1. 'Time-lapse' painting of the great fireball of August 18th, 1783, as seen from Windsor Castle. A number of watercolor paintings and sketches of the event were made by artists Paul and Thomas Sandby.

fireball processions from the typical *Draco Volans*, and the circumstances for such phenomena to occur are only rarely satisfied.

Eyewitness Tiberius Cavallo was known for his interests in magnetism and atmospheric electricity — repeating and refining Franklin's earlier (1752) kite experiments to investigate lightning. Cavallo was also a Fellow of the Royal Society of London, and he provided a description of the procession in the Societies' *Philosophical Transactions*. Cavallo writes, "the weather was calm, agreeably warm, and the sky was serene ... some flashes of lambent light, much like the aurora borealis, were first observed ... which were soon perceived to proceed from a roundish luminous body, nearly as big as the semi-diameter of the Moon ... The whole duration of the meteor was half a minute ... the light was prodigious.... During the phenomenon no noise was heard by any of our company excepting one person, who imagined, to have heard a crackling noise, something like that which is produced by small wood when burning [see Chapter 22]... about ten minutes after the disappearance of the meteor we heard a rumbling noise as if it were thunder at a great distance [see Chapter 21]." The latter sound accounts were not imagined and represent examples of what, in the first case, are called simultaneous (or electrophonic) sounds and in the second sonic booms — these fireball properties will be discussed later. Cavallo also noted that, "it [the initial fireball] parted into several small, each having a tail, and all moving in the same direction, at a small distance from each other, and very little behind the principle body [this is the central part of the painting by the Sandby's]." Nathanial Pigott also wrote to the Society describing his observations from Yorkshire (some 350 km north of Windsor), noting that, "it suggested the idea of a highly brilliant comet, emitting a train or trail, but of a different colour from the ball itself, this last being a most brilliant bluish white, and the tail of a dusky red, the length of which appeared to extend over fifteen or more degrees of the heavens."

The fireball procession of 1783 was widely observed — indeed, it could hardly be missed — and physician Charles Blagden collected and reviewed the available information for the Royal Society of London, his article appearing shortly after that by Cavallo. After reviewing the eyewitness accounts, Blagden moved towards a physical explanation of the phenomenon observed, and adopted the then popular idea that fireballs were some form of electrical fluid action operating in the Earth's upper atmosphere. Indeed, Blagden specifically rejected the idea that fireballs were some form of igniting vapor trail (the Aristotelian model — recall Chapter 2), and he additionally rejected Edmund Halley's cautious 1714 idea that fireballs might be solid bodies entering the Earth's atmosphere from space. The former he rejected according to the prodigious path length over which the procession travelled, and the latter he rejected on the basis that if such objects were solid then surely some of the fragments would fall to the ground intact (here Blagden is expressing the idea that meteorites, falling stones and irons, are terrestrial in origin and not extraterrestrial). "What can these meteors be?" asks Blagden in his report, to which he continues, "The only agent in nature with which we are acquainted, that seems capable of producing such phenomena is electricity." To this, he adds, "the electrical origin of meteors is deduced from their connection with the northern lights [aurora]." In this latter step, Blagden asserts (incorrectly) that all bright fireballs are observed to proceed from the north to the south, traveling along magnetic meridians, and suggests fireballs are really "masses of electric fluid repelled, or bursting from the great collected body of it in the north."

Blagden's model is consistent within the framework of natural philosophy as it was then known, and it drew upon the then new ideas relating to electricity, magnetism, and atmospheric structure. Interestingly, Blagden draws a distinction between fireballs and ordinary shooting stars. Specifically, the latter, he notes, can move in any direction and must accordingly be lower atmospheric phenomena (where, he reasons, the Earth's magnetic influence is not so strong). While the physical model developed by Blagden was consistent with the physical theory of his time, it was the last hurrah for such ideas. Indeed, within 11 years of the appearance of the great fireball procession, Ernst Chladni was to publish the first of his seminal works linking the appearance of fireballs to the fall of meteorites; and within 23 years Jean-Baptiste Biot was to investigate the fall of stones in L'Aigle, concluding that meteorites do, indeed, fall from the sky (recall Section 2.3).

A second grand fireball procession was witnessed from the eastern United States on July 20th, 1860 [2]. Walt Whitman wrote of the experience, "the strange huge meteor procession, dazzling and clear, shooting over our heads. (A moment, a moment long, it sailed . . . then departed, dropped in the night, and was gone)." This procession had a path that stretches from the Great Lakes region, over New York state, and on over the Atlantic, with a ground track of at least some 2000 kilometers. As with the 1783 procession, the fireballs in 1860 moved parallel to the horizon in a stately manner, showing various configurations, fragments, and trails (Figure 20.2). The event was described in the August 4th *Harper's Weekly* (along with several woodcut images) as being "one of the most wonderful of the kind ever witnessed." Lucian Kendall writing from Reading, Pennsylvania, reported that, "four brilliant meteors were seen suddenly to rise [in the west northwest]. The first two were of a brilliant and pale olive, leaving a wake as they fell similar to that of a comet. The other two, a little distant astern, were of a purple color and without the

Figure 20.2. The fireball procession of July 20th, 1860, crossing the Catskill's, as painted by Frederic Church.

wake. The whole thing lasted one and a half minutes, and formed a beautiful appearance." An observer in Brooklyn wrote in more prosaic terms, "the meteor sailed by slowly and horizontally ... the light was indescribable — a mystic pearly tinted radiance filled each moving globe; and they actually seemed intelligent — like eyes! ... it was so long in passing — so stately, silently and slowly — that one had a thousand thoughts." This latter account attests to the startling nature of witnessing a bright fireball event, with an observer struggling to find the right words to describe what they have seen. The last eyewitness accounts describe the fireball procession passing into the east, heading out to sea from the New England coast.

Fireball processions are very rare, only three being reasonably well recorded over the past three hundred years — indeed, they are each approximately spaced by a century. The observations of these events all point to a very shallow entry angle into Earth's atmosphere, and the question that remains is whether the parent meteoroids entered into to temporary Earth orbit before being fully destroyed, or whether they, in fact, returned to interplanetary space. These latter events, where a meteoroid essentially skims through the atmosphere, are somewhat more common than fireball processions, and they are typically called Earth-grazing meteors (or *skimmers*). Perhaps the most well-known Earth-grazing (daylight) fireball event of recent times is that of August 10th, 1972 [3]. This bright fireball followed a path from Utah to Alberta, and spent about a minute travelling through the atmosphere before passing, once again, out into cis-lunar space. Critical to such encounters occurring is the meteoroid's entry angle into Earth's atmosphere: too steep and it will penetrate too deep into the atmosphere and be destroyed (or produce a meteorite); too shallow and it will not penetrate deep enough for a luminous train to be formed.

20.2   What's the angle?

The key observational characteristics of a fireball procession is the parallel nature and streaming flight of its components. The fireballs all move along the same atmospheric path and for most observers (unless the procession passes near overhead) they will swept across the sky in an arc parallel to the local horizon. This result is suggestive of temporary Earth-orbit capture (see below) and also indicative of a very shallow entry angle into the atmosphere. A meteoroid, of course, can enter Earth's atmosphere at a whole range of angles, and the question now is what is the most common angle of meteoroid entry and what is the probability of some specific entry angle to the horizon being realized. Figure 20.3 shows the basic geometry that we are dealing with. In this diagram, the observer is considered to be located at position O, the Earth has a radius R, and the meteoroid path is illustrated by the dashed line. In this analysis we shall ignore any gravitational interaction between the meteoroid and the Earth. The observer at O will see the meteoroid enter the atmosphere at some angle z away from their local zenith. The zenith is the line that passes from the center of the Earth, through the observer's feet and on into the celestial sphere.

Now, the probability P that a meteoroid will hit the Earth within the cross-section area perpendicular to its path is unity (else it missed the Earth all together): hence, $P = \pi R^2 = 1$. The probability with

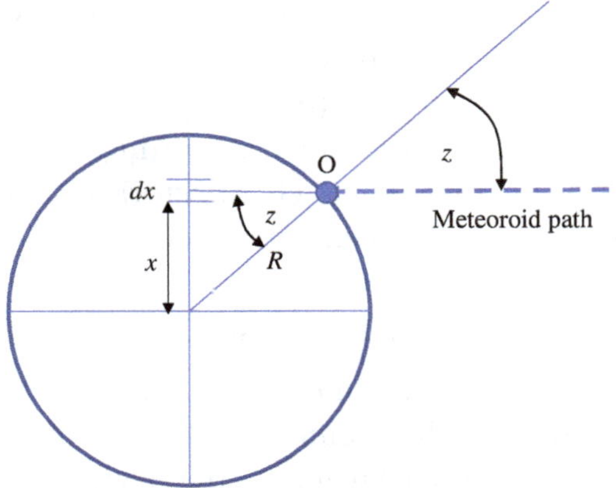

Figure 20.3. The angle of incidence z for a meteoroid approaching the Earth from a random direction. For the observer at O, the angle z corresponds to the so-called zenith angle — literally, the angle away from the zenith.

which a meteoroid will strike a region of width dx a distance x from the center of the Earth, where $0 \le x \le R$, is $dP = 2\pi x\,dx$. From Figure 20.3 it is seen that $x = R \sin z$, and accordingly, $dx = R \cos z\,dz$. Combining known terms, we now have

$$dP = 2\pi x\,dx = 2\pi R^2 \sin z \cos z\,dz = \sin 2z\,dz \qquad (20.1)$$

From Equation (20.1) it follows that the most probable zenith angle of entry is when $z = \pi/4 = 45$ degrees. Hence, an observer is most likely to see a meteoroid enter the atmosphere in a ring around the sky at an altitude of 45 degrees above the horizon. The probability that a meteoroid will enter the atmosphere along the line of an observer's zenith is additionally exactly zero. The probability that a meteoroid will be seen to enter the atmosphere between zenith angles z_1 and $z_2 > z_1$ is given by the integral of Equation (20.1), and this gives:

$$P(z_1 \le z \le z_s) = \int_{z_1}^{z_2} dP = \frac{1}{2}\left(\cos 2z_1 - \cos 2z_2\right) \qquad (20.2)$$

As the zenith angle z approaches $\pi/2$ (or 90 degrees), that is with the meteoroid entering the atmosphere low to an observer's horizon, so the probability approaches zero — in other words, such meteoroid encounters are going to be very rare. The probability that a meteoroid will enter the atmosphere at an angle smaller than 1 degree above the horizon is $P(0 \leq z \leq 1\,\text{deg.}) = 0.0003$, that is something like 1 in 3300 encounters. Accordingly, fireballs entering the atmosphere along grazing impact trajectories (that is with $z \sim 90$ degrees) are not going to be seen very often.

20.3   The Chant Fireball Procession

The last great fireball procession occurred (as I write this page) over 100 years ago, and from all accounts it was an awe-inspiring display (Figure 20.4). It has no official name, but I will use here the term Chant Fireball Procession in honor of Canadian astronomer Clarence Chant who conducted an extensive study and analysis of the event. The procession was first observed from western Saskatchewan in the early evening hours on February 9th, 1913, and its flight was tracked across Manitoba and Ontario (Figure 20.5), and on through New York state, to the island of Bermuda, and eventually over the Atlantic, with the last sightings being reported from ships located off the coast of Brazil. The ground track was at least 15,000 kilometers

Figure 20.4.   The Chant Fireball Procession, as painted by Gustav Hahn, in the skies above Toronto, Canada.

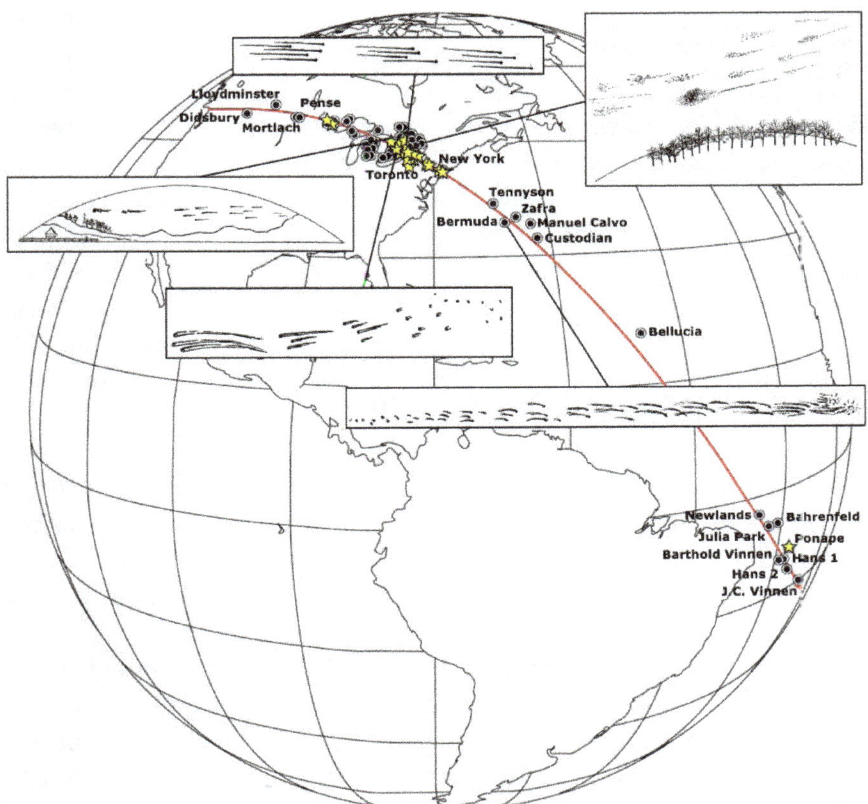

Figure 20.5. The derived ground path of the Chant Fireball Procession. The ground path (red arc) was at least 15,000 kilometers (nearly a quarter of the Earth's circumference), first being seen in western Saskatchewan, Canada, and last being seen from ships located off the coast of Brazil. The various insets show eyewitness drawings of the display at different locations [5].

(Figure 20.5) — a staggering distance for a fireball, and a distance that indicates special Earth encounter conditions.

Eyewitness accounts of the procession were remarkably consistent, and the account given by J. Skidmore, situated in Cobourg, Ontario seems particularly apropos: "they glided along so leisurely and did not seem to be falling as meteors usually do, but kept a straight course about 45°, or a little more, above the horizon. Our first impression was that a fleet of illuminated air-ships of monstrous size were passing. The incandescent fragments themselves formed what to us looked like the illuminations, while the tails seemed to make the frame of the machine. Sometimes there would be just a single collection, forming a single ship; then in a half-minute several collections would pass, looking like ships traveling in company. It took fully 3 minutes to pass. There was no noise; only beauty, beauty!" Indeed, by comparing the various eyewitness reports it is possible to gauge a number of important facts about the procession. Firstly, it was composed of many distinct fireballs, one set of fireballs and fragments appearing after another in a progressive sequence. Secondly, the fireballs moved nearly parallel to the horizon at a near constant (calculated) height of about 50 kilometers. Thirdly, the speed of the various observed fireballs was remarkable slow, with the characteristic velocity being estimated at 8 km/s. It is clear that the Chant Procession was not the result of a single large object entering into the Earth's atmosphere at a shallow angle, but rather it was a display produced by a successive series of encounters with a pre-Earth-encounter fragmented object [4].

The author along with graduate student Mark Comte studied the circumstances of the Chant Fireball Procession in a 2018 research paper [5] and concluded that it was related to the successive capture, into Earth orbit, of a chain of fragments associated with a previously disrupted asteroid — the initial size of the parent asteroid being perhaps just 3 to 4 meters across.

Various indicators support the Earth-orbit capture scenario. Firstly, the meteoroids that made up the successive clusters in the chain of fragments encountered the Earth at a very shallow entry angle, at about 12 degrees to the local horizon, which accounts for the near-parallel-to-the-horizon motion reported by all eyewitnesses. And, secondly, the deduced speed of 8 km/s is exactly that expected of a satellite (temporary moon) orbiting the Earth (not the Sun) at an altitude of 50 kilometers. These encounter conditions can be modeled numerically [5], and the procession can be modeled as the sequential entry of multiple meteoroid clusters that moved through the Earth's atmosphere on grazing-incident trajectories.

That the parent body was likely a small, disrupted stony asteroid is inferred from eyewitness accounts — these accounts indicating that sonic booms were heard in several locations along the ground path. For a meteoroid to produce such sounds (described more fully in Chapter 21) it must penetrate deeply into the atmosphere, to at least 25 kilometers altitude or so. In this manner, while no meteorites were actually found as a consequence of the passage of the procession, it is more than likely that some were; and as discussed in Sections 1.3 and 12.3, the most likely meteorite-producing body is a stony asteroid.

With a stony composition it is estimated that the initial size of the parent body was probably less than 5 meters across, and that this body was disrupted long before the various fragments and fragment clusters (possibly numbering into the several hundreds) encountered the Earth. The question that now arises is what sort of object might the parent body be, and it is suggested that it was one of Earth's mini-moons (or natural satellites).

20.4 Mini-moons

Critical to the production of a fireball procession, or grazing incident (skimmer) fireball, is a shallow entry angle into the Earth's atmosphere. To produce a long ground path the parent meteoroid (or initial fragments) must find a sweet spot between too step an angle and total disruption (possibly producing a meteorite), and too shallow and never actually attaining a height at which ablation (and an associated light phenomenon) can begin. There is no specific restriction on meteoroid entry angle into the Earth's atmosphere, so sooner or later, as indicated by Equation (20.2), very shallow entry angles will occur. Are there circumstances, however, that might enhance not only the shallow entry angle condition, but also provide for a slow entry velocity — this latter effect resulting in temporary Earth-orbit capture? The answer to this appears to be yes, and it draws upon a recently explored linkage between fireball processions and small natural Earth moons — that is, mini-moons (also called, TCOs = temporary captured objects).

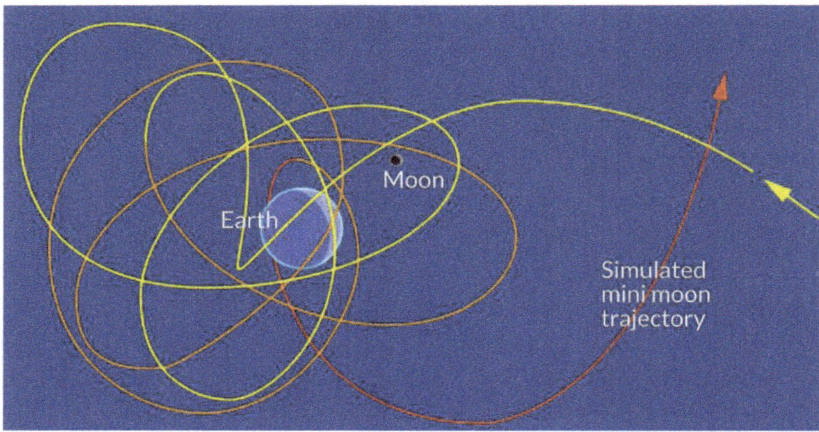

Figure 20.6.   The temporary capture of a mini-moon by the Earth–Moon system. Simulation by Mikael Granvik (University of Helsinki), Jeremie Vaubaillon (Paris Observatory), and Robert Jedicke (University of Hawaii). Image by K. Teramura (University of Hawaii).

The idea that the Earth might have had more than one moon is not new, but the concept of mini-moons refines the idea to the temporary capture of asteroids by the Earth–Moon system. These objects don't orbit the Earth or Moon in a simple fashion; rather, they follow a complicated gravitational dance that temporarily holds them in relatively close Earth proximity (Figure 20.6). A typical mini-moon is only a fleeting visitor, staying in a looping orbit about the Earth and Moon for perhaps a few months to a year or so. It is during such times of *capture*, however, that a closer interaction might take place and an Earth impact occur. The point being that such encounters will typically take place with a low relative velocity, that is only just above the Earth's escape velocity [V_{esc} (Earth) = 11.2 km/s].

Mikael Granvik (University of Helsinki), Jeremie Vaubaillon (Paris Observatory), and Robert Jedicke (University of Hawaii) have calculated the probability that at any given time Earth has more than one moon [6]. They used a computer to simulate the passage of 10 million asteroids past the Earth, and then specifically tracked the trajectories of some 18,000 objects that were captured by

Earth's gravity. From the simulation thus run, Granvik and coworkers concluded that at any moment in time, at least one mini-moon, with a diameter of order one meter, should exist in near Earth proximity. Mini-moons as large as 5 to 10 meters across will additionally be captured every decade or so, with 100-m sized objects being captured on timescales of order 100,000 years. Indeed, Granvik and coworkers argue that about 0.1 percent of meteorites impacting the Earth are derived from the TCO population.

The first confirmed mini-moon, with a size of some 2 to 3 meters, was discovered in 2006 [object 2006 RH120]. With an orbit that brings it close to the Earth–Moon system once every 20 years, it was last in a temporary captured orbit between September 2006 and June 2007. The next close encounter will be in 2028, and currently 2006 RH120 occupies an orbit corresponding to an Amor near-Earth asteroid (see Section 12.1). The second confirmed mini-moon was discovered at the Catalina Sky Survey Observatory on February 19th, 2020. Designated 2020 CD3, this mini-moon is estimated to be about 2 to 4 meters across, and at the time of its discovery was orbiting about the Earth once every 47 days. The first impacting mini-moon candidate (designated EN130114) was recorded by the European Fireball Network cameras on January 13th, 2014. In this case the fireball was consistent with the destruction of a 5-kilogram meteoroid encountering the Earth's atmosphere at a graceful 11.2 km/s.

References

[1] M. Beech. "The great meteor of 18th August 1783." *Journal of the British Astronomical Association*, **99**, 130–134, 1989.

[2] D. Olson, *et al.* "Walt Whitman's 'Year of Meteors.'" *Sky and Telescope*, July 2010.

[3] Z. Ceplecha. "Earth-grazing daylight fireball of August 10, 1972." *Astronomy and Astrophysics*, **283**, 287–288, 1994.

[4] D. Olson and S. Hutcheon. "The great meteor procession of 1913." *Sky and Telescope*, February 2013.

[5] M. Beech and M. Comte. "The Chant Meteor Procession of 1913 — Towards a descriptive model." *American Journal of Astronomy and Astrophysics.* **6**, 31–38, 2018.

[6] M. Granvik, J. Vaubaillon, and R. Jedicke. "The population of natural Earth satellites." *Icarus*, **218**, 262–277 (2018).

Chapter 21

A Roaring Sound and a Mighty Wind

Just after 9 p.m. on April 25th, 1969, a large meteoroid entered Earth's atmosphere over northern France, streaked across the Channel and southern England, over the mountains of northern Wales, and on towards Northern Ireland. It blazed across the sky changing color from fiery white to blue and green. Having passed over Anglesey, Wales, it began to break apart, trailing numerous small fragments. In Londonderry, Northern Ireland, detonations were heard, but these were no part of *the troubles*, rather they were the sonic booms from what was to become the Bovedy and Sprucefield meteorites.

The passage of the fireball was witnessed by hundreds of spectators, and among them was Terence Murtagh (past President of the Armagh Planetarium). "The fireball was brilliant, it was many times brighter than the full Moon and it was shooting across the sky, and little bits of it you could see were falling off," recalled Murtagh in a UTV interview during a 50th anniversary celebration event of the meteorite's fall. "I was rushing out to go to the front door to see where it went when the sonic booms hit. There were three big sonic booms and they rocked the whole house, everything in the house shook like an earthquake." The sonic booms arrived in Belfast some 5 to 6 minutes after the passage of the fireball through the sky and they were described as being "like the steady continuous rolling of thunder." Two meteorites were eventually recovered, one at Sprucefield, 15 kilometers southwest of Belfast, smashing through a roof; and one at Bovedy, 60 kilometers northwest of Belfast, where it made a small impact pit — many more fragments continued onwards, presumably falling into the North Atlantic Ocean.

Far from being the first meteorite to land amidst the rattle of sonic booms, the Bovedy meteorite is particularly remembered for being the first that was instrumentally recorded. Eileen Brown was out recording bird songs in Bangor (a few tens of kilometers to the east of Belfast), when, in the background, for some 5 to 6 seconds, a series of five distinct booms (along with barking dogs excited by the noise) were captured on her tape. Analysis of the tape reveals that the sonic booms produced distinctive low frequency sounds in the range between 30 and 400 hertz. At 30 hertz, the sonic booms have a frequency very close to the limit of human hearing, and are pushing into the low-frequency (long-wavelength) region corresponding to infrasound.

21.1 Shock waves

Any object moving at a speed V greater than the sound speed c_s in the surrounding medium will produce a sonic boom, with the energy of the boom being propagate outwards at the speed of sound. The sound wave will additionally propagate within the confines of the so-called Mach cone for which the half angle is $\theta = \sin^{-1}(c_s/V)$. Typically, the ratio c_s/V is small since the meteoroid velocity is typically 50 to 250 times greater than the speed of sound in the atmosphere. Accordingly, θ is very small and the moving meteoroid can be modeled as a line source for which the shock wave associated with the sonic boom propagates outward in a direction perpendicular to the meteoroid's trajectory (Figure 21.1). If, during the meteoroid's flight, it undergoes catastrophic disruption, then a spherically expanding ballistic shock wave is produced.

Since sonic booms are pressure waves that move through the atmosphere, they not only generate sounds audible to observers, they can cause actual damage to ground structures. The Chelyabinsk event of February 15th, 2013, produced such a powerful set of overpressure waves that thousands of windows were smashed when ground contact was made (at a time several minutes after the passage of the fireball thought the sky), and buildings were structurally damaged — in excess of a thousand people also sought medical

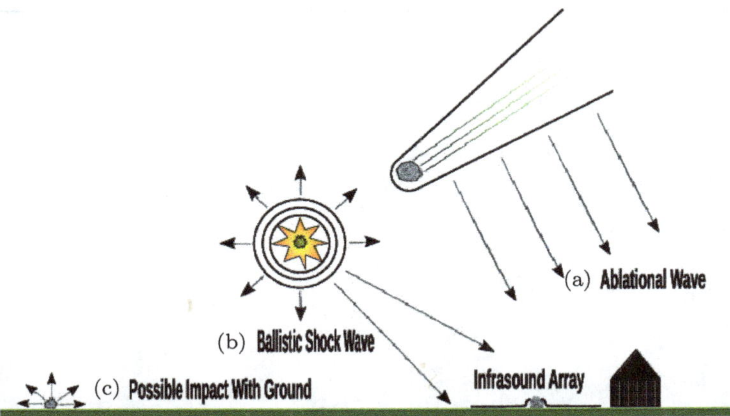

Figure 21.1.   Schematic of a meteoroid's passage through Earth's atmosphere. (a) The meteoroid's path acting as a line source or ablation wave for which the shock propagates outward perpendicularly. (b) Catastrophic disruption of the meteoroid results in the generation of an expanding spherical (ballistic) shock wave. (c) The impact of a meteorite upon the ground can generate ground-coupled seismic waves. The ballistic and ablation shock waves will generate long-wavelength, infrasound waves. Image from T. Ens *et al.* (2012) [1].

attention from cuts caused by flying glass. Figure 21.2 shows the extent of the damage resulting from the Chelyabinsk shock wave after it reached the ground — the region of damage effects extended over some 15,000 square kilometers. The overpressure from the shock wave is estimated to have peaked at about 3 kPa (inner most grey zone in Figure 21.2) and at these pressures, damage to human hearing can and, indeed, did occur.

While damage from the shock wave produced by the disintegration of the Chelyabinsk meteoroid was most severe within a localized zone of some 60 km either side of its ground path, the effects of the shock wave were detectable at much greater distances. In these more distant locations, it was not the shock wave, as such, that was detected but the infrasound waves generated during the disruption event. Infrasound waves are low frequency (or alternatively, long wavelength) pressure waves. Generally, they propagate through the atmosphere at the speed of sound (343 m/s at sea level) with frequencies lower than the threshold of human hearing — that is

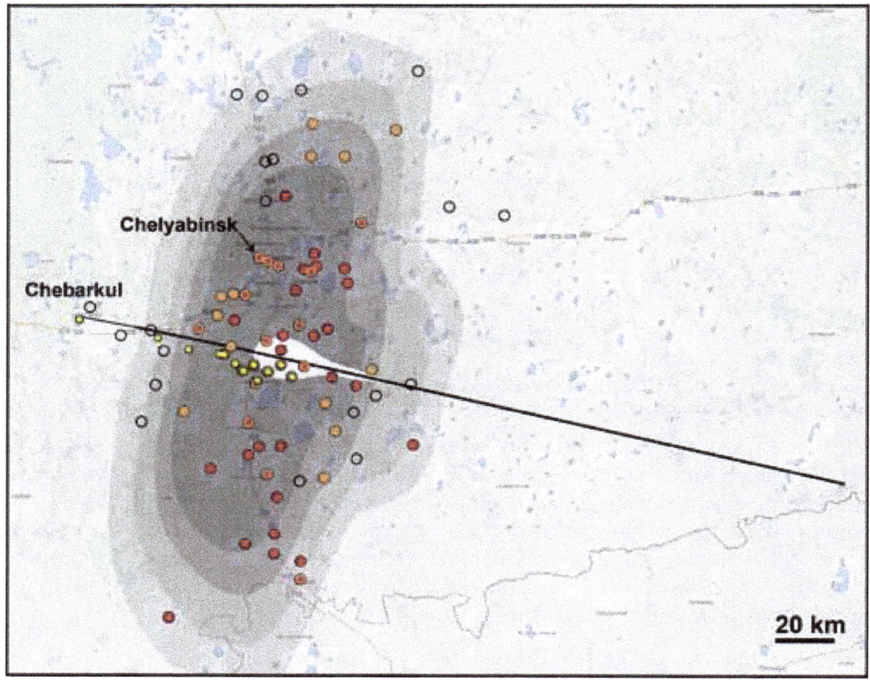

Figure 21.2. Ground path of the Chelyabinsk fireball (moving from right to left) and overlays (grey contours) of predicted overpressure zones. The red and orange dots indicate locations where glass damage was reported. The yellow dots indicate locations where meteorites were recovered. Image courtesy of Olga Popova *et al.* [2].

with frequencies smaller than 20 Hz (wavelengths longer than 17 m). In many ways, infrasound is the sound of silence, but within the silence there is a message. The detection of infrasound requires the use and construction of highly sensitive microbarograph systems (Figure 21.3). These systems can detect the very small pressure variations associated with infrasound waves, and by using multiple systems at the same location, the delay times in the signal detection can be used to determine the azimuth associated with the direction of travel. Having multiple systems located around the globe further allows for the location of the actual generating source to be made.

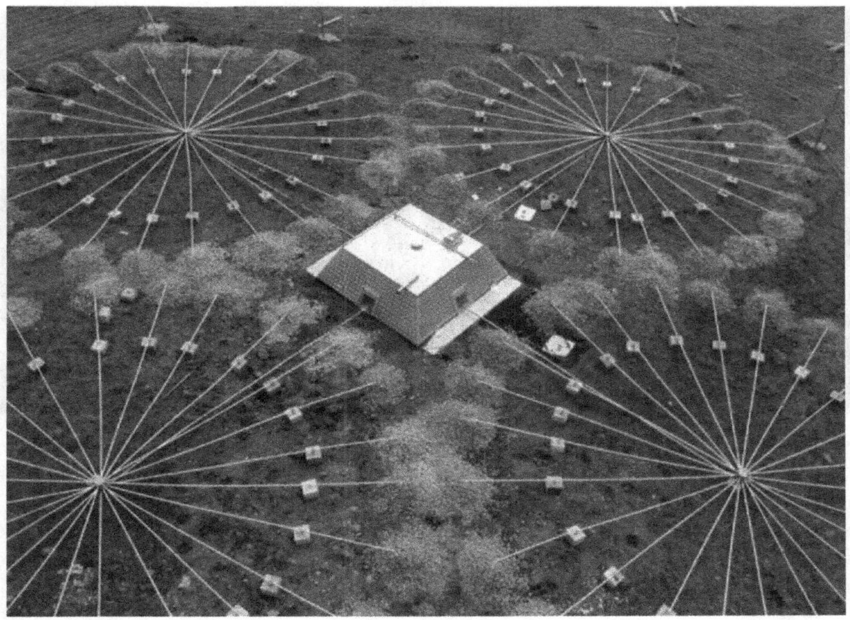

Figure 21.3. The infrasound detection array (IS49) on the island of Tristan da Cunha. A microbarograph for measuring pressure variations is located at the center of each wagon-wheel-like structure. The radial spokes are hollow tubes that help reduce the effects of local wind *noise*. This array forms part of the Comprehensive Test Ban Treaty verification program.

Figure 21.4 shows the infrasound waves produced by the disruption of the Chelyabinsk meteoroid as recorded in Kazakhstan at a range of many hundreds of kilometers away. Indeed, infrasound waves from the Chelyabinsk event were detected around the globe, and as distant as 15,000 km away in Antarctica.

How often might we expect to observe Chelyabinsk-like encounters? Peter Brown (Western University) and coworkers have analyzed the impact data related to numerous bright fireballs detected by well-calibrated space-borne instrumentation [3], and find that the approximate cumulative frequency of impacts from objects larger than D meters across is

$$N\,(>D) = 37 \times D^{-2.7}\,\mathrm{yr}^{-1} \tag{21.1}$$

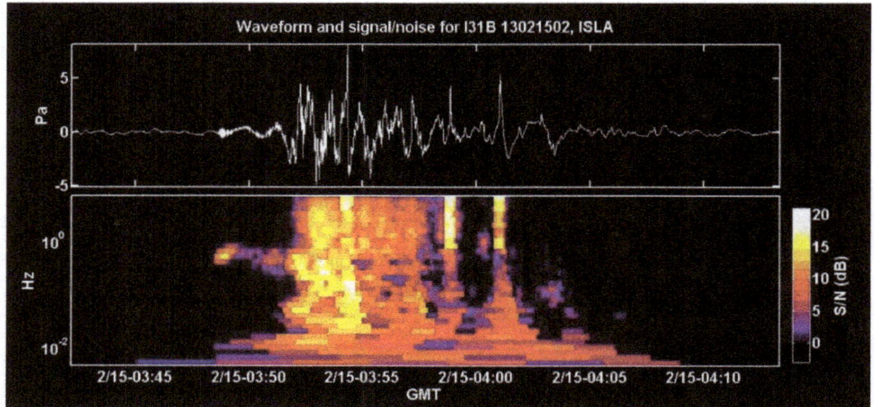

Figure 21.4. Infrasound data for the Chelyabinsk fireball as recorded in Kazakhstan. The lower axis indicates the time, while the upper panel shows the infrasound pressure wave form (measured in pascal), and the lower (spectrogram) panel indicates the frequencies that are present within the signal (in the range window from 10 to 0.01 Hz). The entire event lasted about 15 seconds. Image from the Infrasound Laboratory at the University of Hawaii.

For the Chelyabinsk asteroid, with $D \approx 20\,\mathrm{m}$ ($= 0.02\,\mathrm{km}$), so the likely encounter time interval is of order 88 years.

Figure 21.5 is a composite image relating to the entry of asteroid 2019 MO into the Earth's atmosphere. Unlike Chelyabinsk this asteroid was actually detected about 12 hours prior to impact, although no alert was given. The asteroid was about 4.5 meters in diameter and the energy associated with the impact was estimated to be 3 kilotons of TNT equivalent energy. This latter number is based upon the infrasound signal recorded in Bermuda, some 1000 km away from the impact location, which was to the south of Puerto Rico (see top and bottom of Figure 21.5). The asteroid was optically detected a few hours before impact by the Asteroid-Terrestrial-impact Last Alert System (ATLAS). This system consists of a pair of wide-field cameras, located on two of the Hawaiian islands about 158 km apart. Operated by the University of Hawaii the express purpose of ATLAS is to search for already close, potentially impacting asteroids. To this end the two cameras scan the whole sky several times each night looking for fast moving objects. The system capability is such that

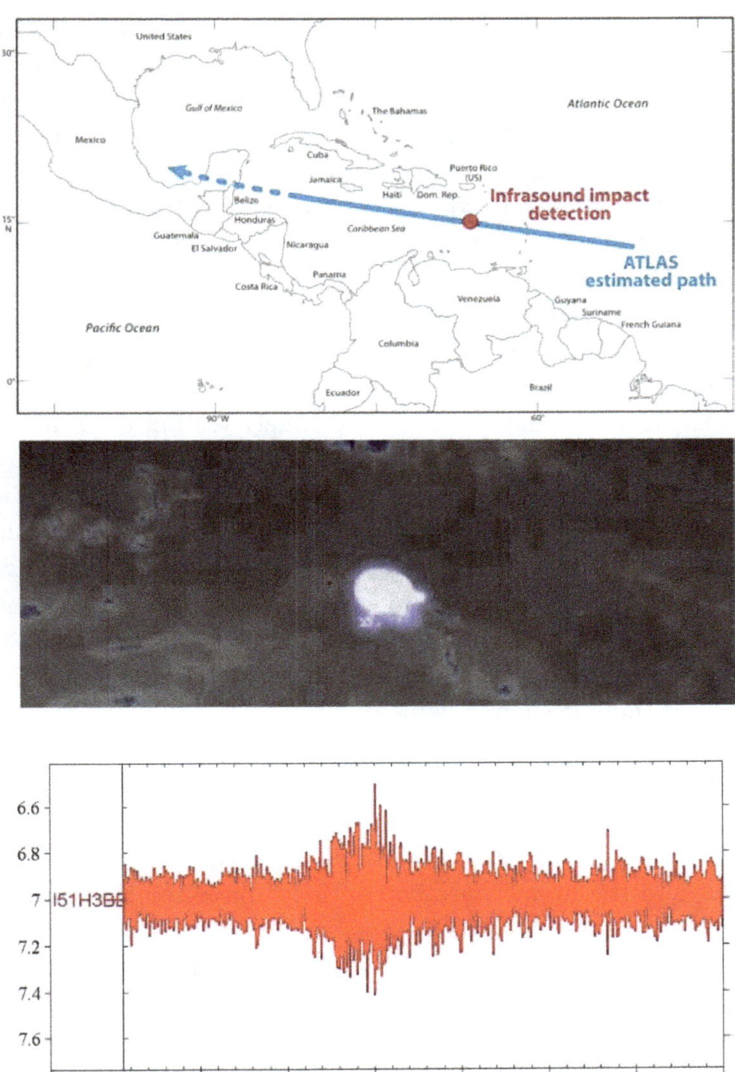

Figure 21.5. Ground path of (top), optical light flash from (middle), and infrasound signal recorded in Bermuda (bottom) due to asteroid 2019 MO entering Earth's atmosphere on June 22nd, 2019. Images courtesy of ATLAS, GOES, and Peter Brown (Western University).

it can detect a 140-m diameter asteroid, on a collision path with the Earth, at a distance equivalent to some 3 weeks travel time before impact. The system can detect 50-m diameter asteroids heading for Earth at a distance providing about a one-week warning time. ATLAS found 2019 MO when it was about 500,000 km from Earth (that is, situated just beyond the Moon's orbit). A weather radar system in Puerto Rico detected the asteroid as it exploded over the Caribbean Sea, and the optical flash of the explosion was detected by the GOES-16 geostationary weather satellite (middle image in Figure 21.5). The upper image shows the estimated ground path of 2019 MO, and while meteorites may well have been deposited during the impact, they are now most likely resting upon the ocean floor.

The orbit deduced for 2019 MO is that of an Alinda asteroid, with a perihelion just interior to the Earth's orbit and an aphelion at about 4 AU from the Sun. The Alinda group of asteroids acquire high orbital eccentricities due to their location close to the 3:1 mean-motion resonance with Jupiter (recall Chapter 3). It has been suggested that 2019 MO might be associated with the June Epsilon Ophiuchids meteor shower, which is normally a low-activity annual meteor shower active between June 21st to 23rd. Interestingly, in 2019 the shower underwent a remarkable outburst of activity, and this begs the question of 2019 MO being a stream member. But caution is required and it could, of course, be that the two events, the outburst and the impact, are two entirely unrelated random events. Interestingly, however, the June Epsilon Ophiuchids have been linked to periodic comet 300P/Catalina, and detailed dynamical studies by Pavol Matlovič and coworkers at Comenius University in Bratislava, Slovakia, suggested that the stream might be no more than 1000 years old [4]. If 2019 MO was a stream member, and hence composed of cometary ices, it was likely only a few tens to hundreds of years old, since the survival time against sublimation for a meter-sized ice fragment in the inner solar system is equivalent to just a few orbits about the Sun.

21.2   The Tunguska event

In terms of the loudest noise and strongest shock wave known to be associated with the entry of a small asteroid into the Earth's atmosphere, that of the Tunguska event in 1908 holds the current record. It was a devastating encounter, and still, more than 110 years later, it is a topic of great speculation and investigation [5]. Over the years it has been suggested that the Tunguska explosion was due to a matter–antimatter annihilation event, the passage of a small black hole through the Earth, the ignition of a vast cloud of methane gas, the crash landing of an alien UFO, divine retribution, and the impact of a comet or asteroid. The enduring mystery of the Tunguska event is that, while clearly a devastating explosion, it left no impact crater (Figure 21.6).

The story of the Tunguska event began at about 7:15 a.m. on the morning of June 30th. The sky was clear, when suddenly, all hell broke lose. According to the nomadic reindeer herders of the Tunguska river region, the heavens literally split apart — the intensity of light was so high that trees were set ablaze and the clothing of some observers was set on fire. People were knocked off their feet, animals were killed, and buildings hundreds of kilometers away were shaken. The cause of this chaos was the catastrophic break up of a 50-meter-sized asteroid (see below, however) at an altitude of about 8 km above the ground. The energy released was colossal, with estimates placing it somewhere between 10–20 megatons of equivalent TNT explosive energy (a thousand times more energy than that of the Hiroshima bomb). About 30 seconds after the blinding flash of light the associated blast wave reached the ground snuffing out the blazing trees. Within seconds of the shock wave's arrival, however, some 80 million trees, over a region of some 2000 square kilometers, were stripped of their branches and lain flat on the ground, their ripped roots pointing backwards to the explosion epicenter (see the inset map in Figure 21.6). A seismic wave, estimated to be of order magnitude 5 on the Richter scale, rippled along the ground for hundreds of kilometers, and over the ensuing hours an atmospheric

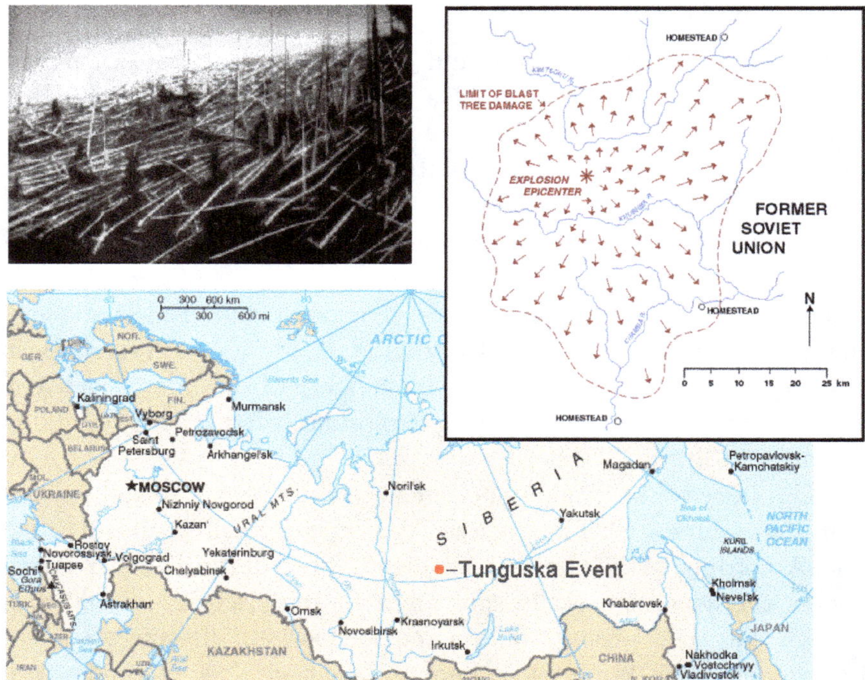

Figure 21.6. The Tunguska event of June 30th, 1908. The image at upper left shows a swath of trees that have been laid over by the blast wave, while the image to the upper right shows the full extent of tree damage — a region extending over some 2150 square kilometers. Note the various directions in which the trees have been toppled can be traced back to the blast wave epicentre.

pressure wave was recorded as circling the entire globe twice. In the days after the event Europe experienced strange 'bright nights,' the glow reportedly being bright enough at midnight, in England, to read a newspaper. This anomalous glow was the result of sunlight being scattered by fine dust particles lofted into the atmosphere by the Tunguska explosion.

The Tunguska region is remote, even to this very day, and it is not at all surprising that it took some 19 years before the first expedition arrived at the site of the explosion. Led by mineralogist Leonid Kulik the first expedition team found a scene of utter devastation — even after two decades, the area resembled a war zone, with those trees

Figure 21.7.   The charred remnants of limb-stripped trees at the epicenter of the Tunguska explosion. One of the many scenes of devastation discovered by the survey expedition led by Leonard Kulik in 1927.

that had not been felled pointing skyward like so many charred fingers (Figure 21.7). It was immediately obvious that some highly remarkable event had taken place, and the question arose as to the cause. Kulik had acquired funding from the Soviet Government on the then reasonable pretext that a large and valuable iron meteorite, located in an impact crater, would be found at the fall location. The problem then, and mystery still, is that no impact crater was found, and indeed, no meteorite fragment has ever been recovered from the deforested region. As soon as the news of what Kulik's expedition had found, or rather, not found, in the region of devastation, speculation as to its potential explanation began.

Straight out of the gate, American astronomer Harlow Shapley and British astronomer F. W. W. Whipple independently suggested that the lack of any crater implied a cometary impactor, and this idea garnered much support for many years. For all this, however, in ensuing years, speculations concerning alternate causes for the Tunguska event began to appear. Indeed, almost every new fad and fashion of the physical sciences produced some alternative explanation. In 1940, for example, Vladimir Rojansky (Union College, NY) brought the then new idea of antimatter into the astronomical realm, suggesting that some meteoroids and cometary nuclei could be composed of contraterrene matter (as antimatter was then called). Lincoln La Paz (University of New Mexico), in 1941, specifically picked up on this idea, and argued that the Tunguska event was possibly caused by the "mutual annihilation of a contraterrene meteoritic mass and terrene material." Interestingly, and reflective of the times, La Paz favored an antimatter explanation because the deduced mass and size of an ordinary matter impactor (set — incorrectly as it turns out — at $\sim 10^{10}$ kg and ~ 200 m across respectively) would have been fully expected to produce a sizeable crater.

In 1978 the Slovak astronomer Lubor Kresak argued that the Tunguska impactor was a fragment detached from comet 2P/Encke — the association being made upon the basis of the time of the event (June) and the occurrence of the daytime β-Taurid meteor shower.[1] This scenario built upon the early suggestions of Shapley and Whipple, but pin points a specific parent comet. Equation (21.1) indicates that objects capable of producing a Tunguska-like blast will occur about once every 1000 years or so, but this number assumes a steady impactor influx rate. If the Tunguska parent body was a chance fragment derived from comet 2P/Encke, then it would be an unexpected, rather than an expected, event. Bill Napier (Armagh Observatory), in a 2019 research paper, has looked at how fragments ejected from the nucleus of comet 2P/Encke might chance to encounter the Earth, and finds that such fragments can produce "brief meteor hurricanes" which might be capable of affecting Earth's climate. Indeed, not only might the Tunguska parent body be fragment from comet 2P/Encke, but Napier suggests that the sudden onset of global cooling, some 13,000 years ago (the so-called Younger Dryas), might have been caused by the atmosphere being dusted by numerous fragments and clouds of dust derived from comet 2P/Encke.

---

[1]Active during daytime hours in the northern hemisphere, this shower is only *visible* via radio and radar studies.

In 1993, fifteen years after Kresak presented his 2P/Encke fragment scenario, Christopher Chyba (NASA Ames), Paul Thomas (NASA Ames), and Kevil Zahnle (University of Wisconsin) argued in a seminal research paper, published in the journal *Nature* [6], that rather than being caused by a low-density cometary impact, the Tunguska impactor was, in fact, a stony asteroid. Bolstering their argument with detailed computer simulations, Chyba and coworkers argued that the stony impactor, initially some 50 meters across, suffered catastrophic deformation and disruption at an altitude of about 10 km. Rejecting the extraterrestrial origin idea entirely, German geophysicist Wolfgang Kundt (University of Bonn) suggested in a 2001 research paper that the Tunguska event was caused by the release and ignition of some 10 million tonnes of sub-terranean methane gas. It would seem that even after a century and more, a full explanation of the Tunguska event remains elusive, and while a stony asteroid impact is the most probable explanation, additional scenarios will no doubt be presented in the future (recall Section 11.1).

A twist to the Tunguska story, and one that potentially supports the asteroid impact scenario, was outlined in 1999. Following an expedition to the Tunguska region in that year, researchers from the University of Bologna suggested that Lake Cheko, located just 8 km away from the blast epicenter, might, in fact, be a water-filled impact crater [7]. There are many lakes in the Tunguska taiga, but Lake Cheko does present some unusual characteristics. At 500 m across, the lake is bowl-shaped in profile — unlike the typical flat-bottomed lakes in the region — and its sediment basement suggests a relatively young age. Magnetic readings taken at the lake surface further suggest that there is a meter-sized fragment of rock located at the lake's deepest point. Not every researcher agrees with the Bologna expedition findings, however, arguing that the sediments in the lake bottom are deeper than would be expected for a recent formation. The story of Lake Cheko will no doubt rumble on for many years yet, and is a mystery that remains to be resolved.

In addition to cometary and asteroid impacts, the Earth is additionally at risk, in terms of climate change and potential mass-extinction events, from the radiation and debris cloud that would be produced by a nearby supernova explosion. Such events must have occurred in the past, and they might well happen again in the future. It is known that the Earth must have encountered material ejected from a supernova in the recent past since radioactive ^{60}Fe has been detected in deep-sea sediment cores. The point being that supernova explosions represent the only known way of generating such iron, and since it has a half-life of 1.5 million years, then its implantation in ocean sediments must have been in relatively recent times (geologically speaking). While no mass-extinction event has been unequivocally linked to a nearby supernova explosions, some researchers have suggested that the end Ordovician-Silurian extinction event (recall Figure 11.9), and distinct climate change episode, some 450 million years ago was the result of such a nearby cataclysm.

While the threat of a supernova explosion might happen on a timescale of once ever 250–300 million years, it is possible that the Earth's atmosphere continuously encounters small, sub-relativistic dust particles ejected from distant supernova. The detection of such particle has recently been investigated by Amir Siraj and Abraham Loeb (both at Harvard University) [8]. Indeed, Siraj and Loeb note that even the fastest meteoroids that the Earth regularly encounters, the Leonid meteoroids with a velocity of 71 km/s, have a speed that is just 2×10^{-4} that of light. Dust ejected into the interstellar medium after a supernova explosion can, however, obtain speeds of 1 to 10 percent that of light, and accordingly if such meteoroids encounter the Earth's atmosphere, they will produce very short-lived, but energetic cylindrical blast waves. Modeling the encounter conditions of dust particles just a few millimeters across, but traveling at 10% the speed of light, Siraj and Loeb find that a readily detectable infrasound signature, lasting just 10^{-4} seconds, should be produced along with a similar duration optical flash. Present day optical and infrasound detectors are not configured to record such short-duration events, but future systems, appropriately configured and setup as a global, all-sky monitoring network, may well be capable of detecting a few sub-relativistic, supernova-derived meteors per year. This result, of course, is predicated on the possibility, not currently proven, that millimeter-sized grains moving at relativistic speeds can survive passage through the interstellar medium over distances corresponding to hundreds of parsecs.

Before his 1941 title bout against Billy Conn, boxer Joe Louis remarked that, "he can run but he can't hide." And Louis was right: catching up with and knocking out his lighter-weight adversary in the 13th round. The same refrain could be leveled against meteorite impact events. The associated fireball might not be seen, or its atmospheric path only vaguely noted, but if there are detonations and something hits the ground and a seismometer is in the area, then their fall will be recorded. In these cases, it is literally the ground motion that is recorded when either the shock wave hits the Earth's surface (so-called acoustic coupling) and/or the meteorite (so-called ground coupling) [9]. Acoustic-coupling-produced seismic waves were first detected in association with the Tunguska event of 1908 — in this case surface propagating ground waves traveling at least as far as 5400 km. Many additional seismic observations of acoustic-coupled ground waves have been recorded since, but only two events are documented in which ground-coupled seismic waves were recorded.

The first of these was related to a bright fireball that was observed near Prince George, British Columbia on August 21st, 1969. The event rushed through the darkened skies just before 11pm, and John Sinclair, in Ainsworth, B.C., explained that an "intense red glow lit the interior of his house, followed by a loud explosion that shock the neighbourhood — like dynamite." This spectacular incident was investigated by Ian Halliday and Alan Blackwell (National Research Council of Canada), and it was soon discovered that several seismic stations in the Prince George region had detected a distinct impact signal. The area, however, is heavily wooded and difficult ground to search, and even after several helicopter flights, and an aircraft photographic survey of the fall location, no impact features, or indeed, meteorites were found. Interpretation of the seismic data indicated that the impact energy was of order 10^9 joules, and Halliday and Blackwell inferred that an iron meteorite having a mass of some 12 tonnes (with a diameter of order 1.5 meters) was involved in the fall. Somewhere in the hills southeast of Tabor Lake[2] lies a good chunk of, as yet unfound, meteoritic iron. Adapting the Gutenberg–Richter magnitude scale [9], a meter-sized meteorite hitting the ground with energy E (in joules) will produce a seismic signal of magnitude:

$$M = 0.67 \, \mathrm{Log}_{10} \, E - 4.53 \qquad (21.2)$$

Under the application of this formula, the Prince George meteorite fall produced a magnitude 1.5 microearthquake seismic signal. The Earth may not have moved in any great sense because of the Prince George impact, but it certainly trembled a little.

The only other example of a ground-coupled impact is that of the Jilin (H5 chondrite) meteorite fall on March 8th, 1976 (recall Figure 7.2). This was an extensive fall of material, but what is taken as being the main mass (with a mass of 1.77 tonnes) created a $\sim$2-meter-wide, $\sim$6-meter-deep plunge pit. Ground-coupled seismic waves were recorded at two stations, and indeed, the impact time of 23:02:36 U.T. is based upon the seismic data.

One location in which the ground coupling of meteoroid impacts has proved particularly useful is on the Moon. In this case, of course, there is no atmosphere with which the meteoroid can interact, and accordingly there is no fireball trail prior to impact. Among the surface instruments left on the Moon during the Apollo Program were a set of seismographs, and these regularly detected meteoroid impacts (as well as Moonquakes). Before the Apollo Lunar Seismic Network (ALSN) was closed down in 1977 (due to lack of funding), a total of 1743 impact events were recorded from meteoroids in the estimated mass range from 0.1 kg to 1 tonne [10]. Figure 21.8 shows the seismogram associated with an impact of May 13th, 1972, and indeed, this impact literally set the Moon ringing, with surface oscillations lasting for at least 80 minutes. Particularly interesting, however, was the ALSN detection of meteoroid impact swarms that

---

[2]The best estimate for the fall location is longitude -122.433, latitude $+53.833$.

13 May, 1972

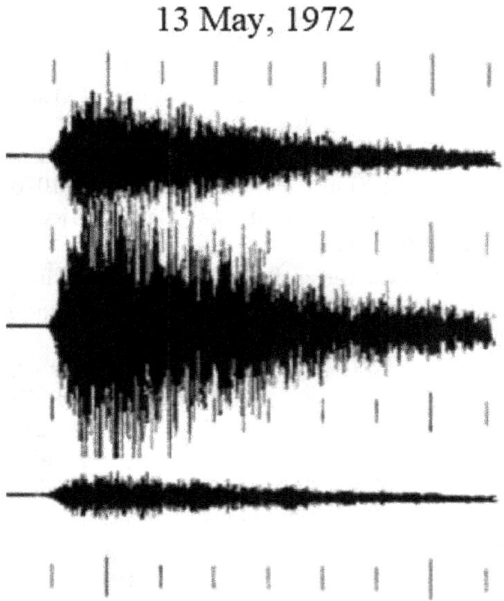

Figure 21.8. Seismogram tracings of a lunar impact recorded on May 13th, 1972. The three traces indicate oscillations in X, Y, Z planes, and the time axis is separated into 10-minute segments (vertical tick marks), starting at 08:50 U.T.

could be associated with the peak times of all the major meteor showers on Earth.

Among the remarkable impact swarm events detected with the ALSN, however, two stood out with respect to further analysis. These two impact clusters were observed in June 1975 and January 1977, and were studied by Jürgen Oberst (German Aerospace Center, Berlin). Oberst suggested that these specific events could be associated with the passage of the Moon through the debris trails associated with the Farmington meteorite (the June event) and the Innisfree meteorite (the January event). The Farmington meteorite fell on June 25th, 1890 and it has the shortest known cosmic-ray-exposure age of all known meteorites, just 20,000 years. Not only this it has been suggested that the Farmington meteorite might be associated with the β-Taurid meteoroid stream, which in turn may have contained the Tunguska event parent body.

References

[1] T. Ens, P. Brown, W. Edwards, and E. Silber. "Infrasound production by bolides: A global statistical study." *Journal of Atmospheric and Solar-Terrestrial Physics*, **80**, 208–229 (2012).

[2] O. Popova *et al.* "Chelyabinsk airburst, damage assessment, meteorite recovery, and characterization." *Science*, **342**, 1069–1073 (2013).

[3] P. Brown *et al.* "The flux of small near-Earth objects colliding with the Earth." *Nature*, **420**, 294–296.

[4] P. Matlovič *et al.* "Characterization of the June epsilon Ophiuchids meteoroid stream and the comet 300P/Catalina." *Astronomy and Astrophysics*, **636**, A122 (2020).

[5] D. Steel. "Tunguska at 100." *Nature*, **453**, 1157–1159 (2008).

[6] C. Chyba, P. Thomas, and K. Zahnle. "The 1908 Tunguska explosion: Atmospheric disruption of a stony asteroid." *Nature*, **361**, 40–44 (1993).

[7] L. Foschini *et al.* "The atmospheric fragmentation of the 1908 Tunguska cosmic body: Reconsidering the possibility of a ground impact." arXiv: 1810.07427. L. Gasperin *et al.* "Magnetic and seismic reflection study of Lake Cheko, a possible impact crater for the 1908 Tunguska event." *Geochemistry, Geophysics Geosystems*, **13**(5), 1525–2027 (2012).

[8] A. Siraj and A. Loeb. "Observational signatures of sub-relativistic meteors." arXiv: 2002.01476.

[9] W. Edwards, D. Eaton, and P. Brown. "Seismic observations of meteors: Coupling theory and observations." *Reviews of Geophysics*, **46**, RG4007 (2008).

[10] Y. Nakamura, G. Latham, and H. Dorman. "Apollo Lunar Seismic Experiment — Final summary." *Journal of Geophysical Research*, **87**, supplement, A117–A123 (1982).

Chapter 22

Electrophonic Sounds

"What is this sound and rumour?" — so begins *The March of the Workers* (1885) by William Morris. Usurping, however, its intended message, the very same sentiments could be used to describe the history of electrophonic (also simultaneous) sounds. Indeed, the existence of such sounds and their association with the atmospheric passage of bright fireballs is steeped in a long history of denial, persistent rumour, physical mystery, and misunderstanding — and much of the mystery, couched, all be it, in terms of numerous physical mechanisms, and a growing body of observational data, still exists to this very day.

The suggestion that an observer might hear a fireball at the same time as seeing it, stretches back to at least the passage of the March 19th, 1714 fireball investigated by Edmund Halley. These simultaneous sounds are distinct from the reverberating, thunder-like sonic booms that accompany the break up and passage of a meteoroid through Earth's lower atmosphere. Rather they are described as being staccato, hissing-like, or crackling sounds. And, indeed, such reports remain largely a mystery to this very day. In his investigation of the 1783 fireball procession Charles Blagden dismissed such simultaneous sight and sound reports as "more doubtful in its nature, and less established by evidence," and not much has changed since — that is, excepting the fact that simultaneous sounds are now accepted as a real observational phenomenon, but have only poorly been recorded instrumentally (poorly, that is, in the ability of the data to be taken as definitive and capable of separating one production mechanism from another).

At stake, of course, is how can one simultaneously hear and see a fireball when the event itself is perhaps several hundreds of kilometers away. Between 10 and 30 kilometers altitude in the Earth's atmosphere, the speed of sound is about $300\,\mathrm{m/s}$ (rising to $330\,\mathrm{m/s}$ at sea level). The sound travel time required to cover a distance of 100 km is therefore about $5^1/_2$ minutes — that is, by the time that a direct sound wave might have travelled to an observer, the light phenomenon associated with the fireball will have long dissipated and passed. If a sound is to be associated with the visible event, therefore, the mechanism responsible for generating the sounds must be operating at a speed comparable to that of light. The problem, of course, is that sound, as a pressure wave, cannot possibly propagate at such speeds — canonically, sound travels a million times slower than light: $c_{\mathrm{sound}}/c_{\mathrm{light}} = 10^{-6}$. For all this, however, the modern interpretation and exploration of simultaneous sounds was put in place in the early 1980s by Colin Keay (University of Newcastle, Australia) [1]. Building upon ideas developed by Vitaly Bronshten (Russian Academy of Sciences, Moscow), Keay championed usage of the term electrophonic sounds over simultaneous sounds, since the important sound generating mechanism is electromagnetic radiation, which, of course, does travel at the speed of light. While Keay's mechanism is robust and well argued, there are still technical issues to be sorted out, and there are also alternative ideas.

22.1   Magnetic spaghetti

The model developed by Keay is built around the way in which meteoroids interact with the atmosphere (that is through ablation), and the manner in which the ionized material following in the wake of the ablating meteoroid interacts with the Earth's magnetic field. Since, Keay argues, the material left behind an ablating meteoroid is ionized, it can, under the right circumstances, trap and distort geomagnetic field lines (Figure 22.1). If these same trapped magnetic field lines are made to wiggle under turbulent wake conditions, then as they relax (once the wake has dissipated) they can generate electromagnetic (EM) waves with a wavelength comparable to the height of the Earth's atmosphere — that is EM waves are

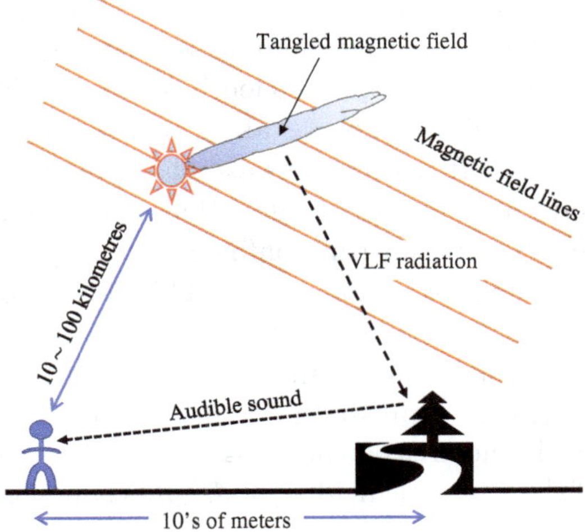

Figure 22.1. Schematic for the Keay–Bronshten *magnetic spaghetti* model for electrophonic sound production. The audible sounds are produced locally to the observer by material (leaf matter, for example) interacting with the VLF radiation that has been generated by the fireball interaction with the Earth's magnetic field.

produced with wavelengths of many kilometers, deep into the very low frequency (VLF) radio region of the EM spectrum. Since the VLF waves will propagate at the speed of light, if their energy can be conducted by objects close to the observer, then in principle sounds can be heard at the same time as the fireball is seen — the sounds are generated at the observer's location (at distance of order of meters), not at the meteoroid's ablation location (which can be situated tens to hundreds of kilometers away). Keay has described the driving mechanism behind the production of electrophonic sounds as the *magnetic spaghetti* model, since it is the trapping and tangling of the geomagnetic field that produces the VLF emission. Having described a mechanism for allowing the production of electrophonic sounds, namely the generation of VLF EM waves, the final, crucial step is the coupling of the EM wave to material in the neighborhood of the observer. The kind of material that can make the sounds required are dried grasses or leaves, frizzy hair, or even glasses being worn by the observer. It is this selectivity in the vibrating medium producing the audible sounds that accounts for the fact that some observers report electrophonic sounds while others do not.

Keay's *magnetic spaghetti* model can be developed to provide onset conditions for when electrophonic sounds might be generated. Firstly, the meteoroid has to be sufficiently low in Earth's atmosphere that mass loss through ablation is taking place — this will result in a trail of ions being left within the wake of the meteoroid. Next, the trail of ions must be able to capture, that is entrain, geomagnetic field lines. The condition for this confinement is expressed according to the magnetic Reynolds number R_m. This dimensionless number is derived according to the ratio of the magnetic inductance and magnetic diffusion of the medium (the plasma of ions in the wake of the fireball), and is expressed in terms of the characteristic flow velocity V, the characteristic length scale L, and the decay time for the diffusion of the magnetic field $\tau_m = L^2/\eta$, where η is the magnetic diffusivity constant.[1] Accordingly,

$$R_m = \frac{V\,\tau_m}{L} \tag{22.1}$$

The larger the value of R_m, so the more tightly can the plasma hold on to the magnetic field lines. It is generally taken that magnetic field lines will be trapped and forced to move accordingly to the motion of a plasma once $R_\mathrm{m} > 10$.

Trapping geomagnetic field lines is one thing, but, as noted by Keay and Bronshten, the field lines also need to be tangled up and distorted for the magnetic spaghetti model to work, and this accordingly requires that the plasma behind the fireball be in a turbulent state. The turbulent flow condition is set by the Reynolds number R_e, which in this case is expressed as the ratio of the inertial forces in a fluid compared to the viscous forces. We have,

$$R_e = \frac{\rho V L}{\mu} \tag{22.2}$$

where ρ is the density and μ is the dynamic viscosity of the fluid. A fluid (plasma) will become turbulent once $R_e > 10^6$. The density

---

[1]The diffusivity constant is further written in terms of the permeability of free space μ_0 and the electrical conductivity σ_0, with $\eta = 1/\mu_0\sigma_0$. In a meteor plasma $\eta \sim 1000\,\mathrm{m}^2/\mathrm{s}$, giving a characteristic diffusion timescale of $\tau_m = 0.001$ seconds when $L \sim 1\,\mathrm{m}$.

term in Equation (22.2) can be substituted for by the exponential atmosphere expression [Equation (6.11)], giving $\rho = \rho_0 \exp(-Z/H)$, here Z is the atmospheric height, $H \approx 8\,\text{km}$ is the atmospheric scale height, and $\rho_0 = 1.225\,\text{kg/m}^3$ is atmospheric density at sea level. Accordingly, the height Z at which the plasma trailing behind an ablating meteoroid is going to encounter a turbulent flow regime can be expressed as:

$$Z = H \ln \left(\frac{\rho_0}{\mu\, R_e}\, V\, L \right) \qquad (22.3)$$

Setting $R_e = 10^6$ (the turbulent flow condition) and taking $\mu \approx 2 \times 10^{-5}\,\text{Ns/m}^2$ (a characteristic value for the Earth's lower atmosphere), the onset height Z for the potential generation of electrophonic sounds can be determined. Inserting constants into Equation (22.3) the onset height condition is $h < Z_{es}$, where, h is the actual altitude of the meteoroid and

$$Z_{es}\,(\text{km}) = 8 \ln (V\,L) + 32.92 \qquad (22.4)$$

where V is now expressed in km/s and L is in meters. Taking a typical fireball speed in the atmosphere to be 15 km/s and a characteristic dimension of $L = 1$ m, so the turbulent onset altitude, and the condition for the possible onset of electrophonic sound generation, is $h\,(\text{km}) < 54.5$.

The condition for turbulence is, as Equation (22.4) reveals, dependent upon the velocity and characteristic size — as these terms decrease so the onset height, and the possibility of electrophonic sound production, also decrease. Recall, however, that not only must condition (22.4) be satisfied, but the magnetic Reynolds number [Equation (22.1)] must additionally be larger than 10 (and preferably much greater than 10). For $L = 1\,\text{m}$, provided the velocity is greater than about $10\,\text{km/s}$, so the magnetic Reynolds number will be greater than 10. A detailed study of the atmospheric flight of several meteorite-dropping fireball events (Pribram, Innisfree, and Lost City) by Keay [2] found that the electrophonic sound production conditions were typically satisfied for between 3 to 5 seconds — a result consistent with the available eyewitness accounts.

22.2   Eyewitness accounts

One of the mysteries still surrounding electrophonic sound generation is the identification of the exact transduction process — this is the process by which the sounds are actually produced. Various surveys suggest that electrophonic sounds can take on many descriptive forms, with eyewitnesses reporting hearing sounds such as a stream of stochastic *pops* or *vuts*, the sound of crackling bacon, leaves rustling, along with fizzing, hissing, rushing, and swishing noises. There is some evidence to suggest that people with long frizzy hair and those people with glasses are more likely to hear electrophonic sounds than observers with short hair and no glasses, but the issue is complicated, and the transduction and sound-generation mechanism behind the production of electrophonic sounds remains largely unexplained. For all this, electrophonic sounds are commonly reported. Indeed, in a survey conducted by the author [3] on the extensive Millman fireball catalog of Canadian fireball reports, gathered between 1962 and 1989, out of the 2131 fireball events described, 97 (about 5 percent of events) clearly indicate the presence of electrophonic sounds. Of the fireballs producing such sounds, some two-thirds were brighter than magnitude -10 (comparable to the full Moon), and about half had durations that lasted between 1 and 5 seconds.

The recent very bright fireball[2] and meteorite-producing event in Chelyabinsk, Russia, was extensively observed and investigated, and a whole host of anomalous effects and sensations were reported by eyewitnesses. Anna Kartashova (Russian Academy of Sciences, Moscow) and coworkers [5] presented a summary of the collected eyewitness accounts, and from 1674 people interviewed some 198 reported hearing electrophonic sounds — that is some 12 percent of eyewitnesses. Most, 76 people, reported hearing a hissing sound, whereas 25 people reported hearing a crackling sound, and some 25 eyewitnesses reported hearing a whistle-like sound. Others (31 observers) reported a sound like a passing aeroplane.

An estimate of the global flux of potential electrophonic-sound-producing fireballs can be calculated from the minimum magnitude

---

[2]At 100-km range, the fireball would have been brighter than the Sun!

limit of -10 derived from the Canadian Fireball Catalog [3]. For a typical velocity of $20\,$km/s, a mass of about $20\,$kg is required to produce a fireball having a peak brightness of magnitude -10. The global flux F_{20} of such meteoroids is about $10{,}000$ per year (see Chapter 18). The number of simultaneous-sound-producing fireballs N_{ss} occurring over an area A in the time interval T will be of order

$$N_{ss} = \frac{1}{2}\left(\frac{A}{A_E}\right) C\, F_{20}\, T \qquad (22.5)$$

where A_E is the surface area of the Earth, C is a cloud obscuration factor (say $1/2$), and the factor of $1/2$ accounts for the fact that most fireballs are seen at night, while many pass unnoticed (and unheard) during daylight hours.

Over an area equivalent to that of Canada, with $A = 9 \times 10^6\,$km^2, the expected number of simultaneous-sound-producing fireball events in a 27-year time interval (the interval covered by the Canadian Fireball Catalog) is $N_{ss} \approx 1200$. Given that this total estimate is about right (to order of magnitude), then only about 8 percent of the potential simultaneous-sound-producing fireball events were witnessed and reported by Canadian observers. If a fireball has to be within $100\,$km of an observer for electrophonic sounds to be heard, then one such event might occur (per observer anywhere on Earth) every $6^1/2$ years. Given, however, an 8 percent detection efficiency, an individual observer would likely have to be out observing every night for 81 years before witnessing a bright fireball with associated simultaneous sound — truly a once in a lifetime event.

The results from the Canadian Fireball Catalog and the Chelyabinsk study all point to the fact that only very bright fireballs are going to produce electrophonic sounds, and this is effectively an energy constraint. The author [5] has investigated the circumstances of the Chelyabinsk fireball and argues that something like 625 watts of radiative energy could be developed via magnetic entanglement, and that accordingly at a range of 100 km from the fireball the energy flux would be of order $6 \times 10^{-8}\,$W/m^2. If the local-to-the-observer transduction process can successfully generate pressure waves with a similar intensity, then acoustic levels of some $50\,$db might result.

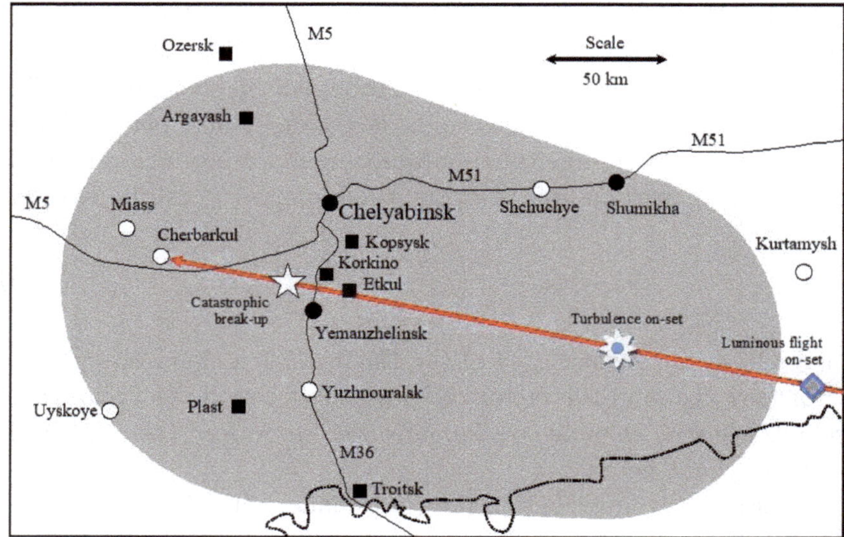

Figure 22.2. Ground path of the Chelyabinsk (2015) fireball (red line) and region (shaded grey) from which electrophonic sounds are predicted to occur [5]. The filled circles correspond to those locations at which electrophonic sounds were definitively reported. The filled square symbols indicate a selection of the locations at which observers reported the feeling of a 'warming heat' during the passage of the fireball. The locations at which luminous flight began, turbulent flow conditions started, and where catastrophic break up occurred are shown as a diamond (far right), an 8-sided star, and a 5-sided star respectively.

Such sound levels are generally described as being equivalent to that of ordinary speech at a few meters away from the listener's ears. Even if the transduction process results in pressure waves with an intensity of just $10\,\mathrm{db}$ — equivalent to an intensity of $10^{-11}\,\mathrm{W/m^2}$, that is a conversion efficiency of 0.01% — then sounds at a level equivalent to that of 'distant rustling leaves' could, in principle, result. The zone within a 100 km distance of the flight path of the Chelyabinsk fireball is illustrated in Figure 22.2, and this zone is reasonably consistent with the locations of those observers who did report hearing electrophonic sounds.

For many years electrophonic sounds were rejected as a real phenomenon, it being reasoned that eyewitness accounts were the result of human physiology — or, less generously, figments of the observer's imagination. This idea of observers experiencing what they expect to experience (rather than what they actually experience) has also extended to accounts of instantly feeling the heat from a fireball, and even smelling a fireball. Both these latter effects were reported by eyewitnesses to the Chelyabinsk fireball [4]. Of the people who submitted reports, some 37 percent reported that they felt a warming effect, and 2 percent of respondents even reported receiving a sunburn. The physical mechanism behind the heat sensation phenomenon is not readily apparent, although it may relate to electrostatic effects upon the skin. Sunburn effects, provided they are truly real and due to being exposed to the fireball, can be accounted for in terms of ultraviolet radiation, but again, the exact process and why only very few people are affected is presently unknown. On more familiar ground, many eyewitnesses in villages located close to the fireball ground path reported a burning or gunpowder-like smell in the air about an hour after the fireball's passage. This phenomenon, and its onset delay, presumably reflect the mixing time for such odors to mix to the ground. Eyewitness accounts of smells and the feeling of heat sensations are not unique to the Chelyabinsk fireball and have been reported (and usually dismissed as illusionary) throughout recorded history.

22.3 Auroral sounds

Another anomalous affect that has also long been dismissed is the attribution of sound generation by the aurora borealis. Keay has suggested a similar mechanism to the *magnetic spaghetti* model might be at play to produce such auroral sounds, but an alternative mechanism has recently been presented [6]. Again, the sound generation is local to the observer, but is dependent upon local atmospheric conditions and the development of electric field gradients (Figure 22.3). Just as with simultaneous sounds from fireballs, the simultaneous auroral sounds are problematic since the region in which the light phenomenon is produced is located between 100 and 300 km altitude, so a local sound source is required, and one such location is that described by the inversion model.

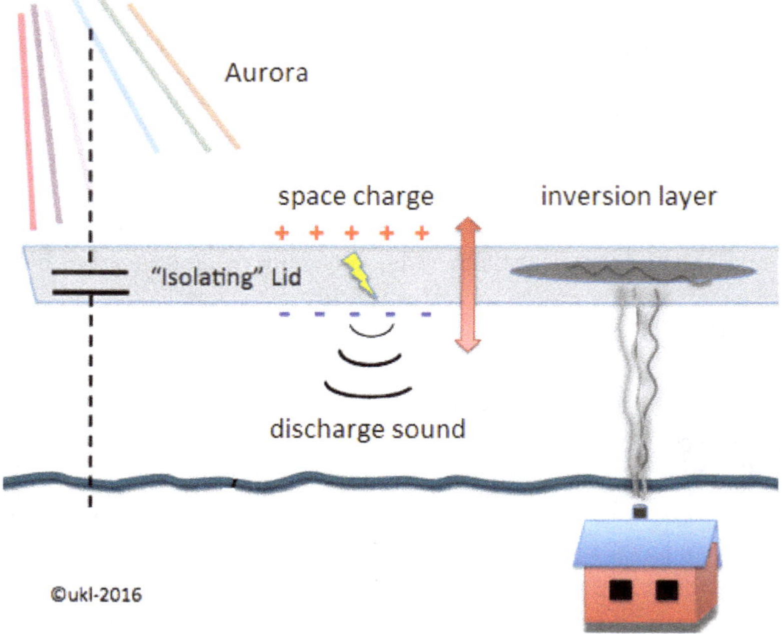

Figure 22.3. Schematic diagram for the inversion layer, space charge, model for auroral sound generation. Image adapted from Uno Laine/Aalto University.

This model for auroral sounds was recently examined experimentally by Unto Laine (Aalto University, Finland) and collaborators and they were able to locate the sound generating zone at an altitude of order 100 meters above the ground. The inversion layer model, as illustrated in Figure 22.3, builds upon the idea that under calm weather conditions a temperature inversion layer can develop anywhere between about 50 and 400 meters above the ground. Warm air moving upwards will carry negatively charged aerosols to the bottom of the inversion layer — this accumulation will cause a space charge to develop across the inversion layer: negative charges building up along its base and positive charges building up along its top layer. Effectively the inversion layer acts like an isolating lid and forms a spatially extended capacitor. Eventually, if the conditions are right, Laine *et al.* argue, an aurora-driven perturbation can activate the charges in the inversion layer enabling them to discharge. It is these locally produced discharge events that generate the sounds heard by observers on the ground.

22.4 Photoacoustic effect

While the *magnetic spaghetti* model has much going for it with respect to explaining simultaneous sounds, the VLF conduction and sound-generation mechanism remains problematic, and some researchers have sought alternative explanations. One recent alternative model for simultaneous sound production has been presented by researchers working at the Sandia National Laboratory in New Mexico, and they have revisited an old idea first developed by Alexander Graham-Bell in the 1880s [7]. Rather than relying on the generation of VLF radiation, the Sandia group explored the photoacoustic effect that Graham-Bell attempted to develop into the photophone (Figure 22.4). This communication device used sunlight

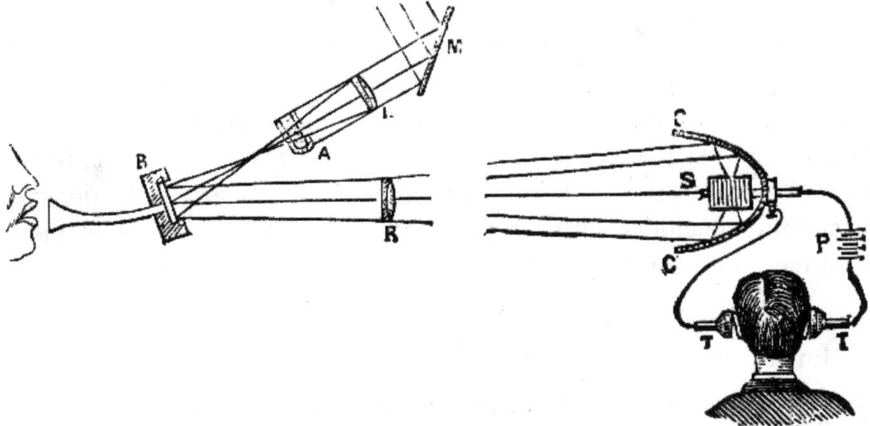

Figure 22.4.   The photophone as envisioned by Alexander Graham-Bell (1880). Sunlight is intercepted (mirror M) and angled so as to illuminate a thin, reflective diaphragm (B) which will vibrate in accord to the words being spoken into a trumpet-like funnel. The reflected light is then collected with a parabolic mirror (CC) at a remote station, where it is brought to the focal point (S) at which a cell of light-sensitive material has been positioned. Accordingly, as the energy flux of the received sunlight varies so the output from the light-sensitive cell will also vary, and this is something that can be monitored.

as the transmitting medium, but applied a thin, reflective diaphragm in the voice apparatus to modulated the energy flux that would be received by a light-sensitive, dielectric (that is normally insulating) material situated at a distant station.

With respect to fireball trails, the Sandia group argued that strong millisecond flares and strong brightness fluctuations that invariably accompany the passage of a fireball can mimic the photoacoustic effect. Specifically, they argue, the modulated light from the fireball will radiatively heat dielectric absorbers such as hair, paint, and cloth. In turn, this miniscule heating effect can produce audible pressure waves in the surrounding air, thereby accounting for simultaneous sounds, and, importantly, the sound pressure waves will vary in response to the fireball's energy output. A detailed calculation by the Sandia group indicates that audible photoacoustic sounds should be produced by all fireballs brighter than about magnitude -10 (that is, at least as bright as the full Moon) [7, 8].

The story (and mystery) of fireball-generated simultaneous sounds has, no doubt, not been fully told, and it is highly likely that more than one mechanism is at play. A current mystery, for example, is the claim that faint, ordinary meteors can also produce a VLF signal and even simultaneous sounds. The problem in this case is one of energetics, and being fully sure that the VLF signals being detected (the VLF background is highly noisy[3]) are truly derived from and coincident with the passage of a specific meteor. The quest to understand the mystery of simultaneous sounds continues.

References

[1] C. S. L. Keay. "Anomalous sounds from the entry of meteor fireballs." *Science*, **210**, 11–15 (1980).

[2] C. S. L. Keay. "Electrophonic sounds from large meteor fireballs." *Meteoritics*, **27**, 144–148 (1992).

[3] M. Beech. "The Millman Fireball Archive II: Sound reports." *Journal of the Royal Astronomical Society of Canada*, **98**, 34–41 (2004).

---

[3]Lightning is one phenomenon that generates strong VLF signals, and there are many additional atmospheric effects that produce so-called whistles and chirps.

[4] A. Kartashov *et al.* "Eye-witness interviews on the Chelyabinsk airburst." *Proceedings of the IMC*, Poznan, Poland. pp.189–192 (2013). O. Popova *et al.* "Chelyabinsk airburst, damage assessment, meteorite recovery and characterization." *Science*, **342**, 1069–1073 (2013).

[5] M. Beech. "Electrophonic sound generation by the Chelyabinsk fireball." *Earth, Moon and Planets*, **113**, 33–41 (2014).

[6] U. Laine. "Auroral Acoustics Project — a progress report with a new hypothesis." *Baltic-Nordic Acoustic Meeting*, June 20–21, 2016, Stockholm, Sweden.

[7] R. Spalding *et al.* "Photoacoustic sounds from meteors." *Scientific Reports*, **7**, 41251 (2017).

[8] C. Sung, P. Brown, and R. Marshall. "A two-year survey for VLF emission from fireballs." *Planetary and Space Science*, **184**, 104872 (2020).

Chapter 23

The Mystery of Ice Meteorites

That large blocks of ice fall out of the sky, to impact the ground and occasionally houses, is evident enough; they are observed to fall and they are collected [1,2], but the central question to be addressed here is, did they enter the Earth's atmosphere from interplanetary space? Indeed, it is this latter requirement that must be satisfied, by definition, for such icy remnants to be considered meteorites. In addition, accepting for the moment that ice meteorites really do fall to Earth, the question of their origin arises — literally, where do they come from? It is certainly true that the solar system contains numerous bodies that have water ice as a major compositional component — for example, cometary nuclei and the moons of Jovian planets. In principle, therefore, one would expect ice meteoroids to exist simply because of the ongoing collisional evolution of objects within the solar system. One can, in fact, be more definite and state that it is a certainty that ice meteoroids must exist; the question, however, is how long they can survive. The outburst and break up of comet 73P/Schwassmann-Wachmann 3 (Figure 23.1) in 2006 provides an example of an event that produced at least 16 icy sub-nuclei, each many tens to hundreds of meters in diameter, and, no doubt, numerous smaller icy meteoroids as well.

23.1 Atmospheric passage

As described in Chapter 6, one of the most important factors in determining the delivery of a meteorite to Earth's surface is its initial encounter speed — the lower the encounter speed, so the better the

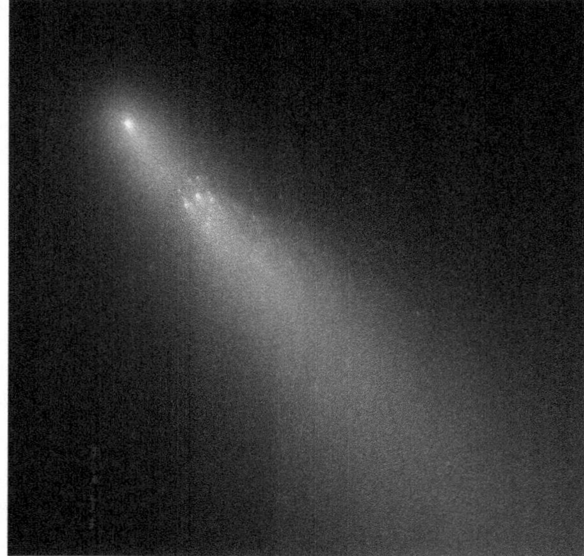

Figure 23.1.   Hubble Space Telescope image of ice fragments produced during the 2006 outburst of comet 73P/Schwassmann-Wachmann 3.

chances of something surviving to reach the ground. With respect to meteoroid streams known to be associated with specific comets, the smallest known Earth encounter speed is the 15 km/s of the occasionally active τ-Herculid meteor shower. The next lowest encounter speeds being those for the π-Puppid meteoroids (18 km/s) and the Draconid meteoroids (20 km/s). These latter two streams being associated with comet 26P/Grigg-Skjellerup and 21P/Giacobini-Zinner respectively. For the moment let us concentrate on the τ-Herculid shower since this stream is known to be associated with comet 73P/Schwassmann-Wachmann 3 [3]. This particular comet is known to have undergone at least two major fragmentation episodes: one in 1995 and again in 2005. While no known ice meteorite, derived from comet 73P/Schwassmann-Wachmann 3 or the τ-Herculid stream, has ever been reported, the comet and the shower provide the best current circumstances for producing such objects (see below).

While ice meteoroids must exist within our solar system, the more important question at this stage is, how long can they survive once ejected into space? Indeed, once any icy nucleus or ice meteoroid approaches within about 2.5 AU of the Sun, then sublimation, the phase transition from solid ice to water vapor, will become important. Detailed numerical calculations [2] indicate that, for a spherical pure water ice object moving in an orbit similar to that of comet 73P/Schwassmann-Wachmann 3, the radius would decrease due to sublimation at a rate of about 1.4 meters per orbit about the Sun (or about 25 cm per year). In other words, a 10-m diameter water ice block would completely disappear within about 4 orbits of the Sun — a timescale of about 20 years. The same sized meteoroid in an orbit similar to that of the Earth, that is, at one astronomical unit from the Sun, would disappear on an even shorter timescale of about 2 years.

Cometary nuclei that move deep into the outer solar system spend much less time in the Sun's sublimation-heating zone, and consequently any ice meteoroids left in their wake will potentially survive longer than a few tens of years. A 10-m diameter ice block with an orbit similar to that of comet C/1861 G1 (Thatcher), the parent comet to the April Lyrid meteor shower, for example, which has an aphelion distance of about 109 AU, might survive for about 2000 years before complete dissipation; but, if so ordained, it would encounter the Earth with an initial speed of 48 km/s — a speed so high that any chance of it surviving atmospheric passage are effectively zero.

The problem with respect to the production of ice meteorites, therefore, is that they must encounter the Earth within just a few years of being ejected from their parent nucleus, and this, dynamically speaking, is unlikely to happen. Furthermore, ice meteoroids must also enter the Earth's atmosphere at a relatively low speed.

The physical characteristics (e.g., the density, specific heat, enthalpy of melting and vaporization) of ice are well known, and consequently the equations that govern the mass loss rate and the deceleration of a meteoroid as it travels through the Earth's atmosphere can be numerically solved for in terms of just two input parameters: initial mass and initial velocity. The lowest possible speed that any object can have at the top of the atmosphere is

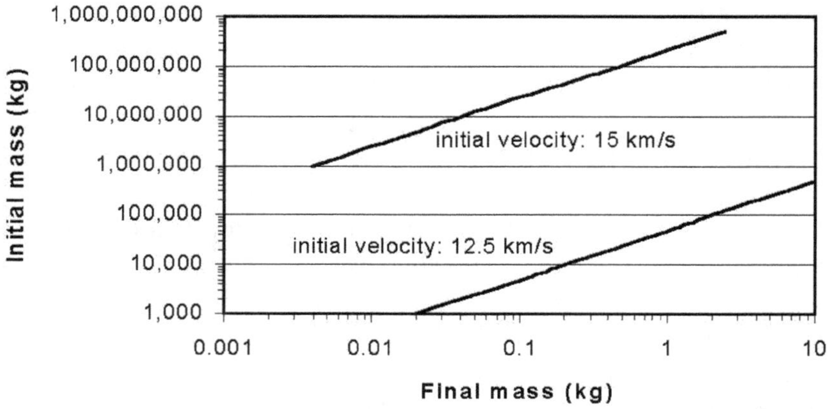

Figure 23.2. Initial and final mass comparisons for water ice meteoroids passing through the Earth's atmosphere for the two initial velocities of 12.5 and 15 km/s. [These calculations — see Chapter 6 — are based upon an ablation coefficient of $\sigma = 1.67 \times 10^{-7}\,\mathrm{s^2/m^2}$, a density of $917\,\mathrm{kg/m^3}$, and an entrance angle of $45°$.]

Earth's escape velocity of 11.2 km/s. Figure 23.2 shows the results of completing a series of ice-meteoroid ablation calculations for encounter velocities of 12.5 km/s and 15 km/s, with the latter being appropriate to a τ-Herculid ice meteoroid.

The calculations leading to Figure 23.2 indicate that when the initial velocity at the top of the atmosphere is 12.5 km/s an ice meteoroid of mass $\sim 50,000$ kg (diameter ≈ 4.8 m) is required to produce a 1-kg ice meteorite on the ground. Such a meteorite (assumed spherical) would be about 12.5 cm across. This size would certainly be noticeable, and markedly different from say a hail stone, but depending upon the weather conditions, not an object likely to survive very long on the ground before melting.

When the initial velocity is 15 km/s, however, even a 1,000,000-kg (diameter ≈ 15 m) ice meteoroid will only produce (in principle) an ice meteorite of a few grams mass on the ground. Several qualification points, however, must be made here. It is clear that no τ-Herculid meteoroid has ever produced a recognized ice meteorite. Moreover, if the Earth did encounter a τ-Herculid fragment of several meters in diameter it would probably produce an air-burst explosion.

Catastrophic fragmentation of all large ice meteoroids in Earth's upper atmosphere is almost inevitable, indeed, the ram pressure due to the oncoming airflow will easily exceed the tensile strength of solid ice or that of a cometary nucleus. The tensile strength of comet D/1993 F2 (Shoemaker-Levy 9), which crashed into Jupiter's upper cloud deck in 1994, for example, was estimated to be about $1000\,Pa$ [5], and while the tensile strength of solid water ice is much higher, falling between 10^6 to $10^7\,Pa$, the ram pressures indicated during atmospheric passage are well in excess of $10^7\,Pa$ for the simulations shown in Figure 23.2.

So, can an ice meteoroid survive atmospheric passage to hit the ground? Well, the answer is perhaps yes — just maybe! If the encounter velocity is not much greater than the Earth's escape velocity then a 5 to 10-m diameter ice meteoroid might just produce a 1 to 10-kg ice meteorite at the Earth's surface (provided that the tensile strength of the ice meteoroid is greater than $\sim 10^7\,Pa$).

So, returning to our question, are there ice meteorites? It seems reasonable to suggest that there are two main factors to argue against their existence on Earth. Firstly, the velocity restriction requires that the meteoroids must encounter the Earth with very low velocities — certainly less than 12–13 km/s. No currently known cometary meteoroid stream, therefore, can produce ice meteorites. This effectively removes from consideration what might otherwise be considered a good source of material for producing such objects. The second reason why ice meteorites must, at best, be exceptionally rare relates to their survival lifetime in space. To get close to the Earth means that an ice meteoroid must become heated, and once this happens lifetimes against sublimation are typically just a few tens to maybe a few hundreds of years. In other words, an ice meteoroid is likely to be 'destroyed' in space long before it might encounter the Earth's atmosphere and thereafter produce an ice meteorite.

At the risk of confounding one mystery with another, it has been occasionally noted that meteorite falls can precipitate distinct smells; most often described as sulfurous, or 'metallic.' Szaniszlo Berczi and Bella Lukacs [6] have picked up on this point and suggested that odors of sulphur and ammonia compounds might in fact be released by 'freshly' fallen ice meteorites, and indeed they further argued that ice meteorites might be identified on and within the extensive Antarctic and Arctic ice fields (where their terrestrial lifetime will be long because of the intense cold). The key distinguishing feature of an ice meteorite (with respect to the surrounding ice) being its water–ammonia composition. To date no candidate objects have been found in Antarctica, where ordinary meteorites have otherwise been found aplenty, but, as ever, there is always the caveat: absence of evidence is not evidence of absence, and, for that matter, unless one is specifically looking, it would be difficult to recognize an ice meteorite in a frozen ice terrain.

23.2 Megacryometeors

Perhaps the best current explanation for those large ice blocks that have been observed (over centuries) to fall from clear skies is that proposed by Jesús Martinez-Frias and coworkers at the Geosciences Institute in Madrid [1, 7, 8]. These researchers have proposed the term megacryometeors to describe the falling ice blocks (Figure 23.3).

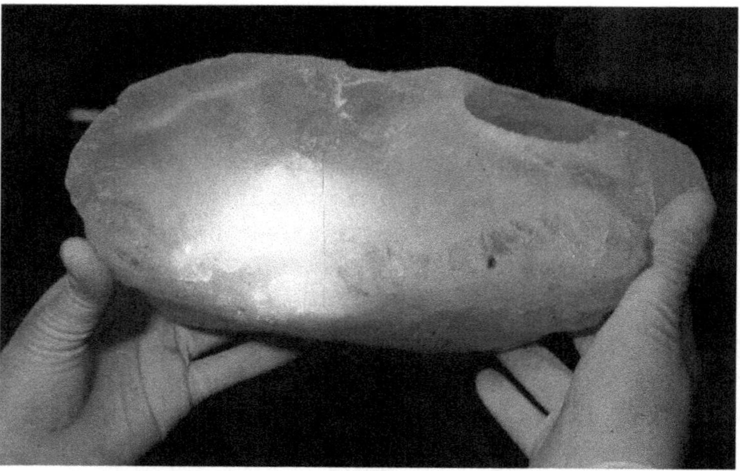

Figure 23.3. A fragment of a megacryometeor that fell in Asturias, Spain in 2000. Image courtesy of Jesús Martinez-Frias: http://tierra.rediris.es/ megacryometeors.

They argue that such objects form under rare, clear-sky conditions, and are produced by a super-sized variant of the nucleation process responsible for the production of ordinary hail. The 'meteor' part of megacryometeors, it should be pointed out, relates to the idea that these objects are considered to be meteorological (that is atmospheric) in origin. It should also be pointed out that these ice falls are distinct in origin from the blue-ice fragments that have occasional fallen from aircraft — the latter being, in fact, discharge from aircraft sanitation systems, the blue color stemming from the presence of disinfectant.

23.3   The Tau Herculids

The τ-Herculid meteor shower is erratic and in most years is barely noticeable. Dedicated observers, over many years, however, have been able to confirm the shower's delicate presence by recording faint, slow-moving shooting stars, radiating from the constellation of Hercules between May 19 and June 19 — this is the time interval when material from the parent comet 73P/Schwassmann-Wachmann 3 will be close to the Earth's orbit and cutting through the ecliptic. The shower reaches its maximum activity on June 9th with the meteors emanating from a radiant point located close to the faint, but still visible to the eye, star τ-Herculis (see Figure 23.4). At its peak rate, typically no more than a few meteors might be seen per hour (under ideal, moonless conditions) from the τ-Herculid shower — and most years no obvious meteors from the shower are seen or reported at all. The shower is far from being well known and its only distinguishing feature, from the better-known meteor showers, is its production of distinctly slow-moving meteors. Rather than generating a fast scratch of light upon the celestial vault, the τ-Herculids perform a gentle (even graceful) illuminated slide.

Arnold Schwassmann and Arno Wachmann, working from Hamburg Observatory, first swept up their 3rd new comet on May 2nd, 1930, and during that return it passed Earth by just 0.062 AU (or about 24 Moon orbital radii). That 73P/Schwassmann-Wachmann 3 might produce a meteor shower was soon realized and a prediction

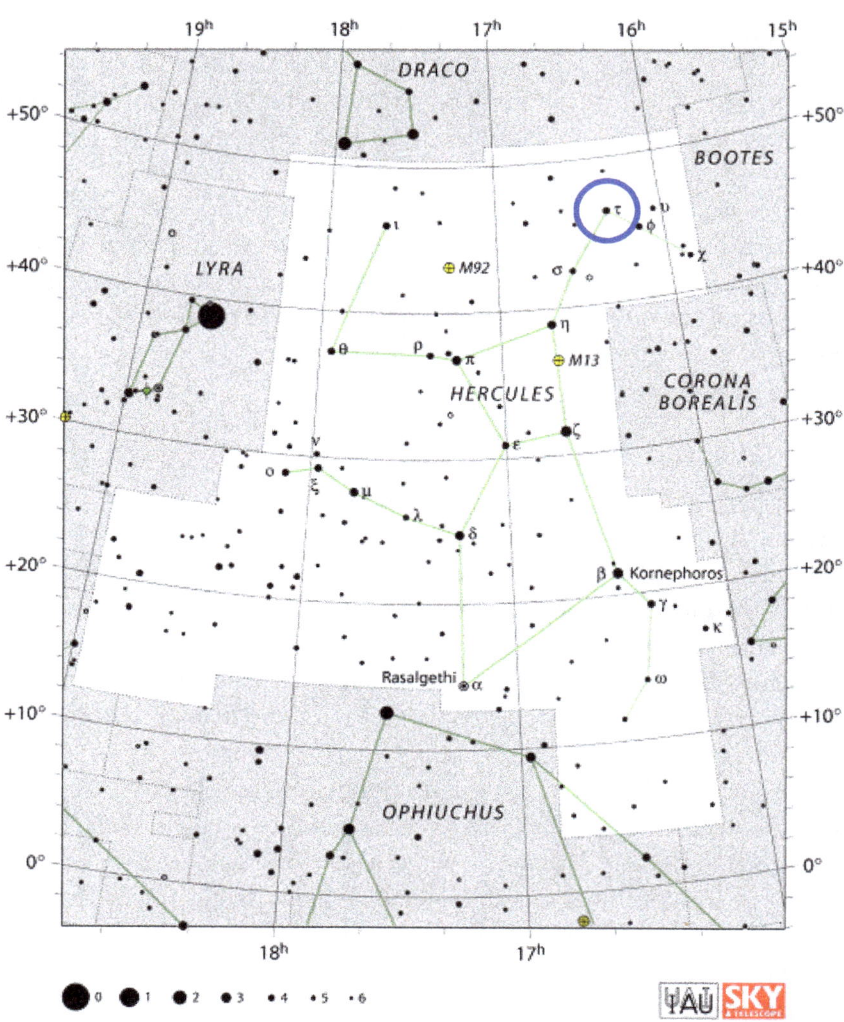

Figure 23.4.   Identification map for the constellation of Hercules. The radiant location for the τ-Herculid meteors is indicated by the circle in the upper right-hand corner. Image courtesy of the IAU and Sky Publishing.

of a radiant location in the constellation of Hercules was made. The first members of the τ-Herculid meteor shower were picked up on May 24th, 1930 by Kaname Nakamura working at the Kwasan Observatory in Japan.

Since its first detection, activity from the τ-Herculids has been intermittent at best, but the major break up event of 73P/Schwassmann-Wachmann 3 in 1995 may have set the stage for better things to come. A detailed analysis of the orbital dynamics by Paul Wiegert (Western University, Canada) and coworkers in 2005 [3] has revealed that very close encounters between the Earth and the debris trail produced by the 1995 break up of 73P/Schwassmann-Wachmann 3 will occur on May 30th, 2022 — the stream distance at this time will be 0.0004 AU (that is, inside of the Moon's orbit). At this particular return, material ejected from the parent comet in 1892 and 1897 will also track across the Earth's orbit. Yet another close stream encounter is predicted to occur in 2049 — this time due to material ejected by the parent comet as it rounded the Sun in 1897, 1903, 1935, and at perihelion returns from 1963 to 1990. In the two super-close encounter years the τ-Herculid shower may leap into life and produce a spectacular meteor outburst, with its slow meteors running across the night sky. Not only this, if there is ever going to be a time when ice meteorites, from a known comet, might chance to be produced, then at the end of May in 2022 and 2049 is the time to lookout.

References

[1] X. Bosch. "Great balls of ice." *Nature*, **297**, 765 (2002).

[2] "Block of ice crashes through roof of house in Bristol." https://www.bbc.com/news/uk-england-bristol-46127798.

[3] P. Wiegert, P. Brown, J. Vaubaillon, and H. Schijns. "The τ-Herculid meteor shower and comet 73P/Schwassmann-Wachmann 3." *Monthly Notices of the Royal Astronomical Society*, **361**, 638 (2005).

[4] M. Beech and S. Nikolova. "The endurance lifetime of ice fragments in cometary streams." *Planetary and Space Science*. **29**, 23–29 (2001).

[5] J. Scotti and H. Melosh. "Estimate of the size of comet Shoemaker-Levy 9 from a tidal breakup model." *Nature*, **365**, 733–735 (1993).

[6] Sz. Berczi and B. Lukacs. "Water-ammonia ice meteorites." *28th Annual Lunar and Planetary Science Conference*, pp. 97–98 (1997).

[7] J. Martinez-Frias *et al.* "Hailstones fall from clear Spanish skies." *Geotimes News*, The American Geological Institute, June, pp. 6–7 (2000).

[8] J. Martinez-Frias *et al.* "Oxygen and hydrogen isotopic signatures of large atmospheric ice conglomerations." *Journal of Atmospheric Chemistry*, **52**, 185–202 (2005).

Chapter 24

Declining Returns

The annual meteorite fall rate (AMFR), literally the number of meteorites seen to fall and then be collected in a given year, is a complicated term to qualify. At the very least it must depend upon the actual distribution of meteorite-producing bodies at one astronomical unit from the Sun, as well as the number and distribution of human beings on the Earth [1, 2]. Not only will human distribution have an effect upon where meteorites are found, but so too can terrain — a landscape dominated by lakes, forests, and shrubland will assuredly make the recovery of any meteorites (falls or finds) nigh on impossible. This being said, landscape can also result in the concentration of meteorite finds (but not falls) in specific areas — as is the case in the Antarctic (Figure 24.1). In these latter cases, the meteorite finds are highly localized (along the Trans-Antarctic Mountain Range), but the position of the initial fall locations will be entirely random on the landscape[1] (Figure 24.2). Not only landscape, however, but the distribution of humans and human infrastructure will affect the meteorite recovery numbers. David Hughes (University of Sheffield) and Robert Dodd (State University of New York at Stony Brook) [3], for example, have drawn attention to the fact

---

[1]While the fall locations are situated at random on the globe, there is a latitude variation effect in play. Indeed, writing in the May 2020 issue of the journal *Geology*, mathematician Geoffrey Evatt and coworkers from the University of Manchester demonstrated that the flux of Antarctic meteorites is smaller by a factor of about 1.5 that experienced at Earth's equator. Evatt and coworkers also used their estimate of the Antarctic meteorite flux to determine a latitude-corrected global mass flux, finding that in the mass range between 0.01 and 1 kg, some 3680 kg of material will fall each year. Furthermore, the global fall rate of meteorites with masses greater than 0.05 kg is found to be 17,600 per year.

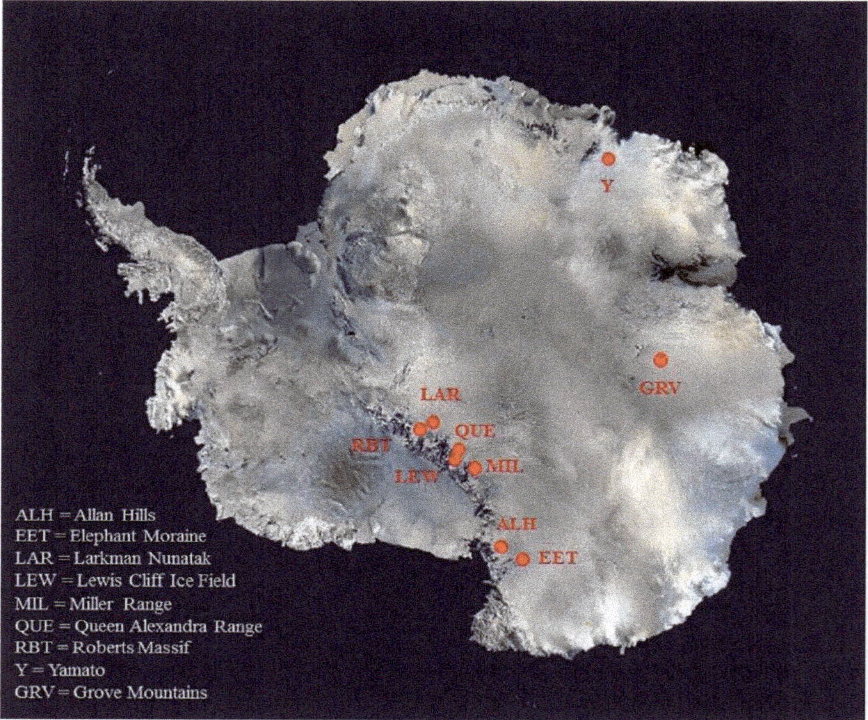

Figure 24.1.   Map of the major meteorite find locations in Antarctica. In this case geography, climate, ice flow conditions, and mountain locations, as well as a sparse human population have entirely dictated where meteorites have been found — the distribution of finds is highly localized, yet the fall locations upon Antarctica will be entirely at random. Image courtesy of NASA.

that a distinct band of meteorite finds that stretches across central Asia follows the path of the Trans-Siberian Railway. Regions were no meteorites have ever been found include the Amazon forest and central Africa, where not only does the landscape work against recovery, but the continuously damp climate will act to weather and erode meteorites rapidly.

24.1   Human population and meteorite fall rate

It has often been suggested that the AMFR should vary in accordance with the world human population, and accordingly a gradual increase

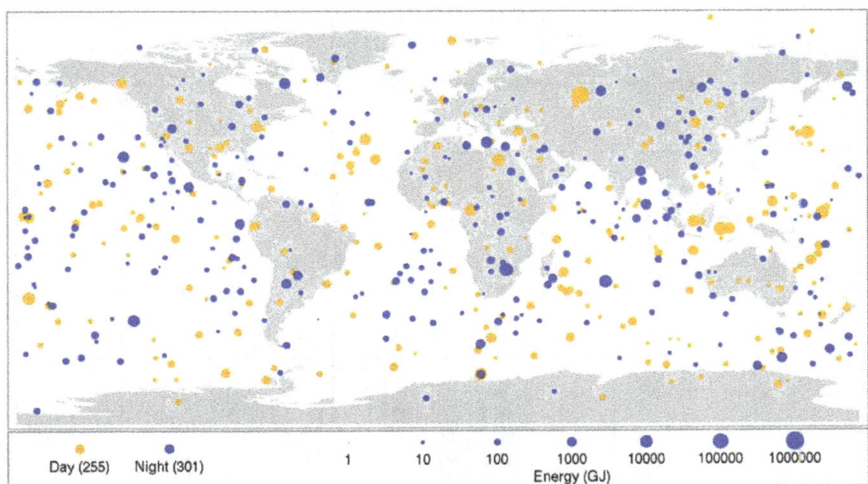

Figure 24.2. Distribution map of bright fireballs (small asteroids impacting Earth's upper atmosphere). Data gathered from 1994 to 2013, with the dot size corresponding to associated energy. The fireball detections, in contrast to meteorite finds and falls, occur essentially at random on the sky. Courtesy Wikipedia Commons/CNEOS.

in the AMFR over time would be expected. This result being predicated upon the simple concept that the more human beings there are in the world, the more observers there must be to notice incoming fireballs and then recover their associated meteorites. Mineralogist Robert Dodd [3] has additionally asked if our collective "outreach toward the planets, which began about 1960, [is] reflected in a higher than normal recovery rate for meteorite falls."

To test the hypothesis of a time increasing AMFR the number of documented meteorite falls per year has been extracted from the 5th edition of the *Catalogue of Meteorites* [4] which covers the time interval from January 1st, 1800, to December 1999. The results of this analysis are shown graphically in Figure 24.3. As might well be expected there is some considerable variation in the AMFR from one year to the next — it was zero, for example, in 1888, but attained a record high value of 17 in 1933, with other record-breaking years occurring in 1868, 1949, and 1976. Why these specific years should stand out is not at all clear. Interestingly, however, there

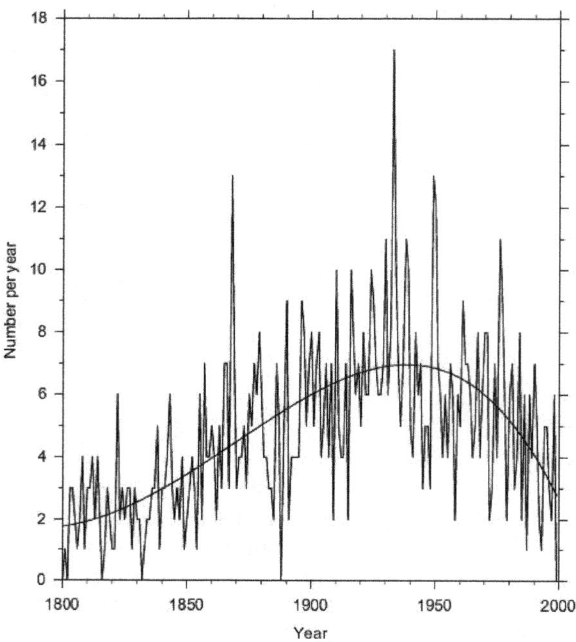

Figure 24.3. The annual meteorite fall rate from January 1800 to December 1999. The smooth curve, which is a 'best fit' fourth-order polynomial to the data, illustrates the background trend in the time variation of the AMFR.

appears to be a reasonably clear background trend in the AMFR data, illustrated by the smooth line drawn in Figure 24.3. From 1800 to circa 1940 the average AMFR showed a steady increase, but from circa 1940 onwards it adopted a steady decline. Rather than increasing in step with world population since circa 1940, the AMFR has, in fact, been steadily falling. This trend is more clearly seen in Figure 24.4 where the 10-year averaged value of the AMFR since 1800 is plotted against world population. From 1800 to 1940 the 10-year averaged AMFR increased almost linearly with world population growth, with every extra 500 million people apparently producing an increase of 2.8 in the average number of meteorites observed to fall. The transitional years, where the increasing trend in the averaged AMFR with world population breaks down, appears to have been between circa 1940 to 1960. Indeed, since circa 1960 the

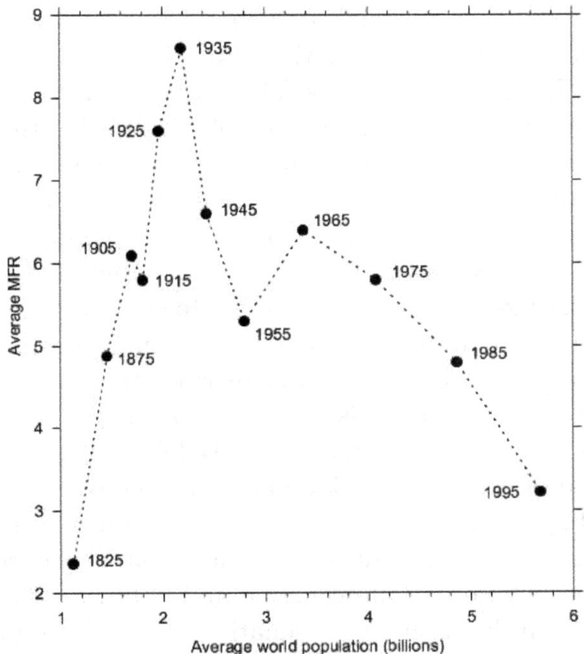

Figure 24.4. Average meteorite fall rate versus world population in billions of people. The averages are decadal after 1900 and per half-century from 1800 to 1900. The data points are labeled according to the average year of the time interval covered. The world population data has been taken from www.census. gov/ipc/www/worldhis.html.

decadal-average AMFR has shown a steady decline of about 0.7 for every additional 500 million people.

As described in several other sections only a very small fraction of the actual meteorite-producing events become observed falls in any given year. Astronomer Ian Halliday and coworkers [5], for example, predict that 5800 meteorite-producing events (with ground masses greater than 0.1 kg) should occur per year on the total land mass of the Earth (recall Chapter 18). This indicates a global fall recovery efficiency of about 0.1% in the modern era. The problem is, of course, that many meteorites fall unobserved during the day, with their associated 'attention-grabbing' fireball going unnoticed due to sunlight conditions. Typically, it would also appear, people in many parts of the world only spend a small fraction of any day outside at nighttime — again missing the all-important fireball sighting. For all this, however, the steady fall in the averaged AMFR since circa 1960 is unexpected. A number of reasons for the downturn can, however, be found. Firstly, human beings are not distributed at random around the globe, in the sense that most people live in highly light-polluted major city centers. So, while the population increases with time it need not mean that the sky coverage increases, and it makes no real difference whether a meteorite fall is seen by one person or by one hundred if they are all in the same location. Secondly the continuing trend towards greater urbanization mitigates against fireball sightings; light and noise pollution along with the 'cocooned,' eyes-locked-on-the-iPhone, indoor, modern lifestyle all tend to reduce the likelihood of fireball detection and hence meteorite recovery.

David Hughes [1] drew attention to the downturn in the meteorite fall rate beginning circa 1940 and argued that it was related to a long time lag between a fall occurring and it being reported in the literature. Given, however, the dramatic growth of the internet and social media platforms over the past 20 years, this suggestion now seems an unlikely explanation for the downturn seen in Figure 24.4. Incredibly, and in spite of a much greater awareness and appreciation of meteorites, the human race today is collectively less efficient at reporting and finding meteorites than it was a hundred years ago. We note in particular that the decadal average AMFR from 1985 to 1995 was, in fact, just one meteorite higher than in the interval from 1800 to 1850!

It might tentatively be concluded that what we are actually seeing with respect to the declining AMFR is a society-based decline in the awareness of their surroundings; people are simply becoming less aware of the night sky — or, indeed, the sky in general. In particular, light pollution in large urban centers has made the night sky less and less accessible to much of the human population. One might speculate that the actual flux of meteorite-producing bodies has truly varied over the time interval of study [2], but there are no obvious mechanisms for producing such short-time-scale effects.

The other possibility, of course, is that we have just been unlucky over the past several decades in locating actual meteorite falls.

24.2 Automatic detection

The future of meteorite recovery is no doubt instrumental, with the human investigator being directed to the fall location by data derived from automatic detection systems working tirelessly 24 hours a day, seven days a week. The historic Prairie Network (PN) and Meteorite Observation and Recovery Program (MORP), in addition to the long running European Camera Network (recall Section 1.1), have paved the way to the development of highly automated systems such as the Desert Fireball Network in Australia. This latter system being able to direct investigators to the fall of the Murrili meteorite, which fell on November 27th, 2015, to within 50 meters of its landing plunge pit [6]. Future surveys may also include the expanded Global Fireball Observatory [7]. At present, this collaboration of various national networks monitors some 0.6 percent of Earth's surface, but aims to expand this to 2 percent coverage during the 2020s. These automatic survey systems, using multiple camera locations, will be able to provide rapid and precise information about falls, and additionally enable detailed orbital data to be deduced.

References

[1] D. W. Hughes. "Meteorite falls and finds: Some statistics." *Meteoritics*, **16**, 269–281 (1981).

[2] K. L. Rasmussen. "Historical accretionary events from 700 BC to AD 1850 — a 1050-year periodicity?" *Quarterly Journal of the Royal Astronomical Society*, **31**, 95–108 (1990).

[3] R. T. Dodd. *Thunderstones and Shooting Stars: The Meaning of Meteorites.* Harvard University Press, Cambridge, Massachusetts, 1986. p. 14.

[4] M. Grady. *Catalogue of Meteorites (5th edition).* CUP, Cambridge, 2000.

[5] I. Halliday, A. Blackwell, and A. Griffin. "Meteorite impacts on humans and buildings." Science, **318**, 317 (1985).

[6] E. Samson *et al.* "Murrili meteorite's fall and recovery from Kati Thanda." arXiv: 2006.07151.

[7] H. Devillepoixa *et al.* "A Global Fireball Observatory." arXiv: 2004.01069.

Chapter 25

Meteorites in Popular Culture

While the sight of a fireball is shocking and awe inspiring, the sight
of a stately comet or a softly gliding shooting star is arguably more
inspiring and romantic. Just like fireballs, however, the comet and
meteor are transitory, they are the oddities of the night sky — the
night sky that otherwise inspires constancy and permanence. This
characteristic of fleetingness and fragility of existence is something
that has not been missed by the poetic eye.

25.1 The poetic muse

Meteors, meteorites, fireballs, *ignius-fatus* (will-o-the-wisps), and
comets have been the subject of much folklore. For some people
meteors are a good-luck light, and that a wish expressed upon seeing
one in flight will be fulfilled. For American First Nations they are the
forerunners of death. To the Australian Aborigines they are ladders
by which the dead climb into the sky. In the Finno-Ugric belief,
meteors are the form in which souls of the dead, or the sleeping,
travel. In Islamic lore they are missiles hurled by angels to keep the
Jinn (spirits and demons) out of heaven. It is from these folklore
origins that the poet has found a vibrant muse: meteors (in all their
forms) are visually spectacular, transitory, and mysterious, they are
streaks of colored light that can dance across the night sky in their
hundreds, or they can be solitary and brooding — they are the souls
of the departed and they are the falling sparks of the sinner. Not
only do meteors provide rich folklore pickings for the poet, but so
too did the Aristotelian explanation for their origin (see Section 2.1)
introduce an earthly element — they are the very vapors of the world,

risen high (indeed, as high as is possible in the Earthly realm) and ignited.

In the Elizabethan age (circa 1600–1700) the sciences had hardly penetrated common consciousness, and few poets were influenced by the new astronomy of Copernicus or Kepler. Poet and playwright William Shakespeare, for example, used mainly ancient astronomical imagery. In *A Midsummer Night's Dream*, Act II, scene I we find the lines:

> *Since once I sat upon a promontory,*
> *And heard a mermaid on a dolphin's back*
> *Uttering such dulcet and harmonious breath,*
> *That the rude sea grew civil at her song.*
> *And certain stars shot madly from their spheres.*
> *To hear the sea-maid's music*

These lines later inspired artist Henry Howard (circa 1801) to produce his *Mermaid Sitting on A Dolphin's Back* (Figure 25.1), where the shooting stars are attached to two areal figures (personification of the winds said to presage and follow the appearance of meteors).

Shakespeare returns to meteor imagery in *Henry IV Part One*, adopting the metaphor for their being omens of fear and bad tidings, with the King warning the Earl of Worcester,

> *Move in that obedient orb again*
> *Where you did give a fair and natural light,*
> *And be no more an exhaled meteor,*
> *A prodigy of fear*

Likewise in *Romeo and Juliet*, Juliet declares to the oncoming dawn that will see Romeo leave her, "it is some meteor that the sun exhales," while in *Venus and Adronis*, Adonis, leaving the garden, gives rise to the simile, "look, how a bright star shooteth from the sky, so glides he in the night from Venus' eye." In Shakespeare's writing the meteor is a metaphor for speed and fear and madness, and his usage relies on an audience familiar with both folkloric attributes and Aristotelian explanations for the phenomenon.

Figure 25.1.   Henry Howard's depiction (circa 1801) of the mermaid on the dolphin, from Shakespeare's *A Midsummer Night's Dream*.

Aware of the growing influence of the newer astronomy, George Chapman, a staunch Christian-Stoic, held that the new learning should be used only for the benefit of moral and spiritual growth, and he often stressed the fragility of human nature. In *The Tears of Peace*, Chapman writes,

And how (like men
Raised to high places) exhalations fall
That would be thought stares.

Here, the meteor acts as a metaphor for the ambitious, social climber — just like a shooting star, they are bright, eye-catching, but short-lived phenomena.

Clear reference to the Aristotelian explanation for meteors can be found in the writings of John Donne. Indeed, Donne can be considered one of the first scientific poets — his science being used to convey imagery, however, rather than matter-of-fact truth. In, *Progress of the Soule*, Donne writes:

> *Twixt heaven, and Earth: shee staies not in the Ayres,*
> *To look what meteors there themselves prepare,*
> *She carries no desire to know, no sense,*
> *Whether th' Ayres middle region be intense,*
> *For th' element of fire, shee doth not know*
> *Whether she past by such places or no.*

A more direct reference to Aristotelian meteorology is found in Donne's *A Fever*:

> *These burning fits but meteors be,*
> *Whose matter in thee is soon spent.*
> *Thy beauty, and all parts, which are thee,*
> *Are unchanging firmament.*

Parallels between meteors and the physical human state were common throughout the 17th century. George Herbert in his poem *The Storme* compares the troubled life on Earth to a meteor shower, "Starres have their stormes even in high degree, as well as wee." Abraham Cowley in his poem *Beauty* further writes,

> *Beauty, whose flames but meteors are,*
> *Short-liv'd and low, though thou wouldst seem a star,*
> *Who darest not thine own descry,*
> *Pretending to dwell richly in the eye,*
> *When thou, alas, dost in the fancy lye.*

The idea of deception, that is, presenting the image that meteors are star-like, but are in fact not stars at all, is further explored by Cowley in his *Ode: Of Wit*, "and sometimes, if the object be too far, we take a falling meteor for a star."

It was commonly believed throughout much of the 17th and 18th centuries that meteors were the portents of strong winds and storms, and John Milton picked up on this connection in his description of the fall of Satan from Heaven in *Paradise Lost*:

> *Thither came Uriel, gliding through the even*
> *On a sunbeam, swift as a shooting star*

In Autumn thwarts the night, when vapors fired
Impress the air, and shows the mariner
From what quarter of his compass to beware
Impetuous winds.

Moving into the later part of the 17th century, the influence of the new scientific philosophies begins to take hold of the common consciousness. This was the time of Newtonianism and the glory of natural philosophy. At first, the new-found rationalism resulted in an optimism which soon began to filter through to the poetic muse. The belief in a benevolent God gave rise to universal harmony, reason, and order. This in turn was reflected in the sentimentalism that flourished in late 18th century poetry. A shift in meteor imagery is also evident at this time. Alexander Pope, who penned the lines, "Nature and Nature's laws lay hid in night. God said, let Newton be, and all was light," was not always so enthusiastic about the new advances brought about by the new natural philosophy, condemning it for ignoring the more important issues of morality and belief, and he chastises in his *The Rape of the Lock*, those that chase after frivolous things: "some less refined, beneath the moon's pale light, pursue the stars that shoot athwart the night." Later, in the same poem, Pope uses meteor imagery under the guise of chasing, like a magpie, bright and shinning things:

A sudden star, it shot through the liquid air,
And drew behind a radiant tail of hair...
The heavens bespangling with disheveled light.
The sylphs behold it kindling as it flies,
And pleased pursue its progress through the skies.

The years of the 18th century saw a harmony of thought between the sciences, religion, and poetry — the sciences were then seen as a new tool for revealing the works of God's divine presence in the world. At this time the scientific poets began to emerge: one such poet being Mark Akenside, who in his *Hymn to Science* writes,

Science! Thou fair effusive ray
From the great source of mental day,
Free, generous, and refined!

Descend with all thy treasures fraught,
Illumine each bewildered thought,
And bless my laboring mind.

As with most love triangles, that between science, religion, and poetry was doomed to be short-lived. The common ground that these kingdoms of thought had held during the middle of the 18th century crumbled as it began to close. This was the time when the sciences became increasingly institutionalized within the university system, with professional scientist becoming teachers and researchers. Science was no longer the plaything of wealthy amateurs or the coffee-house populariser. These developments resulted in a growing alienation of the general public and to a rise in anti-Christian rationalism. The three cultures of science, religion, and poetry were sundered and remain, still to this day, largely polarized magisterium. The growth of romantic poetry was part of the backlash against scientific rationalism — but, for all this, the meteor metaphor held strong. Samuel Rogers in his poem *Byron recollected at Bologna* writes,

He is now at rest
And praise and blame fall on his ear alike,
Now dull in death. Yes, Byron, thou art gone
Gone like a star that thro' the firmament
Shot and was lost, in its eccentric course
Dazzling, perplexing...

Interestingly, here, Rogers uses the term *eccentric* to describe both the path of a meteoroid through the atmosphere and in his description of Byron. John Keats also makes allusions to meteors and meteorites in his *Endymiun,*

For anon, I fled upmounted in that region,
Where falling stars dart their artillery forth,
And eagles struggle with the buffeting north
That balances the heavy meteor-stone;
Felt too, I was not fearful, nor alone,
But lapp'd and lull'd along the dangerous sky.

Indeed, Keats had some scientific training, but was not always enthusiastic abouts its revelations. Keats could argue, for example, that the science of optics had destroyed the beauty and poetry of the rainbow — and this sentiment was expressed hardly a century after James Thomson had written, "how just, how beauteous the refractive law," and Mark Akenside had explained that the rainbow was never more pleasing than when science first explained its form. William Wordsworth perhaps sums up the situation best in his *To the Planet Venus*, where he writes, "science advances in gigantic strides; but are we aught enriched in love and meekness?" In the main part, it would seem, however, that meteor imagery was little influenced by the new ideas introduced by Ernst Chladni (recall Section 2.3) and the occurrence of the intense Leonid meteor storms of 1833, 1866, and 1899 (recall Section 2.4). One exception to this situation, however, was that exhibited by Alfred Lord Tennyson. Indeed, Tennyson was well versed in astronomy and was good friend of Norman Lockyer who had suggested that all luminous bodies in the heavens were composed of swarms of meteoroids and meteoric gas. For all this, however, Tennyson bemoaned that, "art and grace are less and less, science grows and beauty dwindles."

Tennyson's gloom was further intensified by Thomas Hardy, although Hardy generally returned to rural, folklore imagery when describing the heavens.[1] In his *A Procession of Dead Days*, Hardy writes, "a meteor act, that left in its que, a train of sparks my lifetime through." Here the meteor imagery is, once more, employed in the sense of the singular, unexpected, life-changing event. In his *The Second Night*, Hardy returns to the idea of ill fortune following a meteor — when a false lover does not recognize the ghost of his true love who has killed herself upon hearing of his cheating, we are told, "a mad star crossed the sky to the sea, wasting in sparks as

---

[1]The exception being his *Two on a Tower* (published 1882). In this story of two star-crossed lovers, Lady Constantine and astronomer Swithin St. Cleeve, the frailty of human love is set against the stark, unyielding indifference of a cold universe.

it streamed." Later in the narrative, when the false lover meets the ghost for the second time and finally realizes the consequence of his actions, we read:

And she leapt down, heart-hit. Pity she's dead:
So much for the faithful-bent!
I looked, and again a star overhead
Shot through the firmament

Both these passages by Hardy reflect upon the bad fortune associated with meteors as well as the notion that shooting stars are the physical embodiment of the souls of the dead. Henry Wadsworth Longfellow, a contemporary of Hardy, however, picked up a theme of disillusionment in his usage of the term meteor, writing, "Men of genius are often dull and inert in society: as the blazing meteor, when it descends to earth, is only a stone."

With the dawn of the 20th century, a distinct Americanization of the meteor image began to take hold — rather than an emulation to be shunned, it became something to embrace, rather than being a warning against bravado and the chasing of fancies, we now find Jack London writing in his *Credo*,

I would rather be ashes than dust! I would rather that my spark should burn out in a brilliant blaze than it should be stifled by dry-rot. I would rather be a superb meteor, every atom of me in magnificent glow, than a sleepy and permanent planet. The function of man is to live, not to exist. I shall not waste my days trying to prolong them. I shall use my time.[2]

This imagery is certainly more in tune with the 20th and 21st century outlook, where Andy Warhol's "in the future everyone will be world-famous for 15 minutes" is typically considered as being far too short a dwell time in the seemingly mindless spotlight that is social media. Nonetheless, there is a positivity to London's *Credo*: we should all live life to the fullest.

---

[2]Although attributed to Jack London, there is some question as to whether he actually penned the lines.

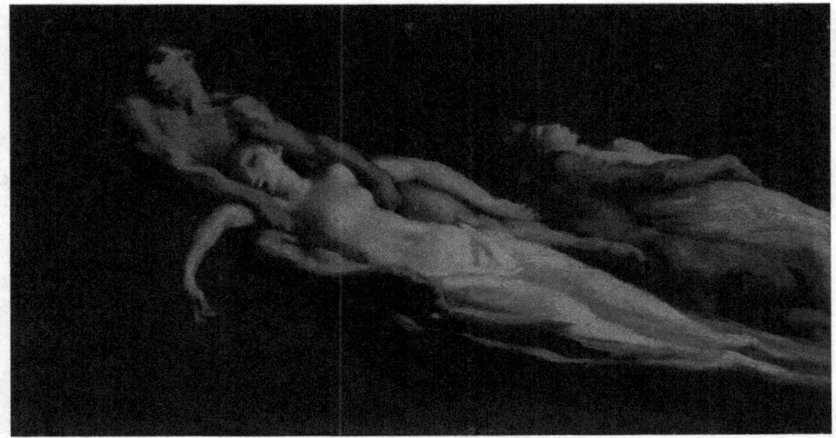

Figure 25.2. *The Shooting Stars* by Jean-François Millet. Oil on canvas: 17.7 cm × 13 cm. Credit: National Museum of Wales, Cardiff.

The fleeting, atmospheric characteristic of meteors has not only inspired the poet but it has inspired the artist. Once again, it is typically used in some metaphorical sense — the portrayal of a meteor indicating something ephemeral, a shimmering moment captured by the artist. Two of the best examples of such imagery were produced by French artist Francoise Millet circa 1847 and circa 1850: the first being entitled *Les Etoiles Filantes* (*The Shooting Stars* — Figure 25.2), and the second *Nuit Etoile* (*The Starry Night* — Figure 25.3). Millet left no notes concerning his inspiration for *Les Etoiles Filantes*, but one is reminded of Andrew Marvell's *The Unfortunate Lovers*, where he writes

Alas, how pleasant are their days
with whom the infant love yet plays!
Sorted by parts, they still are seen
By fountains cool, and shadows green
But soon those flames do loose their light,
Like meteors on a summer's night:
Nor can they to that region climb,
To make impression upon time.

Indeed, there is a passionate sexuality underlying Millet's *Les Etoiles Filantes* — the trailing figures floating in a starlit space

Figure 25.3.   *The Starry Night* by Jean-François Millet. Oil on canvas: 65 cm × 81 cm. Yale University Art Gallery.

betray the desires, struggle, and self-indulgence of new lovers. All these, very human, characteristics being as transient and fickle as the shooting stars in the sky — and sometimes just as destructive.

Millet left no clue as to the inspiration for his *Nuit Etoile*, but its origin coincides with his move to Barbizon, where he turned to the surrounding environment for inspiration. There is a deep sense of the infinite in *Nuit Etoile* and the background and horizon stretch all the way to a star-studded space. The tone of the painting is peaceful, mysterious, and brooding, and the shooting stars give it a sense of moment — a captured moment in the quiet of the universe. The horizon in the image center seems close and finite, suggesting that with just a few more steps the edge of the world will be revealed with the immensity of space opening up beyond. Some written sense of what Millet intended to portray in *Nuit Etoile* can be found in a letter he wrote to a friend in 1856: "if I could only make others feel as I do all the terrors and splendors of the night: if I could but make them hear the songs, the silences and murmurings of the air — one must feel the presence of the infinite." Tennyson expressed similar fears and undertones as Millet in his letter, writing in his poem *God and the Universe,*

> *Will my spark of being wholly vanish*
> *In your deeps and heights?*
> *Must my day be dark by reason, o ye*
> *Heavens, of your boundless nights.*
> *Rush of suns, and roll of systems, and*
> *Your fiery clash of meteorites.*

Millet's *Nuit Etoile* is interesting because he specifically avoids the stereotypical nighttime scene of monotonous, dot-like stars scattered around a romanticised Moon. Rather, his stars pulse with vitality, and the transiency of the shooting stars counteracts the (apparent) agelessness of the stars and the constellations.[3] In this way an element of chaos and surprise is introduced into an otherwise timeless scene.

While Millet's *Les Etoiles Filantes* and *Nuit Etoile* employ aspects of the transitory nature of shooting stars, American artist Frank Tenney Johnson used them in the sense of signaling the approach of bad news. Indeed, in his *An Evil Omen* (1930), the meteor acts to portray the pending destruction of Native American civilization and culture (Figure 25.4). This depiction being in accord with the First Nation's tradition that shooting stars were the forerunners of death.

Accounts of actually observing shooting stars at times of ill omen and disaster can also be found in personal letters and accounts. One account is that by the Reverend M. Davidson to the members of the British Astronomical Association in December 1917, where he described the difficulties of performing meteor work at the Western

---

[3]The constellation of Orion is seen to the upper right corner of Millet's *Nuit Etoile* and the bright star Sirius is depicted to the upper left. Millet is very much in vogue with the philosophy of painting from nature.

Figure 25.4.   *An Evil Omen* by Frank Tenney Johnson. 1930, oil painting,
61 cm × 46 cm.

Front, "there was no adequate means of distinguishing between a
fireball and a star-shell." Another account, riven with pathos, is that
relating to the survivors from the sinking of RMS Titanic on April
14th, 1912. Cold, wet, shivering, and traumatized, survivors reported
seeing numerous shooting stars as they waited to be rescued. Indeed,
these survivors were probably seeing shooting stars associated with
the annual Lyrid meteor shower. Ironically, the radiant point for this
shower is in the constellation of Lyra, named after the harp-like lyre
of Orpheus, and which was capable of producing music so pure it
could even quell the deathly voices of the Sirens.

Moving away from the figurative, sculptor Theodore Roszak
produced a study of a meteor in 1962 that has a strong modern

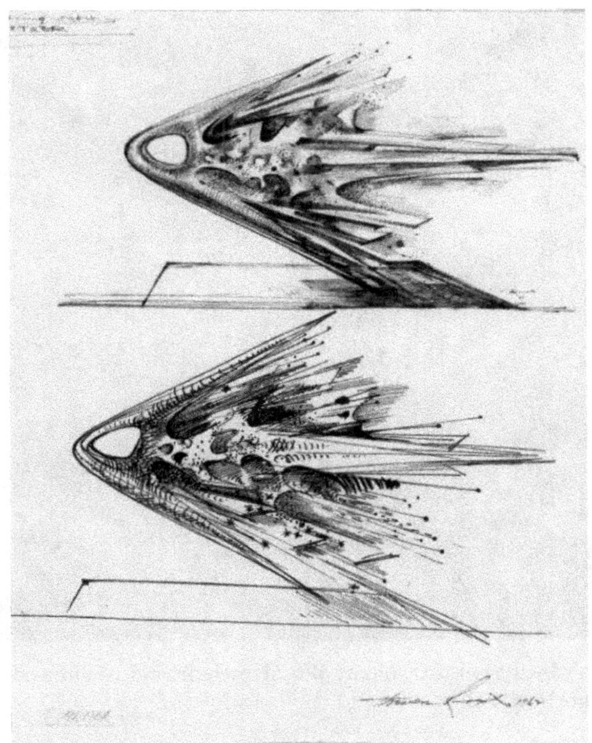

Figure 25.5.   *Study — meteor* by Theodore Roszak, 1962, ink drawing 28.5 cm×
21.5 cm. Collection of Whitney Museum of American Art.

resonance. The depiction, never actually realized as a sculpted work,
invokes a distinct sense of interaction — the meteoroid being melted
as it slams into the Earth's lower atmosphere (Figure 25.5). The
images in Roszak's study are dynamic and full of energy, and they
also reflect a similarity to the Sputnik spacecraft (launched in 1957).

Stamps, just like art movements, provide a glimpse into the
greater zeitgeist of a nation and human creativity. Indeed, stamps are
a visual time capsules in diminutive form, little vignettes of national
and international celebration and recognition. They portray national
treasures and national heroes, and certainly, not to be missed out,
they have long celebrated meteorite falls and discoveries. Perhaps
the ultimate in realism, however, mixing both art and meteoritics,

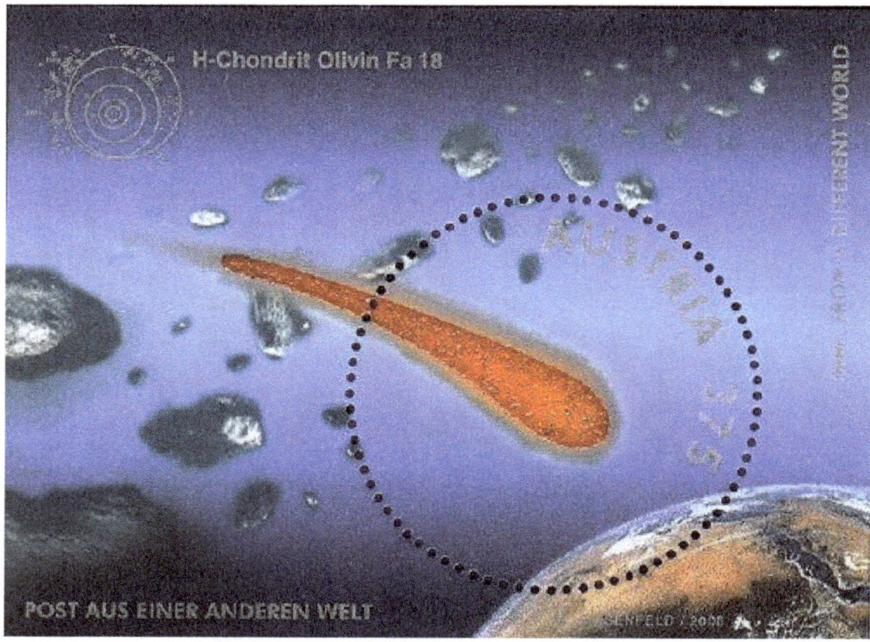

Figure 25.6.   *Mail from a Different World*, meteorite-dust-infused stamp issued by the Austrian Post Office.

is that expressed in a stamp issued by the Austrian Post Office in 2006 (Figure 25.6). In this case actual meteorite dust was mixed with the printer's ink, giving the finished stamp a grainy tactile feel. Some 18 kg of a 19-kg stony meteorite found in Morocco was crushed into a fine powder in order to produce the raw material for the series.[4] Advertised under the title "Mail from a Different World," some 600,000 stamps were produced, each containing 0.03 grams of meteorite dust. The meteor metaphor is many-fold in the Austrian stamp, not only is it otherworldly, but it encapsulates the joy of receiving a missive from a distant friend, and it reflects the business ideal of a speedy delivery.

---

[4]The unused portion of the meteorite is on display at the Vienna Natural History Museum.

25.2 Death from the sky

The story of meteorites would not be complete without reference to its appearance and usage in fictional literature, cartoons, and movies. Indeed, meteors, shooting stars, fireballs, and meteorites have appeared in many literary storylines, but the usage tends to be more metaphoric than physical. The allusion is generally that of something unexpected and dramatic happening — an event that literally turns the story upon its head, throwing everything prior into doubt (Figure 25.7).

Figure 25.7. *Lot and his Daughters*, unknown artist: 1520. The destruction of the twin cities of Sodom and Gomorrah, via meteorites and fireballs, is depicted in the background scene.

Many accounts of death and destruction are presented in the Bible, and the Genesis story concerning the destruction of Sodom and Gomorrah, cities synonymous with impenitent sin, provides a classic example of death raining from the sky. In the biblical account the cities are destroyed by fire and brimstone, but it is an easy step to translate this destruction to the fall of meteorites, or the impact of a comet or asteroid. Moving from Genesis to Revelations, we additionally find the details concerning the opening of the sixth seal[5] — at a time heralding the return of Christ, and the cosmic upheaval that will follow with the unrighteous, who opposed God, experiencing his wrath. Upon the opening of the seal, "there was a great earthquake; and the Sun became black as sackcloth of hair, and the Moon became as blood; and the stars of heaven fell unto Earth." These apocalyptic accounts, intended as a device for reflection upon the consequences of sin, and to ease the real-world suffering being inflicted upon the converted, over time, worked their way into numerous stories and legends, and established the literary idea of destruction of the Earth from space.

In this manner, in modern times, the detection of a comet or an asteroid heading straight for Earth has been used in numerous disaster (if not disastrous) movies and literary storylines. In this usage the comet or asteroid is the would-be harbinger of humanity's doom; the apparent doom, of course, giving rise to the sacrificial hero who will save the day. Characteristically (it would seem) cinematic writers tend to confuse meteor with meteorite, or even meteor with comet or asteroid, but amongst the more recent offerings within the genre of space-disaster movies are *Meteor* (1979, directed by Ronald Neame and starring Sean Connery); *Armageddon* (1998, directed by Michael Bay and starring Bruce Willis), and *Deep Impact* (1998, directed by Mimi Leder, starring Robert Duval).

Meteor is predicated upon the impact destruction of main-belt asteroid Orpheus, an event that sends dozens of small asteroids heading for Earth. The main storyline focuses upon the largest

---

[5]This is literally the sixth seal of the seven that secure the scroll witnessed by John of Patmos in his apocalyptic vision.

fragment, some 10 km across, that would initiate a mass die-off and the destruction of human civilization if it struck the Earth (recall Chapter 11). After many of the smaller fragments strafe the Earth, destroying cities all over the globe, the final extinction event is avoided after the larger fragment is destroyed by a flotilla of missiles. Interestingly, the movie was inspired by and based upon a 1967 MIT graduate student project developed by Paul Sandorff [1]. Called *Project Icarus*, the MIT study looked at ways in which an imagined collision between the Earth and asteroid 1566 Icarus might be avoided.[6] A TV mini-series, also called *Meteor*, was produced by Ernie Barbash in 2009, and in this case the 100-km diameter main-belt asteroid 114 Kassandra is set on a collision course with Earth after being struck by a comet. Once again, a cloud of lesser meteoroids and small asteroid fragments rains death and destruction upon the world, with a flotilla of missiles ultimately deflecting the largest fragment away from the Earth and into the Sun.

The movies *Armageddon* and *Deep Impact* were release rivals, and considered the two possible outcomes of detecting an Earth-intercepting object. *Armageddon* begins with a massive meteor shower destroying the orbiting Space Shuttle *Atlantis*, and the follow-on strafing of New York by meteorites. The storyline centers on a team of "deep-core drillers" who fly to the asteroid and successfully break it apart with a nuclear bomb. The two halves of the asteroid then safely pass by Earth. *Deep Impact*, in contrast, is concerned with the detection of a comet heading for Earth. Once again, the action centers on the failed attempt of a landing crew to destroy the cometary nucleus with a nuclear bomb. Rather than destroying the nucleus, the bomb splits it in two, and these two components continue on to impact the Earth. One component strikes the ocean producing a megatsunami, and the other hits western Canada.

---

[6]The results of a new MIT study, by Sung Wook Paek and coworkers, have recently been published [2]. In this paper an optimization study is presented on how to best deflect an asteroid on an Earth-impact trajectory. Specifically, the study looks at the potential Earth-encounter and mitigation conditions for asteroids 99942 Apophis and 101955 Bennu.

Unlike *Armageddon*, the storyline in *Deep Impact* is largely concerned with humanity (and individual characters) coming to terms with their inevitable death, and it closes with the few survivors of the catastrophe looking to rebuild human civilization.

The contemplation of pending catastrophe, in the knowledge that human civilization is about to end, was also explored by Geoffrey Landis in his short story *The Last Sunset* (published in 1996). Indeed, it is an age-old, unanswerable (until it happens) question: what would you do if you knew that the world was going to end in a week's time? Ray Bradbury considered a similar such question in his short story *Kaleidoscope* (published as part of *The Illustrated Man* in 1951). In this case an astronaut who survives the destruction of his spacecraft finds himself in a slowly decaying orbit about Earth. Realizing that he will not survive re-entry he is forced to anticipate his immanent demise and reflect upon his past life and achievements. With an interesting twist of folkloric attributes, Bradbury has the doomed astronaut realize that, "tomorrow night I'll hit Earth's atmosphere ... and be scattered in ashes all over the continental lands ... I'll burn like a meteor." For the individual astronaut this meteor would be the literal death-flight of his soul; but the story ends with a mother and child seeing the meteor from the ground: "Look, Mom, look! A falling star," and the story ends: "the blazing white star fell down the sky of dusk in Illinois. 'Make a wish,' said the mother. 'Make a wish.'"

While the child at the end of Bradbury's *Kaleidoscope* looks to make a wish upon the observed falling star, other observers, at other times, and in different cultures, have viewed the sight of a meteor in very different ways. In Shakespeare's *Julius Caesar* we find the well-known lines from Calpurnia, who has had a dream of comets in the sky, "when beggars die there are no comets seen. The heavens themselves blaze forth the death of princes." The comet, and even the dream of a comet is here taken as a premonition of death. In modern times there is even a medically accepted phobia for the fear of being hit by a meteorite — meteorophobia[7] — and one can additionally

---

[7]Not to be confused with metrophobia, which is a fear of poetry.

take out insurance against property damage from a meteorite strike. The fall of the Ensisheim meteorite in 1492 sufficiently impressed the Holy Roman Emperor, Maximilian I of Austria, that he took it as a good omen for success in his wars with France and Turkey. He additionally ordered that a large fragment of the meteorite (some 56 kg in mass) should be placed in the church at Ensisheim — there it was fixed to the wall with iron clamps so that it would not depart the church in the same violent way that it had arrived from the sky.[8]

The ultimate in meteorophobia, however, is probably the folktale of Henny Penny (also Chicken Little and Chicken Licken). There are many variants of this particular tale, which builds upon the invocation of hysteria, and/or the inducement of fear and panic that a disaster is about to unfold. Indeed, Henny Penny, the hapless chicken of the story, turns the fall of an acorn into the irrational fear that the sky itself is falling down. The moral of the story largely depends upon which of the many variants that is being told, but the essential idea is that one should not believe all that one hears,[9] or that, in the face of adversity don't be a coward (or a chicken) but rather adopt a stance of courage and fortitude. Bringing the Henny Penny story into the modern era, Australian artist Luke Jurevicus and the Chocolate Liberation Front production company developed the animated character of Figaro Pho who, poor soul, has every known phobia — two episodes of the cartoon series are dedicated to meteorophobia.

Many books have considered the end-of-days scenario, where either a large asteroid or a cometary nucleus is found on an orbit carrying it towards an Earth impact. In *Lucifer's Hammer* (published in 1977) by Jerry Pournelle and Larry Niven, the story is predicated upon the Earth being struck by a cometary nucleus, and follows the chaos and war that is precipitated by this event.

---

[8]The meteorite remained in the church until 1793, and is now housed in the Musée de La Régence in Ensisheim.

[9]This same sentiment applies, even to this very day, and certainly with respect to the nonsense, down-right lies, and misinformation that is routinely published and perpetuated on social media networks.

Arthur C. Clarke's *Rendezvous with Rama* (published in 1972) also begins with an asteroid slamming into Earth, and the actions that follow. In Clarke's case, however, it is the foundation of Spaceguard that takes place — this organization being charged with the task of finding such rogue objects before they strike Earth. The storyline evolves, however, with the Spaceguard survey finding Rama — an alien spaceship. Clarke further developed the idea of asteroid impacts upon Earth, and specifically ways in which they might be deflected, in *The Hammer of God* published in 1993. Rather than using missiles and nuclear bombs, the stock-in-trade of the action movies, Clarke advocated the use of rocket thrusters attached to the surface of the asteroid as the Earth-avoidance mechanism.

Stephen Baxter explores human hubris in his dystopian-future book *Titan* (published in 1997), with a storyline that sees China, striving to place an asteroid in orbit about the Earth (to be used as a weapons station), accidently, instead, have it hit the Earth and thereby engender a mass extinction event. Jack Williamson considers a similar storyline in his book *Terraforming Earth* (published in 2001). In this case, an asteroid strikes the Earth destroying the ecosystem and human civilization along with it, with the only surviving humans being a cohort of clones raised within an automated Moon base. It becomes the task of many clone generations to begin the process of making Earth habitable again — a task, in fact, that does not go as originally planned. This book reflects a certain irony since it was Williamson who coined the term terraforming — that is make Earth-like — in his 1942 short story *Collision Orbit*. In this early work, Williamson envisions engineer Jim Drake working to terraform a 2-km sized asteroid by sinking shafts into its interior, thereby allowing for living space, and mining operations that will ultimately produce a surrounding atmosphere. This latter possibility of the asteroid holding on to an atmosphere is not physically possible (see Chapter 5), but the concept of terraforming, and especially the terraforming of Mars, is a subject that has seen great interest and research effort in the last half-century.

While the direct impact of an asteroid, cometary nucleus, or meteorite into the Earth conveys a real sense of power and kinetic destruction, numerous authors have explored the more subtle, less explosive way in which an extraterrestrial object might influence life on Earth. Drawing upon the 1910 return of Halley's Comet, H. G. Wells in his novel *In the Days of the Comet* (published in 1906) introduces the idea that the chemical make-up of Earth's atmosphere is changed after its passage through the tail of a great comet. Rather than being destructive, however, Wells introduces a utopian change, with the invigorated atmosphere inducing a state of human happiness and a state of peace on Earth. The Earth did, in fact, pass through the tail of Halley's Comet on the night of May 19th, 1910, but rather than invoking a sense of awe and inspiration, the human reaction was generally one of panic. The panic was induced as a result of the detection of cyanogen in the comet's coma — this being a deadly gas. Prophets of doom soon began their siren cry: comet pills were sold, comet-proof rooms were constructed (in which one was literally sealed inside), and submarine voyages were advertised (in this case the idea was to be underwater when the comet tail encounter took place). Needless to say, no comet-related deaths occurred. In contrast to Wells, John Wyndham in his 1951, *The Day of the Triffids*, invoked the mass disaster of human blindness being initiated through the observation of a meteor shower (bright green meteors produced as a result of the Earth passing through the tail of an unnamed comet), with the additional apocalyptic result that aggressive, bioengineered plants capable of motion (the Triffids) were able to escape their confines and begin to take over the world.

H. P. Lovecraft explored the possibility of deadly terrestrial contamination in his short story *The Colour Out of Space* (first published in 1927). This story is centered on the fall of a meteorite in Arkham, Massachusetts, at a location called the "blasted heath." The meteorite, however, has exceptional qualities, it was "oddly soft ... almost plastic ... it was a metal though beyond doubt." Not only did the meteorite have a strange, otherworldly structure, it had "a color impossible to describe." After examination of the strangely malleable material by several scientists, it was concluded that, "it was nothing of this Earth, but a piece of the great outside; and as such dowered with outside properties and obedient to outside laws."

The meteorite fell next to a water well located at a remote farm, and soon after the meteorite's arrival strange animal adaptations were noted and the vegetation grew in an oddly grotesque manner. The owners of the farm went mad and died. Lovecraft reveals through the unnamed narrator that the well had been poisoned by the colored vapour, and from there it "sucked the life out of living things." Ultimately the vapour returned to space, climbing into the sky as an "unrecognizable chromaticism." *The Colour Out of Space* not only draws upon aspects of the horror story, but it explores the not entirely irrational fear that there are things out in space that are potentially lethal to living things on Earth, and which are entirely beyond human comprehension and description.

Lovecraft's idea has been adapted on numerous occasions by many authors, including Steven King in his 1978 book *Tommyknockers*, and by Flann O'Brian (one of the pseudonyms of Brian O'Nolan) in his wonderfully strange novel *The Third Policeman* (published posthumously in 1967). In the latter case the author explores ideas of eternity, limits, and infinities, all mixed together in a cacophony of absurd irrationalities, including a colour that, "no man carries inside his head" and which, upon seeing, will drive the mind insane. Andrew Ross in his (2019) book *The Meteorite* considers the same basic idea as Lovecraft, but on a grander scale. Lovecraft's story has also been adapted to the television and motion picture screen on numerous occasions, most recently in a 2019 film directed by Richard Stanley, staring Nicolas Cage. Stanley's story begins with a meteorite falling in Russia, but in this case, it is not the physical impact that is the problem but the alien virus carried by the meteorite (and others that fall worldwide) — the virus induces a state of dementia which sees human civilization begin to slip into a state of apathetic decay.

Clive Cussler and Craig Dirgo adopted both extraterrestrial and terrorist-induced disaster themes in their 2004 book *Sacred Stone*. One of the subplots in this work is that concerning an ancient radioactive meteorite, called the sacred stone, which additionally contains an extraterrestrial, oxygen-destroying microbe. The storyline follows two groups of fanatics: one group of fanatic Islamist is intent upon stealing the meteorite and turning it into a *dirty* nuclear bomb, while the other, an anti-Islamist group, is determined to use it to destroy the Black Stone contained in the Kaaba.

Jules Verne also described a meteorite search in his *The Meteor Hunt* (also called *The Chase of the Golden Meteor* — Figure 25.8), published posthumously in 1908. In Verne's case a rivalry between

Figure 25.8. The golden meteor descends through Earth's atmosphere — here illustrated by French artist George Roux (who produced illustrations for many of Verne's texts). Although the meteor is shown above a crowded city vista, Verne actually describes the meteorite as landing in western Greenland, invoking a connection with the Cape York meteorites, then just recently sold to the American Museum of Natural History.

two American meteorite hunters is followed, both of whom are seeking to make their name through the collection of the meteorite (especially so when it is deduced that the meteorite is made of solid gold). *The Meteor Hunt* is not a particularly good read, and while the science behind it is not well thought through, the narrative is largely concerned with greed and human relationships — there is also some interesting commentary concerning the inevitable collapse of the gold market[10] if such a golden meteorite were ever found — in fact, in Verne's story, the meteorite is lost, after sliding into the sea and exploding.

More recently, Dan Brown made the discovery of a meteorite the center point of his novel *Deception Point* (published in 2001). In this case a fraud is initiated around a meteorite in which it is claimed that evidence for extraterrestrial life has been found. In a twist to the theme concerning the discovery of extraterrestrial life in a meteorite, Jerry Dawson's story *Meteor* (published in 2018) begins with a mission sent out to deflect an asteroid that will be passing uncomfortably close to Earth. Unfortunately, the mission causes a large meteorite to crash to Earth, landing on a small and remote Pacific island. The meteorite, however, contains deadly alien lifeforms and people soon begin to die. A similar, but less deadly, story is that presented in Hergé's 1942 graphic novel *The Shooting Star*. In this case the meteorite's fall location, in the high Arctic,[11] is tracked down by Tintin, Hergé's hero character, and it is found that the meteorite accelerates the growth process, producing a super-sized tree from a thrown away apple core, exploding mushrooms, and a monstrous spider.

The somewhat strangely titled *Fifteen Hundred Miles an Hour* by Charles Dixon (published in 1895) uses a meteorite fall to deliver an astounding message from Mars. The meteorite descends during the onset of a sand storm over North Africa, and Dixon writes in highly

---

[10]Astrophysicist Fred Hoyle speculated on this very same topic in his short story *Element 79* published in 1968. Element 79, of course, being gold.

[11]One of the ships in the race to recover the meteorite is called *Peary*, invoking a connection to Robert Peary and his infamous theft of the Cape York meteorites.

descriptive terms, "my eyes were almost blinded by a brilliant light, brighter than the flame from an incandescent lamp, and a thousand times as larger, which seemed to shoot from space. At the same awful moment the very dome of heaven seemed cracked asunder by a loud report ... the earth shook and trembled to its utmost foundations. ... I, for several moments, was as one stone-blind." Dixon's description rings so true that one suspects that at some time or another he must have witnessed such an event. The text continues, however, with the finding of the meteorite, "a large round mass of dark-looking substance... it was a meteorolite [*sic*] of unusual dimensions." It was upon examining the meteorite that a rectangular piece of iron was found, and within this was a tightly packed pile of papers. These papers, it transpired, were from "The City of Edos, Planet Mars, or Gathma." The rest of Dixon's book is concerned with an expedition to Mars to investigate the source of the meteorite's concealed letters. Travelling in an electrically powered spaceship, *Sirius*, the title of the book is explained in that 1500 miles per hour is the speed of the ship — at this speed the travel time to Mars when at its closest to Earth is $2^{1}/_{2}$ years.

Not only have authors of fiction considered the direct effects of collisions due to asteroids, comets, and meteorites, along with the contamination of Earth by alien microbes, some authors have centered their narratives around the idea of strange meteor radiation (outside of light generation, there is no such radiation effect, of course). Wyndham's *The Day of the Triffids* is predicated upon the idea of people going blind through the observation of meteors, but perhaps the most interesting usage of the radiation-effect mechanism is that developed by Philip José Farmer under the general heading of the Wold Newton Universe. The Wold Newton family of characters was first described by Farmer in his 1972 book, *Tarzan Alive: A Definitive Biography of Lord Greystoke*, and it is based around the actual fall of the Wold Cottage meteorite on December 13th, 1795. Wold Newton is the name of the nearest town to the meteorite's fall location, and Farmer imagined that some 18 people, seated in two coaches that chanced to be close to the fall location, had their DNA genetically altered by radiation emitted by the meteoroid as it fell towards the ground. This mutation resulted in the descendants of the coach party members siring the most famous detectives, villains, scientists, and explorers of the 19th and 20th century. Characters linked to the Wold Newton Family include, Sherlock Holmes, Professor Challenger and Professor Moriarty, Tarzan (Lord Greystoke), Philias Fogg, Sam Spade, Doc Savage, The Shadow, and even James Bond. Many authors have developed Farmer's theme, but the origins and history of *The Family* are nicely brought together in Win Scott Eckert's short story, *The Wild Huntsman*, as found in *Tales of the Wold Newton Universe* (published in 2013). The meteor radiation effect has been used as a mutation device in several graphic novels, including the reinvented Marvel Comics character Green Lantern, whose ring and lantern are explained as extraterrestrial artifacts. In DC Comics, *The Atom* (Dr. Ray Palmer) additionally obtained his powers of shrinkage and mass control after finding a meteorite made of white dwarf matter.

That a meteorite might have a *memory* or hold a record of its past was explored by J. R. R. Tolkien in his *The Notion Club Papers*. This story, written as Tolkien was struggling to finish his better-known *Lord of the Rings*, considers aspects of time travel and communication through time and ancestry — "travelling on a dream-meteor." Indeed, Tolkien writes (as ever) in a beautiful manner, and the character in the narrative that can deep search past *being* and past *imprints* describes the meteorite's journey: "I caught sight of blazing fire ... A mountain corroded into a boulder in a few seconds of agonizing flame. But above, or between, or perhaps through all the rest, I knew endlessness."

25.3 Meteors and mushrooms

The human mind strives for order and seeks to find cause and effect connections between events. This trait has obvious survival benefits, but it can also lead to false conclusions and/or associations. Many folklore beliefs fit into this latter category of false association, and indeed, there is a long history of incorrectly linking gelatinous fungi with the residue of meteors [3]. The logic of this connection is certainly understandable, given that the appearance of these phenomena was not understood until relatively recent times — both appear at random times, in random locations, and given the Aristotelian premise that meteors are a consequence of the chemical transformation of burning vapors in the atmosphere, then why not speculate about the residue that might result and what it might look like.

William Caxton offers an early hint to this meteor–mushroom connection, reasoning in his *The Mirror of the World*, the first English language encyclopedia (translated from a 1464 French manuscript) published in 1480. Here, Caxton informs us that very bright meteors can sometimes fall to the ground and that if an observer is lucky enough to find the fall location then, "they find none other thing but a litil asshes or like thing som leef of a tree roten, that were weet." The meteor–mushroom connection can be traced back even further than Caxton, however, and Plutarch in the 2nd century A.D. informs us in his *Symposiacs* (book IV) that mushrooms are created whenever thunderous lightning hits the ground. Erasmus Darwin reiterated this connection in his epic poem *Botanic Garden* (published in 1790), writing that, "so from dark clouds the playful lightning springs, rives the firm Oak, or prints the fairy ring." A fairy ring is a characteristically circular structure, associated with fungal growth, found in open grassland. In the Aristotelian theory (recall Chapter 2) meteors and lightning were considered to be similar phenomena, differing only in their atmospheric heights and speeds of combustion. In folklore belief, it was long held that lightning could additionally produce a *thunderstone* if it chanced to strike a common rock. Given a starting point based upon Aristotle's *Meteorologica*, it is perhaps not so surprising that an association eventually evolve between mushrooms (fungi) and meteors, and while the connections can be logically developed, it does not, of course, have any correspondence with physical reality.

For all this, however, more than one hundred years after the appearance of Caxton's *Mirror*, poet John Donne can readily challenge the readers of his *Song* to:

Goe, and catche a falling star,
Get with child a mandrake root,
Tell, me where all past years are,
Or who cleft the Divels foot.

The list of tasks outlined by Donne in his *Song* are all impossibilities, but the notion of catching a falling star is not so much in that it cannot be done in the sense that there is no tangible object to catch hold of, but in the impossibility of knowing when and where to be in order to catch a falling star. In a second illusion to the meteor–mushroom connection, Donne additionally informs us in his poem *Epithalamion X* that:

As he that sees a starre fall, runs apace,
And finds a gellie in the place,
So doth the bridegroom hast as much
Being told this starre is faln, and finds such.

Here Donne uses the term gellie[12] in relation to the residue of the meteor, and in a metaphorical sense to the inevitable decay of beauty.

In a reverse sense, Robert Heath echoes Donne and Caxton, and writes in his poem *Clarastella*:

So have I seen bright falling stars in show
Quench in dark gellies here below
When they false meteors did (descended) spie
A truer light in Stella's eye.

Here the meteors fall and decay in the apparent shame of beholding a greater beauty. These same sentiments are further explored by Abraham Cowley in his poem *Reason*:

So stars appear to drop to us from the skie,
And gild the passage as they fly,
But when they fall, and meet th' opposing ground,
What but a sordid slime is found?

---

[12] *Gellie* is the forerunner of the modern-day jelly, and is derived from the Latin verb *gelare*, meaning to freeze or coagulate.

Figure 25.9.   The gelatinous fungus *Tremella Mesenterica*. Image courtesy of Wikipedia Commons.

Once again, the meteor metaphor applies to an apparently bright and transient phenomenon (or person) that, upon closer inspection, turns out to be neither luminous nor beautiful.

A clear folkloric, rather than poetic, link between meteors and mushrooms has been traced with respect to the gelatinous fungus *Tremella Mesenterica* (Figure 25.9). Commonly called the Yellow-Brain fungus, or Witches' Butter fungus today, during the 18th and 19th century its common name varied between *fairy-butter* (due to its distinct yellow color), *star-jelly*, and *star-shot*. Carl Linnaeus also noted that the Swedish name for this particular fungus is *sky-fall*. Indeed, a letter addressed to Sir Hans Sloane (the foremost collector and botanist of his time) of the Royal Society in 1733, and read to the society's Fellows, described the observation of a bright fireball from Fleet in the shire of Dorset, and the subsequent fruitless search for "those jellies [*Tremella M.*] which are supposed to owe their beings to such meteors."

Another interesting account is given by John Ramsbottom in his book *Mushrooms and Toadstools* (Collins, London, 1953), where a folk tale from the Italian Tyrol is recoded: here, it is supposed that on the nights of August 10th and November 11th a winged dragon passes over the fields scorching the ground below with his fiery tail. Wherever the dragon passes the grass is burnt and it is said that nothing will grow in such locations for 7 years, after which time a circle of mushrooms will appear. This appears to be a wonderful hybrid of many stories, linking meteors to a flying dragon and mushrooms. Additionally, the specific nights upon which the dragon flies correspond to the peak activity times of the Perseid and Leonid meteor showers. Since the annual occurrence of these showers was only realized towards the close of the 19th century, this particular version of a more ancient folk tale must be of a relatively recent genesis.

References

[1] Project Icarus: see https://www.thespacereview.com/article/175/1 (accessed February 2020).

[2] S. W. Paek, O. de Weck, J. Hoffman, R. Binzel, and D. Miller. "Optimization and decision-making framework for multi-staged asteroid deflection campaigns under epistemic uncertainties." *Acta Astronautica*, **167**, 23–41 (2020).

[3] H. Belcher and E. Swale. "Catch a falling star." *Folklore*, **95**, 210–220 (1984).

Appendix: Solving Kepler's Problem

The first step in determining the location of an object, within its orbit, at time T is to calculate its mean anomaly M based upon the known time of a perihelion passage T_0. Accordingly, given an orbital period P,

$$M = (T - T_0)(2\pi/P) \tag{A.1}$$

Equation (3.9) can now be solved for the eccentric anomaly E at time T. In order to do this an iteration scheme needs to be developed. This can be done by taking a first guess at the value of E and then refining the guess, by repeatedly using Equation (3.9), until a converged, final value is found. Accordingly, guess some value for E_0 — this can be as simple as setting $E_0 = 1.0$. Remember at this stage that the eccentric anomaly is expressed in radians in Equation (3.9) so it can have a value anywhere between 0 and 2π. The Newton–Raphson (NR) iteration scheme is a convenient numerical technique for finding a solution to Equation (3.9) once an E_0 value has been set. The NR scheme specifically looks for iterated solutions to the equation:

$$E_{j+1} = E_j - \frac{E_j - e\sin E_j - M}{1 - e\cos E_j} \tag{A.2}$$

where $j = 0, 1, 2, 3, \ldots$ The iteration scheme (A.2) can be readily and rapidly solved to some pre-set precession $E_{j+1} - E_j < \varepsilon$ on a computer. Having found the value of the eccentric anomaly

E corresponding to the mean anomaly M at time T, the heliocentric distance r can be determined via Equation (3.2). If the true anomaly f is required at time T, this can be found via the identity

$$\cos f = \frac{\cos E - e}{1 - e \cos E} \tag{A.3}$$

Once the eccentric anomaly has been evaluated for the given mean anomaly M, the position of the object within its orbit (that is its r and f values) at time T is fully specified.

The orbital velocity can additionally be evaluated from Equation (3.7), which, when set in the units of km/s for the velocity and astronomical units for the heliocentric distance and semi-major axis, gives:

$$V\left(\frac{km}{s}\right) = 29.86 \left(\frac{2}{r} - \frac{1}{a}\right)^{1/2} \tag{A.4}$$

By way of example let us look at the orbit of the Pribram meteorite (a meteorite with a terrestrial residency age the same as that of the author — the author actually being 86 days older). Several fragments of the Pribram meteorite exist, but they all arrived at 8 p.m. in the evening on April 7th, 1959. The orbital elements deduced for the parent body of the meteorites are given in Table 3.1, with $(a, e, \omega) = (2.4, 0.67, 241.75)$. Immediately, we have from Kepler's third law (Equation (3.1)) that the orbital period of the meteorite (prior to impact) was $P = (2.4)^{3/2} = 3.7186$ years $= 1358.02$ days. The descending (N_{des}) and ascending (N_{as}) node distances are provided for by Equation (3.2) with $f = 61.75$ and $f = 241.75$ degrees respectively, giving $r_{\mathrm{des}} = 1.0042$ AU and $r_{\mathrm{as}} = 1.9369$ AU. The meteorite accordingly encountered the Earth while passing through its descending node, and Equation (A.4) reveals that its (heliocentric) velocity at that time was 37.47 km/s. The location of the Pribram parent body in its orbit about the Sun, at 20-day time intervals, is shown in Figure A.1. The figure indicates, as expected

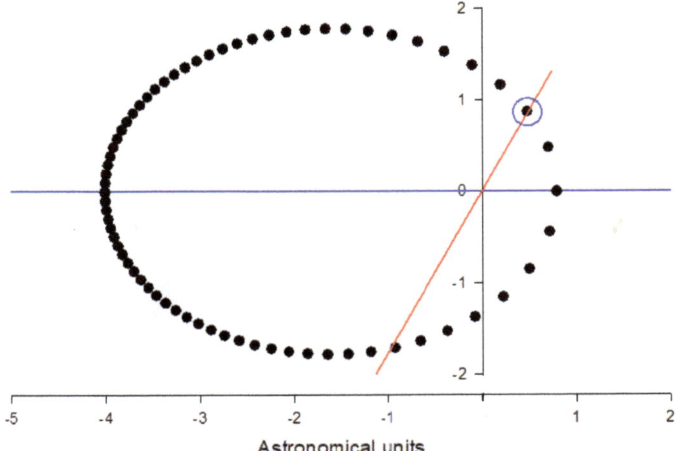

Figure A.1. Orbit of the Pribram parent meteoroid prior to impact with the Earth. The meteoroid position (black dots) is indicated at 20-day time intervals. The redline represents the line of nodes, and the blue circle indicates the location of the descending node and the point at which Earth intercept took place — the meteoroid, in fact, hit Earth's atmosphere just 40 days after perihelion passage.

from Kepler's second law, that the meteoroid is travelling more rapidly and hence moves further along its orbit per 20-day time interval when it is close to perihelion. At aphelion the motion along the orbit per 20-day time interval is relatively small.

General Index

Object Index

Asteroid Index

Comet Index

Crater Index,

Interstellar Objects

Kuiper Belt Objects

Meteor Shower Index

Meteorite Index

Star Index

9 789811 224911